S0-BLI-341

ADVANCES IN METAL-ORGANIC CHEMISTRY

Volume 4 • 1995

ADVANCES IN METAL-ORGANIC CHEMISTRY

Editor: LANNY S. LIEBESKIND
Department of Chemistry
Emory University
Atlanta, Georgia

VOLUME 4 • 1995

Greenwich, Connecticut *London, England*

55 Old Post Road, No. 2
Greenwich, Connecticut 06836

JAI PRESS LTD.
The Courtyard
28 High Street
Hampton Hill, Middlesex TW12 1PD
England

ISBN: 1-55938-709-2

Manufactured in the United States of America

CONTENTS

LIST OF CONTRIBUTORS

James R. Behling — Department of Chemical Sciences
Searle Laboratories
Skokie, Illinois

Paul W. Collins — Department of Chemical Sciences
Searle Laboratories
Skokie, Illinois

René Grée — Laboratoire de Synthèses et Activations de Biomolecules
Ecole Nationale Supérieure de Chimie de Rennes
Rennes, France

Masanobu Hidai — Department of Synthetic Chemistry
Faculty of Engineering
The University of Tokyo
Tokyo, Japan

Youichi Ishii — Department of Synthetic Chemistry
Faculty of Engineering
The University of Tokyo
Tokyo, Japan

Jean Paul Lellouche — C.E.N. Saclay
Service des Molécules Marquées
Gif Sur Yvette, France

Bruce H. Lipshutz — Department of Chemistry
University of California
Santa Barbara, California

John S. Ng — Department of Chemical Sciences
Searle Laboratories
Skokie, Illinois

James H. Rigby
Department of Chemistry
Wayne State University
Detroit, Michigan

PREFACE

Continuing with the precedence set in the three previous volumes of *Advances in Metal-Organic Chemistry*, Volume 4 maintains an international flavor to the contributions and covers a variety of synthetically useful transition metal-based processes that are both catalytic and stoichiometric in the metal.

The first two chapters of the current volume present different aspects of some of the more important organocopper chemistry to be developed in recent years. Chapter 1, "Recent Progress in Higher Order Cyanocuprate Chemistry," written with an entertaining and personal flair unique to Professor Lipshutz, gives the reader insight into the endeavors of the Lipshutz group to understand and develop synthetically versatile higher order cuprate reagents. This is followed in Chapter 2 with an account by Dr. James Behling and his colleagues of a very significant industrial application of organocopper chemistry, "The Evolution of a Commercially Feasible Prostaglandin Synthesis" that was carried out at Searle Laboratories. Together these two chapters highlight the possibilities that organotransition metal chemistry brings to both academic and industrial settings.

In Chapter 3, Professor James Rigby provides the reader with a thorough account of his group's exploration of "Transition Metal-Promoted Higher Order Cycloaddition Reactions." The reader is presented with a clear message of some of the unique and synthetically useful transformations that can be developed when utilizing the templating ability of transition metals. Professors René Grée and Jean Paul Lellouche have written in Chapter 4 what I believe is a seminal contribution on "Acyclic Diene Tricarbonyliron Complexes in Organic Synthesis." This scholarly article from one of the premier research groups studying the synthetic potential of iron diene π-complexes leaves no stone unturned and possess an information content typically seen in review articles appearing in *Organic Reactions*.

Completing Volume 4, Professor Masanobu Hidai in Chapter 5 describes his group's studies of "Novel Carbonylation Reactions Catalyzed by Transition Metal Complexes," which covers the cyclocarbonylation of allylic acetates and related substrates (a novel method for the synthesis of aromatic and heteroaromatic compounds) as well as Professor Hidai's studies of unique homogeneous multimetallic catalysts for carbonylation reactions.

Lanny S. Liebeskind
Editor

RECENT PROGRESS IN HIGHER ORDER CYANOCUPRATE CHEMISTRY

Bruce H. Lipshutz

Advances in Metal-Organic Chemistry
Volume 4, pages 1–64.

ISBN: 1-55938-709-2

I. DO "HIGHER ORDER" CYANOCUPRATES EXIST?

The title of the *JACS* communication to the editor in front of me read: "Higher Order Cyanocuprates: Are They Real?" Ce n'est pas possible! Apparently, on the strength of both ^1H and ^{13}C NMR data, the claim put forth was that what we had originally formulated as "$R_2Cu(CN)Li_2$", i.e., a Cu(I) dianionic salt [$R_2Cu(CN)^{-2}$ $2Li^+$], implying that the nitrile ligand is attached to copper, is *not* a "higher order" (H.O.) cyanocuprate; there is no such species.[1] Rather, just as treatment of a copper halide, CuX (X = I, Br), with two equivalents of an organolithium (RLi) in THF leads to "R_2CuLi + LiX,"[2] so does the analogous expression involving copper cyanide (Eq. 1).

$$2RLi + CuCN \rightarrow R_2CuLi + LiCN \qquad (1)$$

To the readership, especially those who have followed and/or benefited experimentally from the evolutionary development of these reagents,[3] it must have come as somewhat of a surprise. Indeed, I can vividly recall the comments of a UCLA colleague relayed to me upon first glance at this article: "Gee, I guess Bruce was wrong." But then he turned the page and there was our rebuttal article,[4] which not only provided new spectroscopic results strongly supporting the presence of the CN ligand on the copper center, but also pointed out (albeit briefly in a reference) how the NMR experiments leading to such a conclusion had been misleading.

In essence, the argument against the existence of H.O. cyanocuprates goes as follows.[1] The ^1H NMR (500 MHz) spectrum of "Me_2CuLi" (+ LiI) in THF, prepared from 2MeLi + CuI, which gives a single absorption at δ–1.57, is very close to that observed using the 2MeLi + CuCN alternative formulation (δ–1.60). Further support comes from their ^{13}C NMR spectra, in which singlets are observed also in the same region ($\Delta\delta$ ca. 0.2 ppm) for the methyl signals. Most damaging, however, is the position of the signal due to the CN ligand which, contrary to expectation in going from a "lower order" (L.O.) cyanocuprate MeCu(CN)Li to a more electron-rich H.O. cuprate $Me_2Cu(CN)Li_2$, moves *downfield*. To "prove" that the similarities in chemical shifts were not coincidence, Me_2CuLi in THF was admixed with a LiCN in HMPA solution, which leads to only a small, if any (presumed solvent-dependent) change in the ^{13}C NMR position of the CN carbon (Scheme 1). Moreover, $Me_2Cu(CN)Li_2$ in THF containing HMPA showed essentially the same chemical

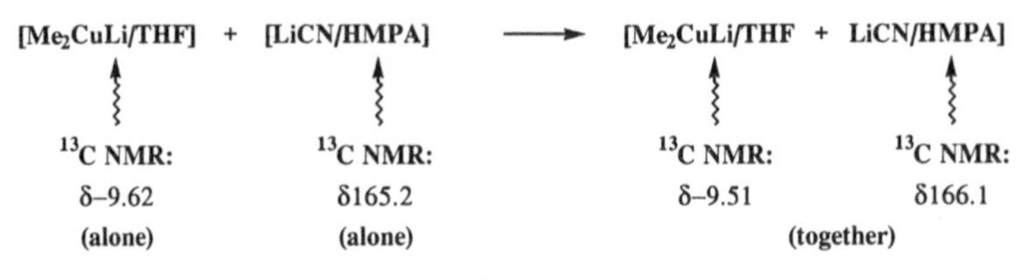

Scheme 1.

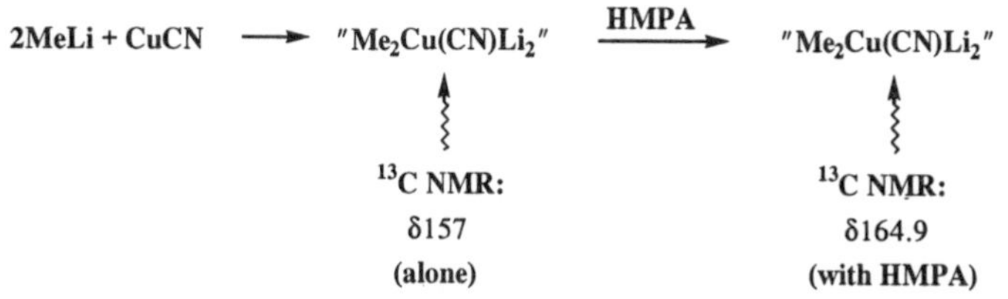

$$2MeLi + CuCN \longrightarrow \text{"}Me_2Cu(CN)Li_2\text{"} \xrightarrow{HMPA} \text{"}Me_2Cu(CN)Li_2\text{"}$$

^{13}C NMR: δ157 (alone) — ^{13}C NMR: δ164.9 (with HMPA)

Scheme 2.

shift (δ164.9) for the CN signal (Scheme 2) as that seen for LiCN in HMPA alone (δ165.2) or for Me_2CuLi + LiCN in THF/HMPA (δ166.1; compare Schemes 1 and 2). Thus, given these data, along with identical results for related cuprates derived from EtLi and PhLi, there was only one logical conclusion: irrespective of the distinguishable reactivity patterns between cuprates derived from CuI/CuBr versus those of reagents from CuCN,[5] the difference was in the LiX formed as by-product of cuprate formation (Scheme 3).

So what's wrong with this analysis? This was the pressure-packed question which faced us; fortunately, I had two extremely talented co-workers, Edmund Ellsworth and Sunaina Sharma, both highly experienced in not only copper chemistry but especially in low temperature NMR analyses of cuprate reagents, available to respond to this challenge. With the benefit of hindsight of course, it is now easy to see the culprit: HMPA. As it turns out, the amount of HMPA added to the various cuprate solutions is extremely influential in terms of chemical shift movements for the nitrile ligand, and points to a critical component of any spectroscopic study on these organometallic reagents: control experiments. Thus, what proved to be the fatal flaw in this otherwise reasonable and seemingly cogent argument was the fact that *the chemical shift of the cyanide group in the ^{13}C NMR spectrum of either cuprate mixture (i.e., prepared according to either Scheme 1 or Scheme 2) could be altered as a function of the amount of HMPA in solution.* Chemical shifts between 158 ppm (0% HMPA) and >164 ppm (>50% HMPA) could be arrived at on demand! This was clearly established by a series of spectra involving sequential increases in the relative proportions of HMPA to THF, and as illustrated in Figure 1, the CN peak can be positioned, e.g., at 159 ppm using 10% HMPA in THF. As a bonus, this series actually allowed us to observe a second cyano peak *upfield* of that seen for

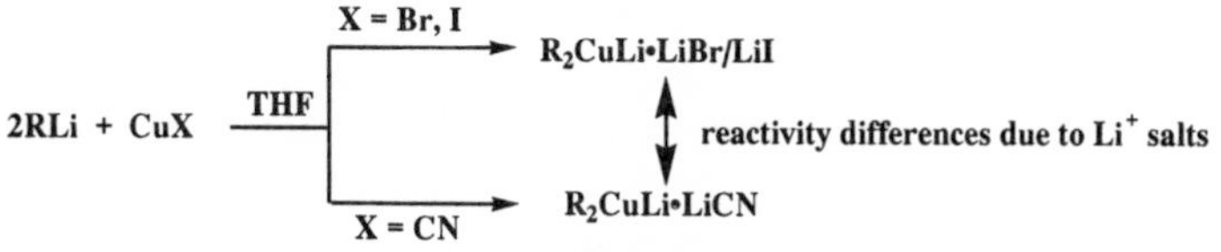

$$2RLi + CuX \xrightarrow{THF} \begin{cases} \xrightarrow{X = Br,\ I} R_2CuLi\cdot LiBr/LiI \\ \xrightarrow{X = CN} R_2CuLi\cdot LiCN \end{cases}$$

reactivity differences due to Li^+ salts

Scheme 3.

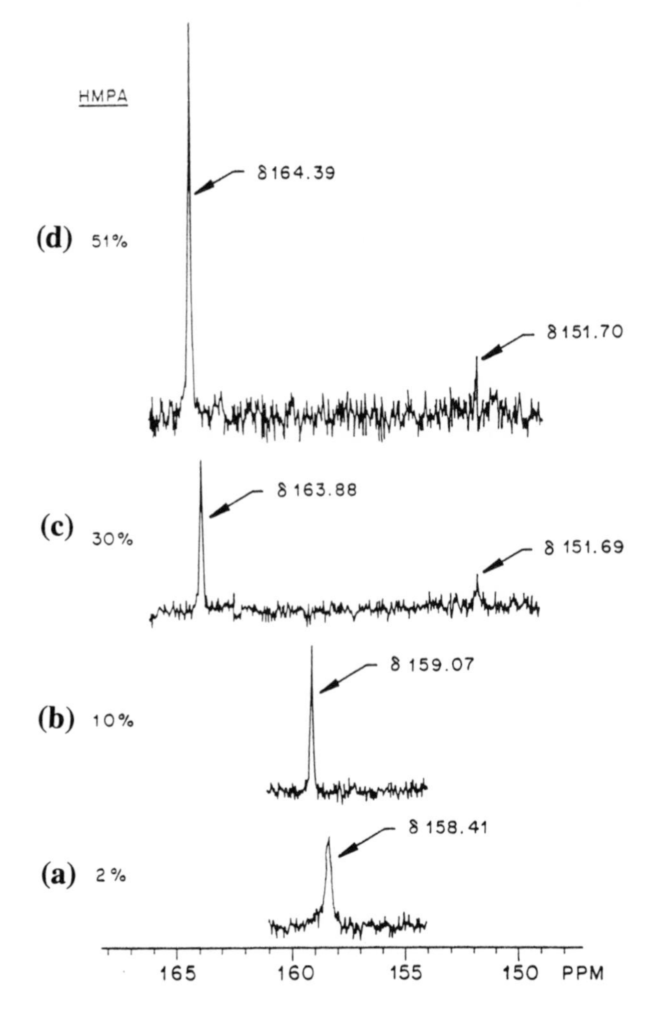

Figure 1. ^{13}C NMR spectrum (downfield region) of $Me_2Cu(CN)Li_2$ in THF/HMPA at –40 °C

the H.O. species $Me_2Cu(CN)Li_2$ (δ158 in THF alone), exactly as predicted (*vide supra*).[1] These two peaks, shown in Figure 2 (not coincidentally), are also seen starting with the Me_2CuLi + LiCN combination using the same concentrations of HMPA added to $Me_2Cu(CN)Li_2$ seen above [compare Figure 1(c) and Figure 2]. Is there an alternative explanation as to how LiCN in THF/HMPA can give rise to these two signals, along with the *complete disappearance of the peak at 166+ ppm due to LiCN,* other than to conclude that the CN ligand *prefers* to be on copper? The same conclusion is reached by carrying out IR experiments on these identical

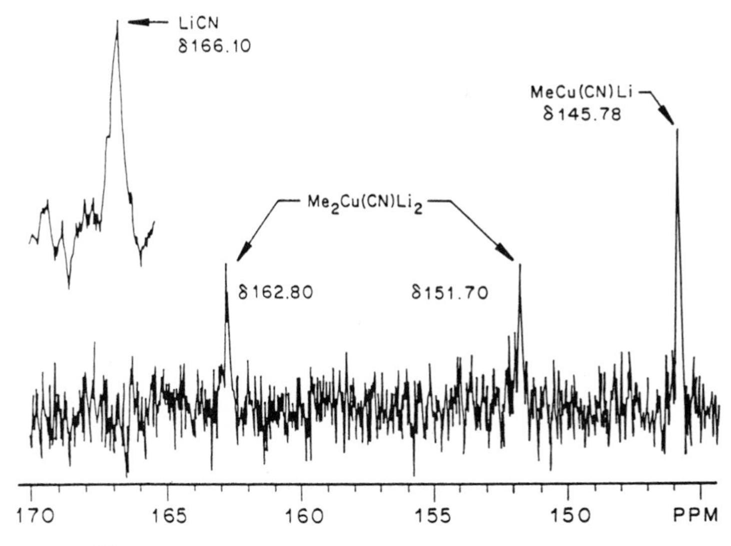

Figure 2. ^{13}C NMR spectrum of Me_2CHLi in THF + LiCN/HMPA at –40 °C

gemisches, in the same solvent system, scrutinizing the nitrile stretching frequencies. Clearly, this latter technique, in and of itself, permitted us to rule out two of the three possibilities (**1**, **2**, and **3**), as only **2** fits the data.

$$\underset{\mathbf{1}}{R_2CuLi{\cdot}LiCN} \quad \underset{\mathbf{2}}{R_2Cu(CN)Li_2} \quad \underset{\mathbf{3}}{R_2Cu^{-}Li_2CN^{+}}$$

So the bell ending "round 1" sounded, and I thought after hearing that same UCLA colleague declare, now having read our follow-up *JACS* paper,[4] "Gee, I guess Bruce was right after all", that maybe the contest was over. Well,... I've been wrong before.[6] But while the fervor forced us to critically examine and physically prove what we felt had to be the case all along, it fortunately did little to change our principle focus of developing new synthetic methods involving copper chemistry.

II. "KINETIC" H. O. CUPRATES: THE LINK BETWEEN SPECTROSCOPY AND SYNTHETIC METHODOLOGY

Somewhat after the "Bertz Affair" had simmered down, we speculated that it might be possible to learn about the innate characteristics of H.O. and L.O. cuprates by finding oxidative conditions which would lead to their controlled "decomposition". In other words, if two different ligands on copper (say, R^1 and R^2, below) could be coupled to afford R^1–R^2 without significant competition from the symmetrical products (R^1–R^1 and R^2–R^2),[7] then a novel route, e.g., to unsymmetrical biaryls might result.[8] Since there would be no externally added substrate (i.e., the reagent **4** or **5** is the substrate, and vice versa), the resulting ratios of products and overall efficiency of the process could be used to directly compare the H.O. cuprate version (Eq. 2) with the L.O. analog (Eq. 3), again hopefully providing evidence of distinction between the reagents themselves.[9]

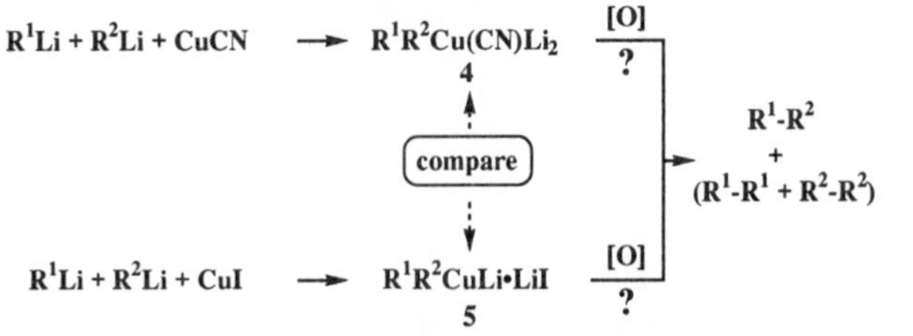

$$R^1Li + R^2Li + CuCN \longrightarrow \underset{\mathbf{4}}{R^1R^2Cu(CN)Li_2} \xrightarrow[?]{[O]} R^1\text{-}R^2 + (R^1\text{-}R^1 + R^2\text{-}R^2) \tag{2}$$

compare

$$R^1Li + R^2Li + CuI \longrightarrow \underset{\mathbf{5}}{R^1R^2CuLi{\cdot}LiI} \xrightarrow[?]{[O]} \tag{3}$$

This, at least, was the game plan as first proposed to Konstantin Siegmann, a then recent postdoctoral arrival from Vananzi's group at the ETH in Zurich. It wasn't long before we realized that cooling solutions of $ArAr'Cu(CN)Li_2$ (**6**), prepared in the usual manner,[10] to temperatures down around –125 °C did nothing to improve an otherwise statistical mix of three products upon treatment with an oxidizing agent. Rather, it was the manner in which the reagent was formed that represented the key to success. Thus, while treatment of diaryl cuprate **6** with $^3O_2$ at –125 °C led to three biaryls in a roughly 1:2:1 ratio, prior generation of Ar′Cu(CN)Li, cooling to –125 °C and then addition of ArLi (ostensibly giving the same species **6′**, at least in terms of stoichiometry), and subsequent oxidation afforded >93% of the unsymmetrical biaryl (Scheme 4).

Our study on the generality of this atypical biaryl coupling, where two anionic (substituted) benzene ligands produce one carbon–carbon bond, is typified by the examples illustrated in Table 1.[11] It is noteworthy that both the initially examined electron-rich aryl moieties employed by Siegmann (entries 1 and 3), and electron-

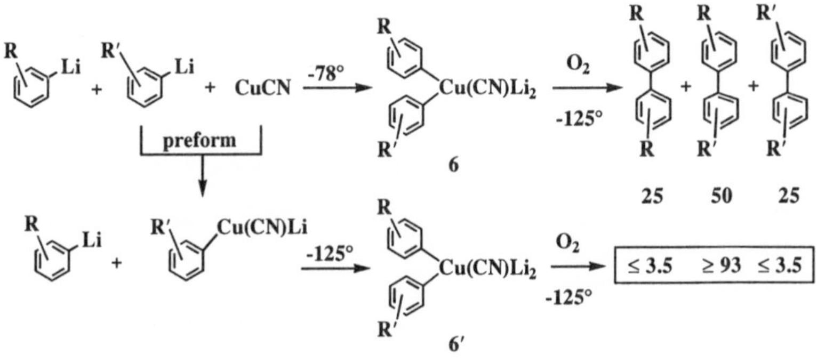

Scheme 4.

Table 1. Representative, $^3O_2$-Induced Couplings of Kinetically Generated, Mixed Diaryl H.O. Cyanocuprates at –125 °C in 2-Methyl-THF

Entry	*Cuprate*	*Coupling Product*	*Yield (%)*
1	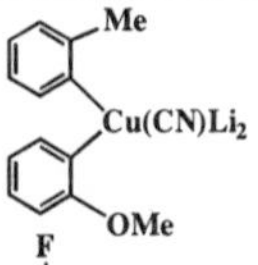	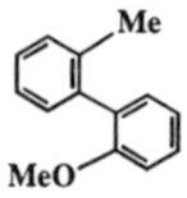	81
2	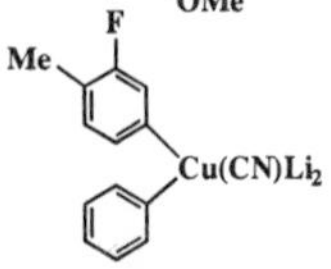	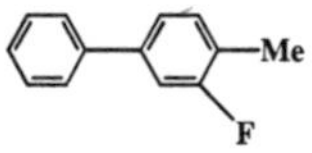	90
3	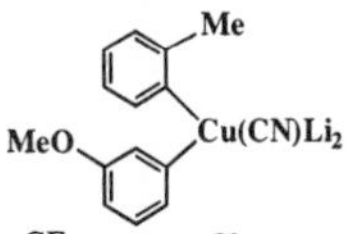	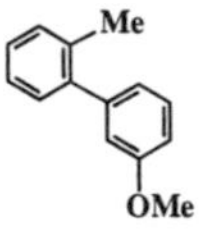	84
4	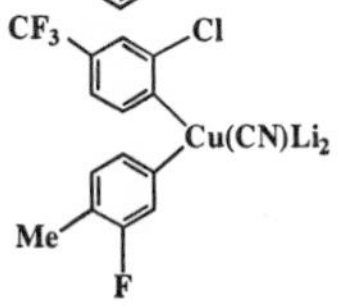	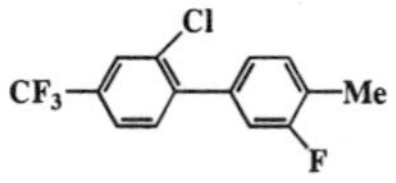	78

poor aromatics later contributed by graduate student Emiliano Garcia (entries 2 and 4), give similar results, both in terms of efficiency and ratios.

Table 1, entry 3:[11] An oven-dried 50-mL three-necked round bottom flask equipped with a 3-way stopcock (for Argon/vacuum), a thermometer adapter fitted with a gas dispersion tube (Aldrich, fritted glass porosity 40–60 m), and a rubber septum was gently flame dried *in vacuo*, then allowed to cool to room temperature under argon, the process being repeated one additional time. Copper cyanide (90 mg, 1 mmol) was then added and the flask was once again gently flame-dried under vacuum and allowed to cool to room temperature under argon.

In another oven and flame-dried 10-mL round bottom flask was added 6 mL of dry 2-methyl THF followed by 0.12 mL (1.0 mmol) of *o*-iodotoluene. The flask was cooled to –78 °C and 2.1 mmol of *t*-BuLi were added dropwise and the resulting solution stirred at this temperature for 30 minutes. Identical treatment of *m*-iodoanisole (0.12 mL, 1.0 mmol) afforded the second aryl-lithium reagent. The flask containing the CuCN slurry was cooled to –78 °C to which was introduced the pre-cooled (–78 °C) toluyllithium and the mixture was allowed to warm to room temperature forming a homogeneous

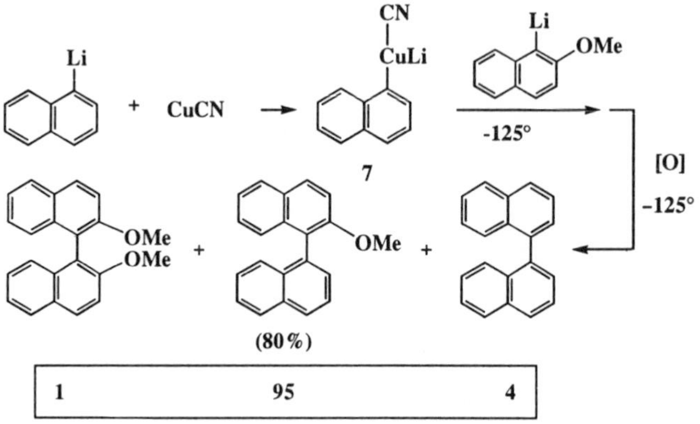

Scheme 5.

solution. The resulting L.O. cyanocuprate was then re-cooled to –125 °C (pentane/N_2) and equilibrated at this temperature for ca. 10 min. The anisyllithium reagent, cooled to –78 °C, was then added dropwise via cannula. After the addition was complete the reaction was stirred at –125 °C for another 10 minutes. TMEDA (0.7 mL) was then added dropwise and the reaction was stirred for another 10 minutes. The argon flow was then stopped and dry O_2 (passed through a –78 °C trap) was bubbled through the reaction mixture whereupon the solution darkened after a few minutes. Oxygen flow was continued for 1 h after which the reaction vessel was briefly evacuated and the argon flow reestablished. The reaction was then quenched with methanol/conc aq. $NaHSO_3$ solution. The reaction mixture was allowed to warm to room temperature, acidified with conc HCl, extracted (Et_2O), and the extracts washed with brine. The organic phase was dried (Na_2SO_4), filtered, and the solvent removed under vacuum. Flash chromatography (silica gel) using hexanes/dichloromethane (9:1) afforded 0.167 g (84%) of the desired product as a colorless oil.

The Garcia contribution to the project did not end at this stage; rather, extension of this procedure to the binaphthyl manifold was investigated next. The location of eventual lithiation of this ring system was initially considered, since the 1-position would likely be the more sterically demanding of the two alternatives presumably due to the in plane *peri*-hydrogen at C8. If our preliminary indications with couplings of, e.g., an *o*-anisyl with an *o*-toluyl group (cf. Table 1, entry 1) were truly reflective of the tolerance of steric factors, lithiation of naphthalene at either C1 or C2 should be of no consequence. This is indeed the case, for Garcia found that treatment of the H.O. cuprate, prepared from the L.O. cyanocuprate, e.g., **7**, and 1-lithio-2-methoxynaphthalene, under our usual low temperature conditions[11]

Table 2. Binaphthyl Couplings of Kinetic Cyanocuprates

Entry	*Cuprate*	*Coupling Product*	*Yield (%)*
1	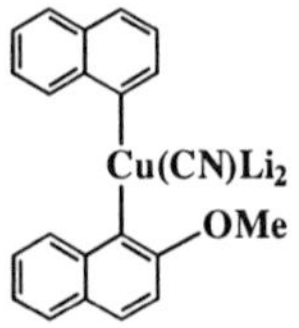 $Cu(CN)Li_2$, OMe	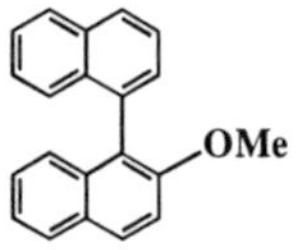 OMe	80
2	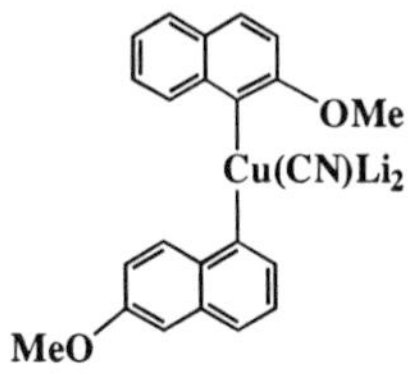 OMe, $Cu(CN)Li_2$, MeO	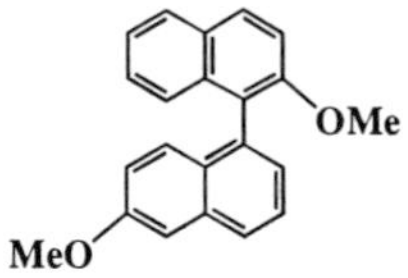 OMe, MeO	81
3	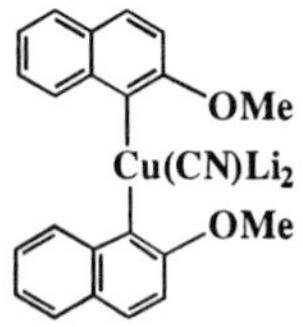 OMe, $Cu(CN)Li_2$, OMe	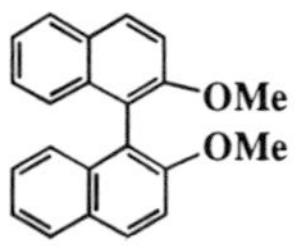 OMe, OMe	81

affords the desired unsymmetrical binaphthyl in 80% yield, with the initially obtained ratio as illustrated above (Scheme 5).[12] A few additional examples have been carried out as well (Table 2), each of which deserves comment.[13] Those involving the 1-metallo-2-methoxynaphthalene ligand (entries 1,2) represent especially noteworthy examples in that they each possess what is tantamount to a 2,6-disubstituted aryl group, a substitution pattern notoriously problematic in traditional Pd(0)-based couplings.[14] The symmetrical system (entry 3) represents the extreme case of *two* highly hindered ligands of the 2,6-disubstituted variety being coupled with equal facility. Potentially more significant would be the demonstration, given the judicious choice of chiral auxiliary, that the intramolecular version could lead to the 2,2′-binaphthyl series of predetermined stereochemistry (Scheme 6).[16]

The prospects for branching beyond benzene and benzene-like nuclei to other aromatics seemed bright. A test case by Garcia employing the combination of a 1-lithionaphthalene and 2-lithiothiophene resulted in the controlled coupling to biaryl **8** in good yield.[13a] With the torch then passed to postdoctoral student Frank Kayser, a recent recruit from the Reetz group while in Marburg, heteroaromatic couplings are now being investigated in greater detail. For example, a benzthiazole has been attached to a thiophene group under our usual conditions (Scheme 7).[13b]

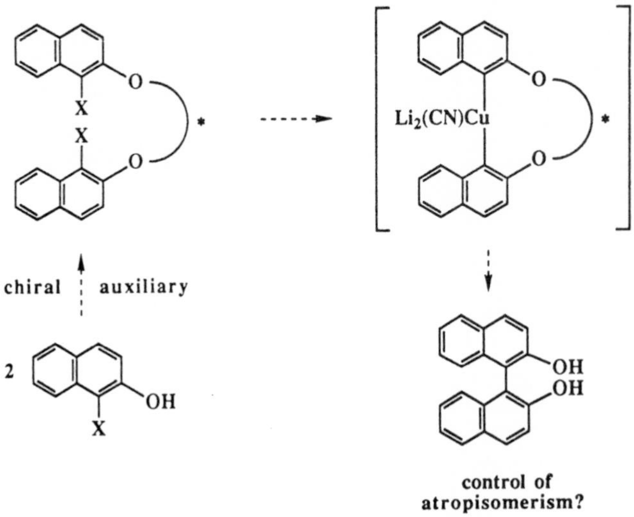
X
X
O
O
Li2(CN)Cu
chiral
auxiliary
2
OH
OH
OH
control of
atropisomerism?

Scheme 6.

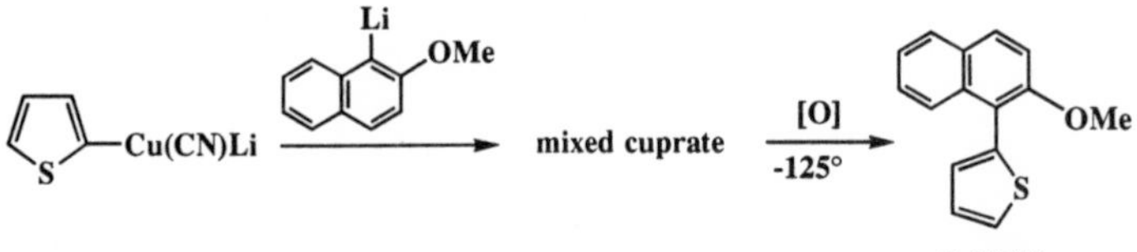
Cu(CN)Li
Li
OMe
mixed cuprate
[O]
-125°
OMe
S
S
8 (80%)

While the synthetic front was advancing, we felt obligated to once again subscribe to our group's philosophy of using physical techniques (mainly NMR spectroscopy) to attempt to understand our new synthetic methodology.[16] Recording NMR spectra in the –120 to –130 °C temperature range is beyond the specifications of our General Electric GN500 probe, and so other arrangements had to be made. Fortunately, just around this time, a conference in Sweden on recent developments in organocopper chemistry was taking shape, spearheaded by Martin Nilsson (Chalmers University). With independent collaborators Christina Ullenius and Thomas Olsson, this Swedish school had established itself as world leaders in very low temperature NMR studies of copper reagents,[17] and so it was off to Sweden for Emiliano, an assignment he accepted with a smile. Our goal, of course, was

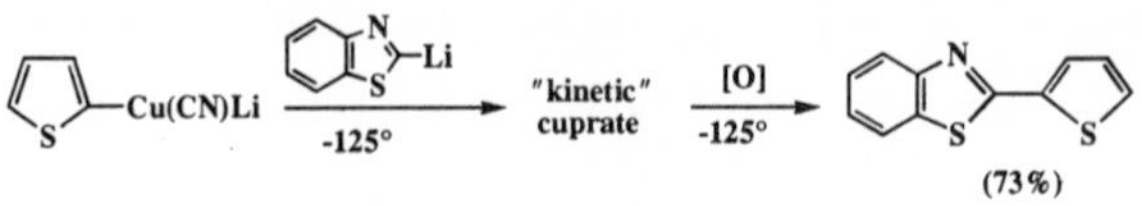
Cu(CN)Li
Li
-125°
"kinetic"
cuprate
[O]
-125°
(73%)

Scheme 7.

straightforward: to demonstrate that a H.O. diaryl cuprate, Ar′ArCu(CN)Li_2, prepared under our "kinetic" conditions gives rise to a ^{13}C NMR spectrum which is neither maintained upon warming to −78 °C or above, nor realizable upon recooling to −130 °C. Just one or two, perhaps three "simple" experiments of this sort would be all that is needed to establish the concept of a "kinetic" reagent, in full harmony with the synthetic work.

Our choice of cuprate was that derived from phenyllithium, 2-lithiothiophene, and CuCN [i.e., Ph(2-Th)Cu(CN)Li_2]. The expectation was that a nine-line pattern in the ^{13}C NMR spectrum at −130 °C would be seen; four lines due to each aryl moiety plus the CN ligand. In recording the spectrum for this reagent, Garcia and Olsson in fact did see nine peaks, as illustrated in Figure 3a. Upon warming this sample, the ligands should reorganize, giving rise to a more complicated spectrum. As shown in Figure 3b, this is precisely what happens. The key portion of the experiment, however, remained. That is, upon recooling this same sample, all that had to happen to the spectrum was . . . nothing! If it just remained as is, given the "thermodynamic-like" conditions to which it had already been exposed, the *prima facie* evidence that a true kinetic species had been found at −130 °C would be in hand. Well, enter Mr. Murphy! Incredibly, and to this day, inexplicably, the spectrum observed at −45 °C (Figure 3b) reverted completely to that seen initially at −130 °C (Figure 3c): bizarre, but true, and not just for this particular cuprate. Two others were studied, [(Ph)(*p*-MeO-Ph)Cu(CN)Li_2 and (*p*-F-Ph)(*p*-MeO-Ph)Cu(CN)Li_2], with the same results observed. No wonder assignments which combine synthetic methods and low temperature NMR work in our group have oftentimes been dubbed "projects from hell". I can see why, but in reality, with enough experimentation, it has usually been possible to gain sufficient insight and ultimately make the correlation—at least this is what I keep telling myself! This time, though, the synthetic side looks very much at odds with the spectral data.

Whatever the explanation behind these ^{13}C NMR spectral data, it should be recalled that the impetus, in part, behind this discovery concerned the eventual comparison between the oxidations of Ar′ArCu(CN)Li_2 and a copper halide-derived diaryl cuprate Ar′ArCuLi. Siegmann also carried out this study using cuprates prepared from both CuI solubilized in THF with LiBr, and also by LiI. Oxidative treatment of these analogous L.O. reagents afforded the results summarized in Table 3.[18] Not only are the ratios (normalized to 100%) nowhere close to those obtained via H.O. cyanocuprates, but the desired unsymmetrical biaryl is not even the major isomer produced under these standardized conditions. We concluded, therefore, that the bottom line is obvious: $R_2Cu(CN)Li_2 \neq R_2CuLi{\cdot}LiCN$.

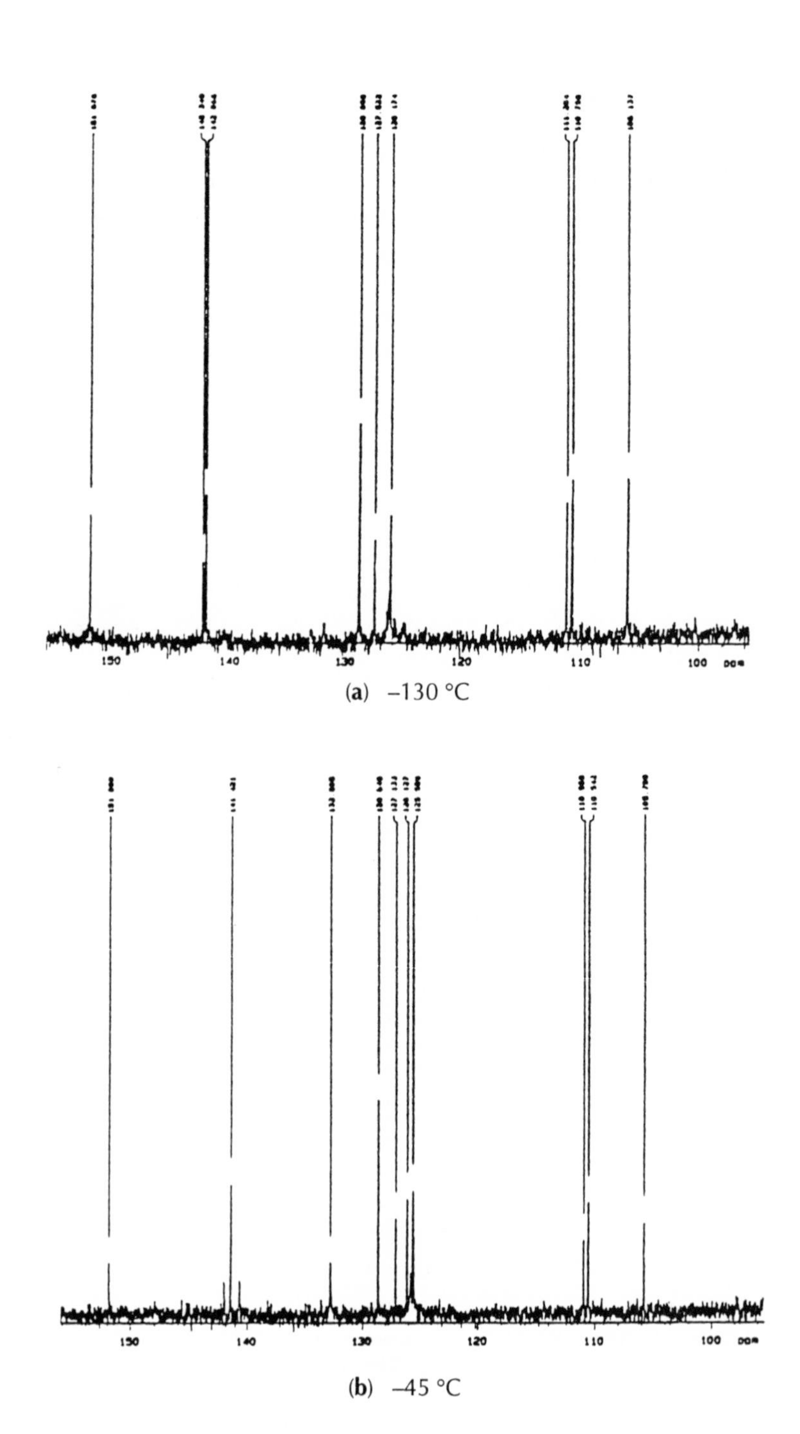

Figure 3. ^{13}C NMR spectra of $Ph(2\text{-}Th)Cu(CN)Li_2$ in 2-Me THF.

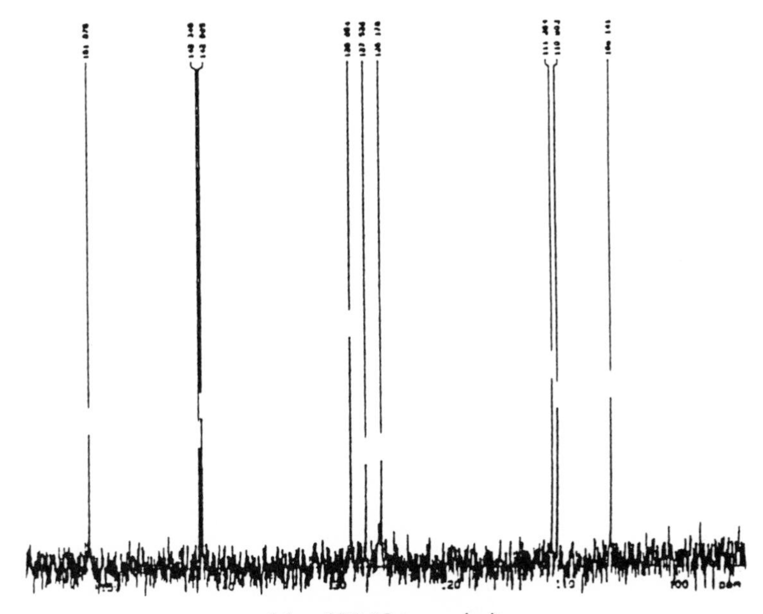

(c) −130 °C (recooled)
Figure 3. (Continued)

Table 3. Comparison Oxidation of Lower Order Diaryl Cuprates

Li / OMe (aryllithium) + Li (aryllithium) + CuX $\xrightarrow[\text{conditions}]{\text{'kinetic'}}$ ArAr'CuLi $\xrightarrow[\text{-125°}]{^3O_2}$ biaryls

	Biaryl Products (%)		
		OMe	OMe, MeO
CuI•LiBr:	**66**	**19**	**15**
CuI•LiI:	**56**	**24**	**20**
CuCN	**3.5**	**93**	**3.5**

⇧ **desired, unsymmetrical biaryl**

III. TRANSMETALLATIONS OF SI AND SN INTERMEDIATES WITH H.O. CUPRATES

While the scope of the biaryl couplings continues to be elucidated, that particular project represents only a small fraction of the organometallic-based methodology of interest to our group. Most of the emphasis has swung over to transmetallation phenomena, viewed schematically below, involving silicon, tin, aluminum, and a heavy dose of ongoing organozirconium chemistry. Our guideline behind the choice of metals requires that each have its own characteristic synthetic chemistry which proceeds through an organometallic intermediate susceptible to ligand exchange with a source of Cu(I) [e.g., a trivial H.O. cuprate (Figure 4)].

Transmetallations in the silicon and tin manifold have already been shown to occur on Si–Sn,[19] Sn–Sn,[20] and Sn–H[21] bonds due to their susceptibility to attack by $R_2Cu(CN)Li_2$, thereby yielding highly useful mixed silyl- or stannylcuprates. The Si–Sn bond is particularly interesting in that, depending upon the steric demands of the groups on each metal in $R_3Sn–SiR'_3$, either $(R'_3Si)RCu(CN)Li_2$ or $(R_3Sn)R'Cu(CN)Li_2$ can be realized (Scheme 8).[19] Thus, with the trimethylstannyl moiety on a bulky, trisubstituted silicon, $R'_2Cu(CN)Li_2$ (R′ = Me, *n*-Bu) attacks tin to give rise to a mixed alkyl silyl cuprate **9** which selectively transfers the silyl ligand (as shown by Fleming)[22a–c] to an electrophile (cf. Eq. 4). Alternatively, with a trimethylsilyl group attached to a bulky tributylstannyl residue, the same H.O. cuprate now attacks at silicon to form the corresponding mixed stannyl cuprate **10**, which also is well behaved and selectively delivers the Bu_3Sn ligand to the electrophile (cf. Eq 5).[19]

3-Methyl-3-tributylstannylcyclopentan-1-one (Eq. 5).[19] To a two-neck flask equipped with a three-way stopcock that is connected to an argon tank and a vacuum pump is added CuCN (0.0745 g, 0.832 mmol). It is gently flame dried under vacuum followed by an argon purge; this evacuation and argon reentry process being repeated three times. Addition of THF (1.0 mL) to the CuCN produces a slurry which is cooled to –78 °C at which time MeLi in

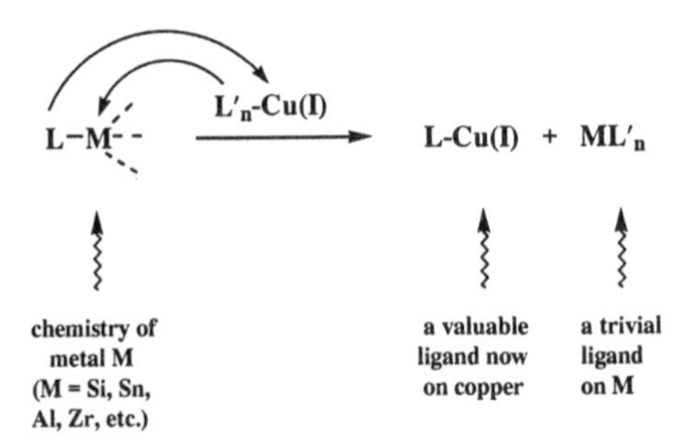

Figure 4.

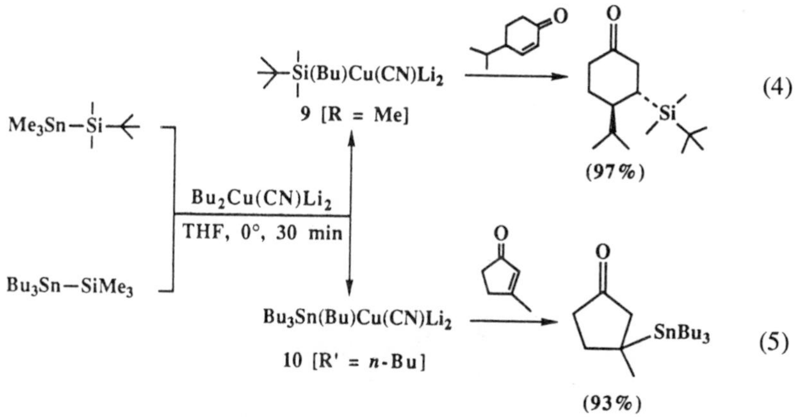

(4)

(5)

ether (1.24 mL, 1.74 mmol) is added dropwise. After alkyllithium addition, the reaction is allowed to warm via removal of the ice bath and stirred until a pale yellow homogeneous solution results (approximately 3–5 min). While warming, it is important that all the CuCN be dissolved which may require swirling of the flask. The flask is recooled to –78 °C and (trimethylsilyl)-tributylstannane (0.63 mL, 1.82 mmol) is added via syringe. The reaction is warmed to 0 °C for 30 min to invoke transmetallation and then recooled to –78 °C. The 3-methylcyclopentenone (0.06 mL, 0.606 mmol) is added neat via syringe to the reaction and this is stirred for 10 min. The reaction is quenched at –78 °C via addition of 10% conc. NH_4OH/saturated aqueous NH_4Cl and allowed to warm to room temperature. The mixture is extracted with Et_2O and the combined extracts are washed with brine and dried over Na_2SO_4. Evaporation *in vacuo* and chromatography (silica gel; 1% TEA, 19% Et_2O, 80% hexanes) yields the stannane (0.218 g, 93%) as a clear oil.

What we have not accomplished as yet is a transmetallation route to the simplest mixed silyl cuprate $(Me_3Si)RCu(CN)Li_2$, **11**. The homocuprate (**11**, R = Me_3Si) is

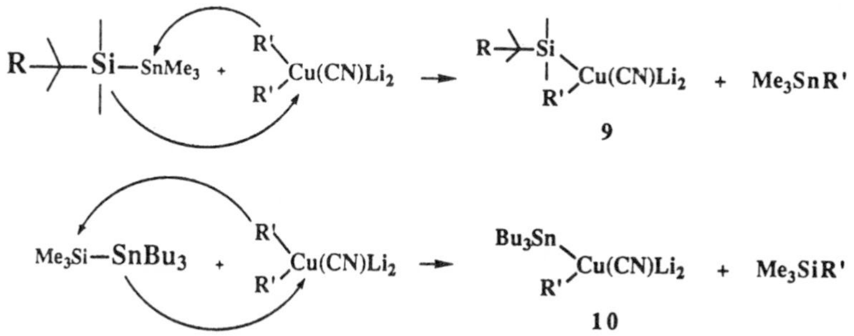

Scheme 8.

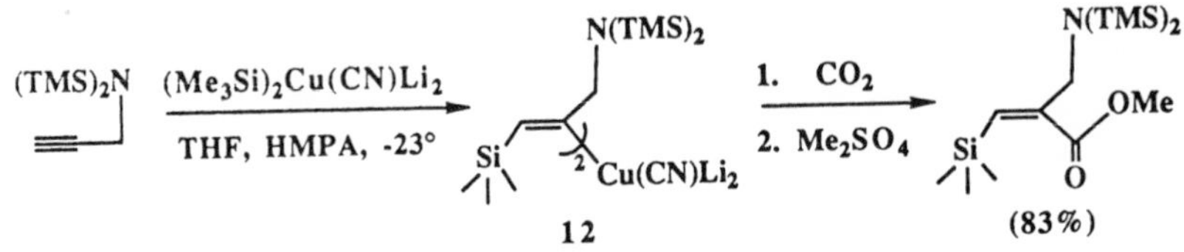

(TMS)$_2$N ... $\xrightarrow[\text{THF, HMPA, -23°}]{(Me_3Si)_2Cu(CN)Li_2}$ 12 [N(TMS)$_2$; Si; Cu(CN)Li$_2$] $\xrightarrow[\text{2. } Me_2SO_4]{\text{1. } CO_2}$ N(TMS)$_2$, OMe, Si, O (83%)

Scheme 9.

normally made from $2Me_3SiLi$ and CuCN,[23] and has been found useful (e.g., by the Ricci group in silylcuprations of propargyl amines). The regiochemistry of these additions follows that observed with the analogous $PhMe_2SiLi$-derived cuprate promoted by Fleming,[22b] but is opposite to that normally found in carbocuprations as advanced by Corriu.[24] The intermediate vinylic cuprates **12** can be trapped by various coupling agents such as X_2, H_2O, allylic/vinylic halides, CO_2, and acid chlorides (e.g., see Scheme 9).[25]

Formation of **12** *followed by alkylation: standard procedure.*[25] **Silylcupration**. To a solution of (trimethylsilyl)lithium (1.50 mmol) in THF, prepared according to the method of Hudrlik,[26] was added 70 mg (0.80 mmol) of CuCN at –23 °C. After stirring for 20 min, 300 mg (1.50 mmol) of the propargyl amine was added with a syringe and the temperature held at –23 °C for 30 min. **Alkylation**. The appropriate alkyl halide (1.50 mmol) was added at –23 °C to the solution of the vinylcopper reagent **12** and then the reaction mixture was allowed to warm to room temperature overnight. After hydrolytic workup and bulb-to-bulb distillation, the product, e.g., of reaction with allyl bromide, was isolated (bp 100 °C at 5 mm Hg, 85% by GC) as a colorless liquid.

During a visit to the ETH in 1989, I was told by Seebach and Amberg about their most unusual and unexpected discovery involving a net β-trisubstituted silicon delivery to an α,β-unsaturated lactone of particular interest to their group. When a 1,3-dioxan-4-one such as **13** is treated with $Bu_2Cu(CN)Li_2$ in the presence of a silyl chloride, R_3SiCl, the product obtained results from the 1,4-addition, formally, of

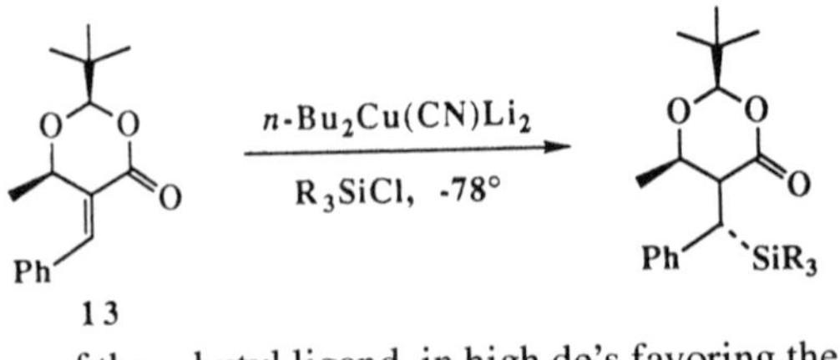

13 [O, O, O, Ph] $\xrightarrow[R_3SiCl,\ -78°]{n\text{-}Bu_2Cu(CN)Li_2}$ [O, O, O, Ph, SiR_3] (6)

"R_3Si^-" rather than of the *n*-butyl ligand, in high de's favoring the 1′S isomer (Eq. 6).[27] The silyl group may be Me_3Si, Et_3Si, *t*-$BuMe_2Si$, or Me_2PhSi, with yields usually above 70%. The process turns out to be general for enones bearing a phenyl moiety at the β-carbon, as educts **14–17** all afford similar adducts. It is not at all

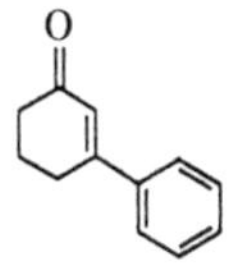

14 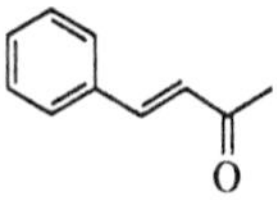 **15** 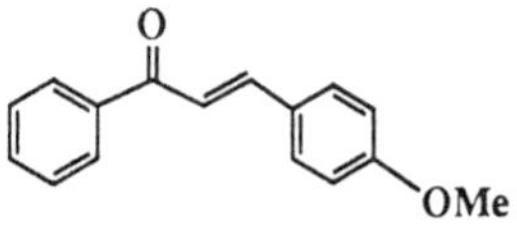 **16** 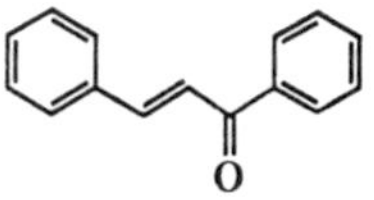 **17**

clear how such an inversion in character at silicon is brought about by the dibutyl H.O. cyanocuprate at –78 °C, although NMR studies on this combination have been done previously in our group and may provide some early clues. In essence, we found that mixing $Me_2Cu(CN)Li_2$ with TMSCl at temperatures as low as –100 °C leads rather quickly to the L.O. species Me_2CuLi, due to abstraction of the nitrile ligand by Me_3SiCl, thereby generating an equivalent of Me_3SiCN.[16c] Because these Swiss researchers employed >1 equivalent of TMSCl versus H.O. cuprate,[27a] both TMS-CN (1 eq) and residual TMSCl were present during this reaction (Scheme 10). Just what the role of TMSCN is in this chemistry must await further experimentation, now underway in our laboratories.

To arrive at $Me_3Sn–SiMe_2R$ (R = *t*-Bu, thexyl), it was necessary for Debbie Reuter to develop a route to Me_3SnH,[28] which upon lithiation with LDA[29] and quenching with RMe_2SiCl affords the desired silylstannanes as fairly stable, water-white liquids (Scheme 11).

It was appreciated, however, that the association of highly toxic Me_3SnH[30] with almost any procedure, even where its isolation is unnecessary,[28] would detract from the usefulness of this methodology. This was especially important in the preparation of trimethylstannyl cuprates, where the choices were limited to Me_3SnH and the expensive $Me_3Sn–SnMe_3$ (Scheme 12). The latter material, nonetheless, had been reported by Oehlschlager[20] to readily transmetallate with H.O. cuprates. An alter-

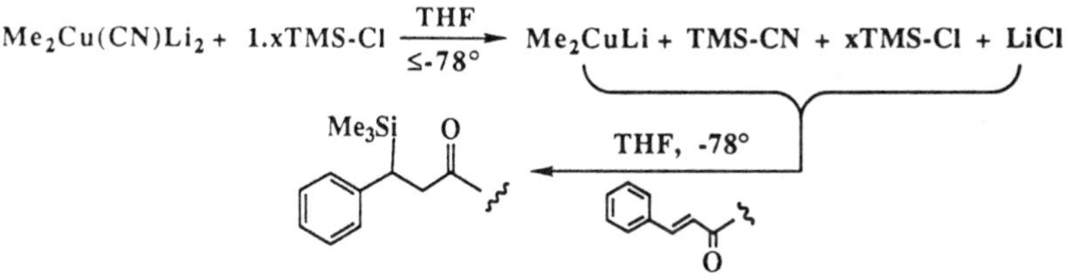

$Me_2Cu(CN)Li_2$ + 1.xTMS-Cl $\xrightarrow[\leq -78°]{THF}$ Me_2CuLi + TMS-CN + xTMS-Cl + LiCl

THF, -78°

Scheme 10.

Me_3SnCl $\xrightarrow[\text{triglyme, } 60\text{-}68°]{LAH}$ [Me_3SnH] $\xrightarrow[\text{THF, } -78°]{LDA}$ Me_3SnLi $\xrightarrow{RMe_2SiCl}$ $Me_3Sn—SiMe_2R$

[R = *t*-Bu; R = thexyl]

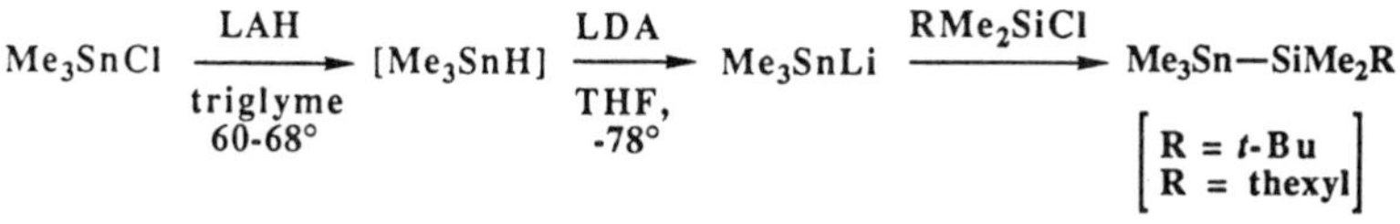

Scheme 11.

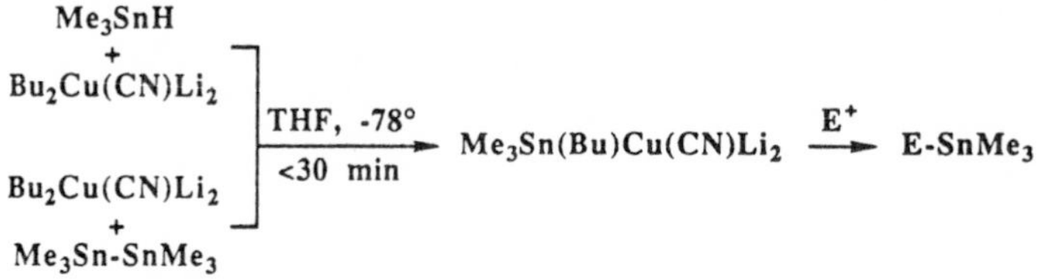

$$\left.\begin{array}{c} Me_3SnH \\ + \\ Bu_2Cu(CN)Li_2 \\ \\ Bu_2Cu(CN)Li_2 \\ + \\ Me_3Sn\text{-}SnMe_3 \end{array}\right] \xrightarrow[<30\ min]{THF,\ -78°} Me_3Sn(Bu)Cu(CN)Li_2 \xrightarrow{E^+} E\text{-}SnMe_3$$

Scheme 12.

native route was devised by Sharma where the source of Me_3Sn would be the more tolerable (commercially available) Me_3SnCl.[31a] In addition, Reuter had adapted an Argus Chemical Company patent[31b] to prepare Me_3SnCl in quantity from the bulk chemical Me_2SnCl_2, generously supplied to us by Otto Loeffler and Mike Fisch, both from this company. Thus, from inexpensive $Me_3Si–SiMe_3$, the Still protocol gives rise to Me_3SiLi,[23b] which is trapped *in situ* with Me_3SnCl (Scheme 13). The resulting $Me_3Si–SnMe_3$ is highly susceptible to attack by $R_2Cu(CN)Li_2$ at silicon, as anticipated based on earlier, seminal work of Chenard.[32] Introduction of the electrophile to the (presumed) cuprate $Me_3Sn(R)Cu(CN)Li_2$ resulting from transmetallation affords products incorporating the desired Me_3Sn group (Scheme 13). Noteworthy are the points that: (1) the entire sequence can be effected in a single flask operation; (2) the $Me_3Sn–SiMe_3$ solution prepared in THF/HMPA is stable at room temperature under argon for months; and (3) 100% of the tin utilized in cuprate formation is employed in the reaction as the selectivity of transfer for Me_3Sn over R is very high, to be expected based on a contribution from the Piers group (where R in the mixed cuprate is 2-thienyl).[33]

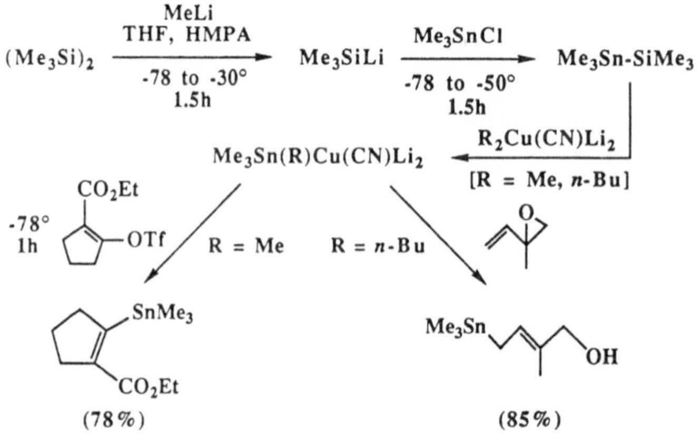

$$(Me_3Si)_2 \xrightarrow[-78\ to\ -30°,\ 1.5h]{MeLi,\ THF,\ HMPA} Me_3SiLi \xrightarrow[-78\ to\ -50°,\ 1.5h]{Me_3SnCl} Me_3Sn\text{-}SiMe_3$$

$$Me_3Sn\text{-}SiMe_3 \xrightarrow[[R\ =\ Me,\ n\text{-}Bu]]{R_2Cu(CN)Li_2} Me_3Sn(R)Cu(CN)Li_2$$

R = Me: 2-(ethoxycarbonyl)cyclopent-1-enyl triflate (CO_2Et, OTf), -78°, 1h → 2-($SnMe_3$)cyclopentene-1-CO_2Et (78%)

R = n-Bu: isoprene monoxide → Me_3Sn-allylic alcohol (OH) (85%)

Scheme 13.

Stannylcupration of 3-butyne-1-ol (homopropargyl alcohol).[31] Ethereal methyllithium (6.6 mL, 1.5 M) was added to a solution of hexamethyldisilane (2.0 mL, 10.0 mmol) in 24 mL of THF/HMPA (3:1 by volume) at –78 °C under argon. The resulting deep red solution was stirred for 1 h while allowing it to warm to –30 °C. The reaction was cooled to –78 °C after which Me_3SnCl (1.99g, 10.0 mmol) in 2 mL of THF was added. The reaction mixture was further stirred for 1.5 h while warming to –50 °C. In a separate flask, $Bu_2Cu(CN)Li_2$ (10.0 mmol, prepared from 8.7 mL of 2.3 M *n*-BuLi and 0.89 g of CuCN) in 10 mL of THF was prepared at –45 °C. After stirring for 0.5 h, this solution was transferred via cannula to a solution of $Me_3SnSiMe_3$ maintained at –78 °C. The resulting lemon yellow solution was warmed to –50 °C and stirred for 1 h to ensure complete transmetallation. 3-Butyn-1-ol (0.63g, 9.0 mmol) was then added neat via syringe followed by 5 mL of MeOH. The reaction immediately turned red in color. After 30 min the bath was removed and the solution warmed to room temperature. Usual extractive workup followed by chromatography on silica gel (hexanes:ethyl acetate, 8:1 as eluant) gave 1.56 g (74%) of 4-hydroxy-2-trimethylstannyl-1-butene and 0.17g (8.0%) of 4-hydroxy-1-trimethylstannyl-1-butene. GC analysis revealed a purity of >95% for both isomers.

Higher order stannylcuprates $(R_3Sn)_2Cu(CN)Li_2$ and $R_3Sn(R')Cu(CN)Li_2$ have been used by other groups of late to effect stannylcupration reactions of acetylenes,[34] enynes,[35] propargyl amines,[36] and allenes.[37] Treatment of acetylene itself with $Bu_3Sn(Me)Cu(CN)Li_2$ at –78 °C leads to the *syn* addition product **18** which undergoes a variety of subsequent cuprate couplings. Pulido and Fleming et. al. have introduced electrophiles such as TMSCl, Bu_3SnCl, ethylene oxide, and cyclohexenone to give excellent yields of the corresponding vinylstannanes (Scheme 14).[34a]

Both the Normant[38] and Quintard[39] groups have added $Bu_3Sn(Bu)Cu(CN)Li_2$ across acetylenic acetal **19** with the indicated regiochemistry shown below. Protic quenching affords the vinyl stannane **21** in excellent yields. Intermediate **20** can be

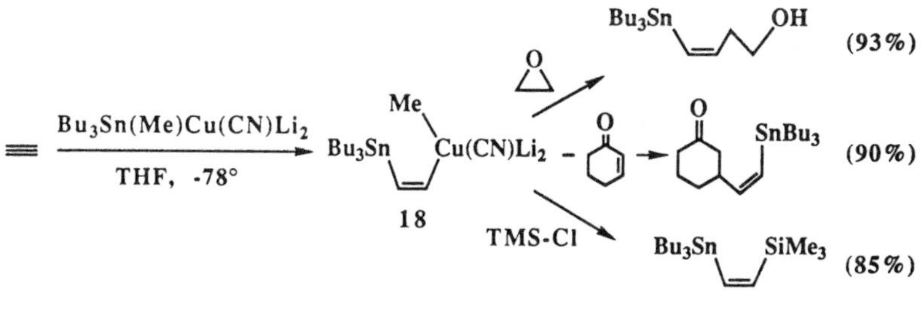

Scheme 14.

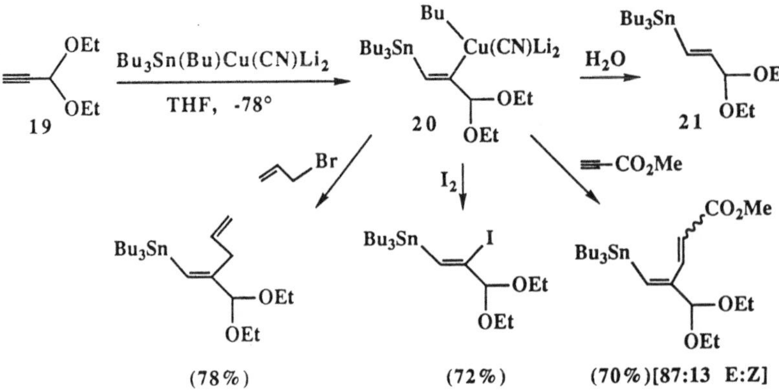

Scheme 15.

converted to products of further complexity upon treatment with other electrophiles such as I_2, methyl propiolate, and allyl bromide (Scheme 15).

Conjugated enynes have been examined by Aksela and Oehlschlager,[35] their reactions with homo- or mixed H.O. stannylcuprates again raising questions of regioselectivity. Both hydrocarbon cases, **22** and **23**, could be controlled to give >97% of the 1-stannylbutadienes **24** and **25**, respectively (Scheme 16). Enyne **26**, however, gave an 82:18 ratio of products (Scheme 17). Interestingly, the relative

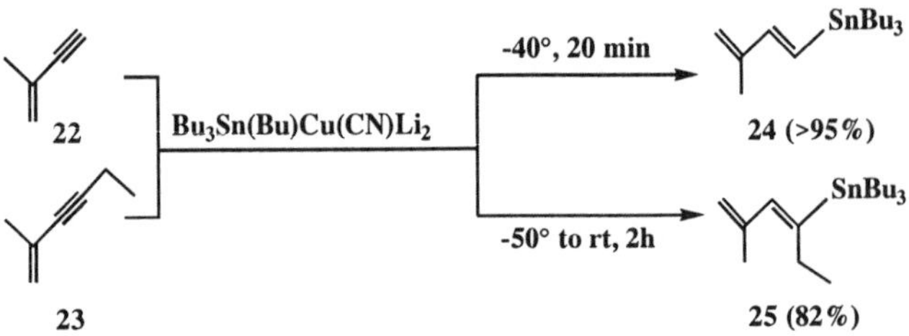

Scheme 16.

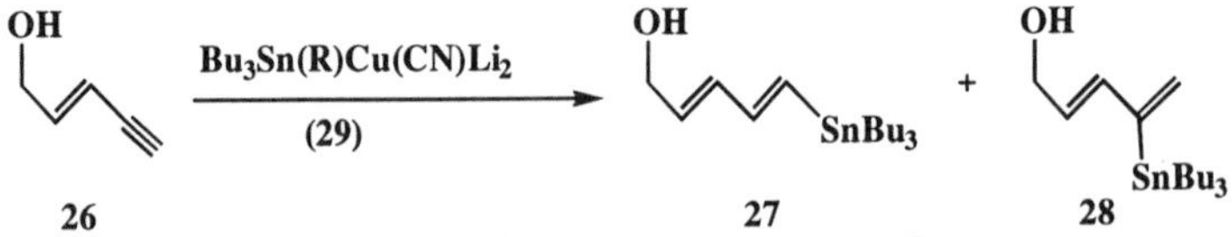

(R) in x	Conditions	27	28
n-Bu	-50°, 30 min	82	18
t-Bu	-60°, 3h	19	81

Scheme 17.

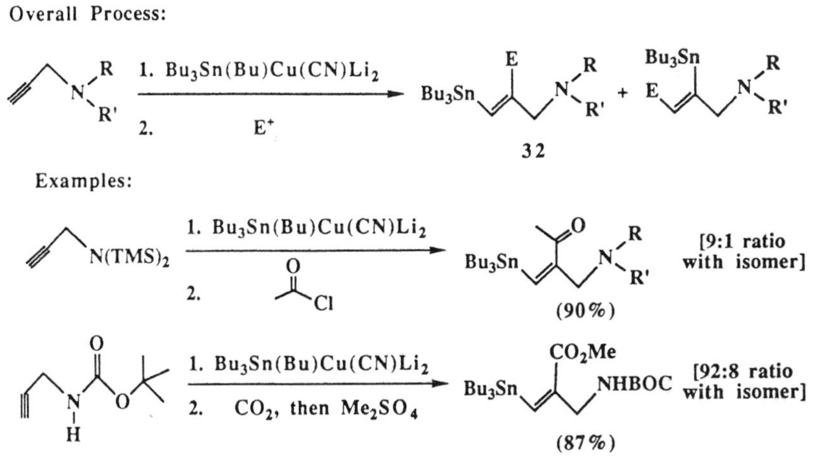

Overall Process:

1. $Bu_3Sn(Bu)Cu(CN)Li_2$
2. E^+

Bu_3Sn E R N R' **32** + Bu_3Sn E R N R'

Examples:

$N(TMS)_2$ 1. $Bu_3Sn(Bu)Cu(CN)Li_2$ 2. Cl → Bu_3Sn O R N R' (90%) [9:1 ratio with isomer]

O N H O 1. $Bu_3Sn(Bu)Cu(CN)Li_2$ 2. CO_2, then Me_2SO_4 → Bu_3Sn CO_2Me NHBOC (87%) [92:8 ratio with isomer]

Scheme 18.

amounts of isomers **27** and **28** could be completely turned around using a bulkier cuprate (**29**, R = *t*-Bu).[35]

As a follow-up study to silylcuprations of propargyl amines (*vide supra*),[25] Ricci, et. al. have addressed related stannylcupration–electrophilic trapping reactions on both the bis-TMS and N-BOC derivatives **30** and **31**, respectively.[36] Additions of $Bu_3Sn(Bu)Cu(CN)Li_2$ to **30** and **31** are usually highly regioselective, again placing the Bu_3Sn moiety at the terminal locale. Representative manipulations are shown in Scheme 18. The resulting vinylstannanes bearing the allylic amino group in protected form (i.e., **32**) are excellent "building blocks" suitable for further processing, in particular via palladium couplings.[40] One case in point surrounds the generation of the dienyl amine grouping (cf. conversion of **33** to **34**, Scheme 19)[41] found in a number of natural products (e.g., virginiamycin, **35**;[42] Figure 5).

$N(TMS)_2$ **30** O N H O **31**

BOC N H 1. 29 [R=*n*-Bu] 2. MeI → Bu_3Sn Me HNBOC **33** → Ph I, $PdCl_2(CH_3CN)_2$ (5 mol %) → Ph Me HNBOC **34** (89%)

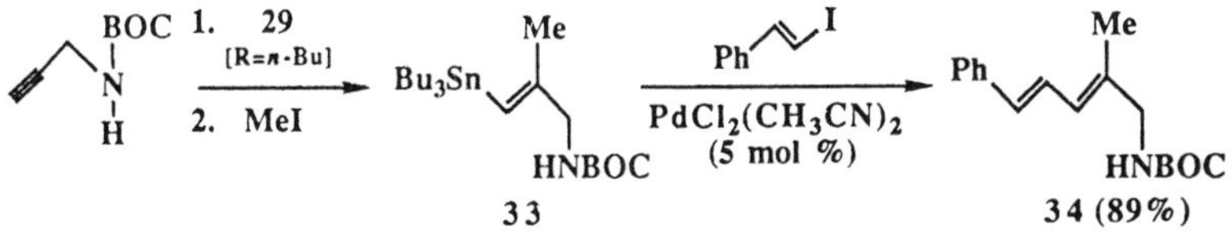

Scheme 19.

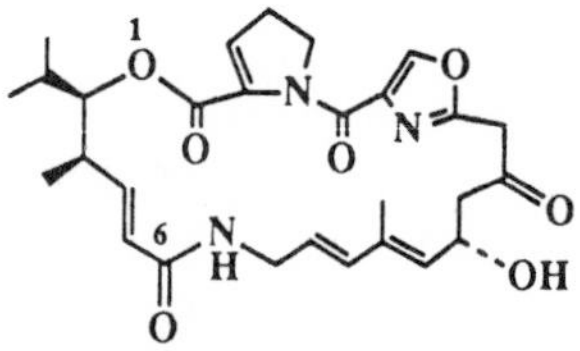

35, virginiamycin M_1

Figure 5.

Stannylcuprations of allenes have also been studied by Pulido and Fleming using the homocuprate $(Bu_3Sn)_2Cu(CN)Li_2$, **36**.[37] While the substitution pattern around the allene can determine whether vinylic or allylic stannanes result from these additions, allene itself presents a very informative example of the events which are occurring between substrate and reagent. Its exposure to **36** at –100 °C gives the kinetically formed vinylcuprate **37**, which can be quenched at this temperature (e.g., with Br_2 or acetyl chloride) to give **38**. Warming to 0 °C over one hour, however, gives rise to the thermodynamically preferred allylic cuprate **39**, which reacts with several electrophiles to produce substituted vinylstannanes **40** (Scheme 20).

Both silyl- and stannylcuprates have been looked at spectroscopically. Prior to Sharma's arrival at UCSB, she had carried out low temperature NMR studies on a number of homocuprates[43] and mixed reagents,[44] of both the L.O. and H.O. variety, as part of her Ph.D. thesis work with Cam Oehlschlager at Simon Fraser University. This Canadian group found that silyl cuprates, such as those derived from $PhMe_2SiLi$ plus a Cu(I) salt, do not necessarily follow the same patterns noted with alkyl cuprates.[16a] Thus, from ^1H, ^7Li, ^{13}C, and ^{29}Si experiments in THF, the

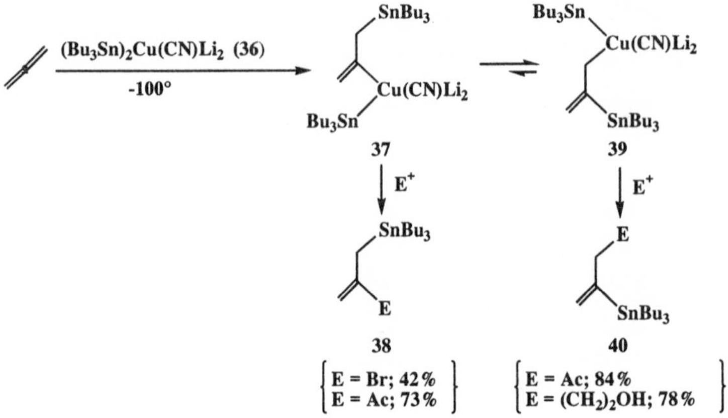

$(Bu_3Sn)_2Cu(CN)Li_2$ (36), -100°

37: $SnBu_3$, $Cu(CN)Li_2$, Bu_3Sn

39: Bu_3Sn, $Cu(CN)Li_2$, $SnBu_3$

E^+

38: $SnBu_3$, E — {E = Br; 42%, E = Ac; 73%}

40: E, $SnBu_3$ — {E = Ac; 84%, E = $(CH_2)_2OH$; 78%}

Scheme 20.

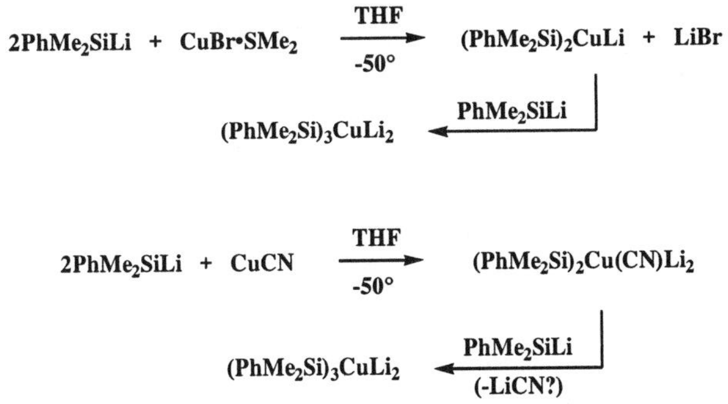

$$2PhMe_2SiLi + CuBr{\cdot}SMe_2 \xrightarrow[-50°]{THF} (PhMe_2Si)_2CuLi + LiBr$$

$$(PhMe_2Si)_2CuLi \xrightarrow{PhMe_2SiLi} (PhMe_2Si)_3CuLi_2$$

$$2PhMe_2SiLi + CuCN \xrightarrow[-50°]{THF} (PhMe_2Si)_2Cu(CN)Li_2$$

$$(PhMe_2Si)_2Cu(CN)Li_2 \xrightarrow[(-LiCN?)]{PhMe_2SiLi} (PhMe_2Si)_3CuLi_2$$

Scheme 21.

combination of $2PhMe_2SiLi$ + CuX (X = Br, I) gives rise to a L.O. cuprate which upon further exposure to $PhMe_2SiLi$ converts to the higher order species $(PhMe_2Si)_3CuLi_2$. Such is not the case for the 3RLi + CuX ratio (R = alkyl) in this medium, although Ashby has reported[45] its existence, observable at −136 °C, in Me_2O, and Bertz has observed Ph_3CuLi_2 in dimethyl sulfide.[46] Remarkably, while the $2PhMe_2SiLi$ + CuCN mixture gives the H.O. cyanocuprate as expected, addition of a third equivalent of the silyllithium gives the same $(PhMe_2Si)_3CuLi_2$, implying breakage of the Cu–CN bond (Scheme 21).[43]

Spectral data on the more highly mixed cuprate $PhMe_2Si(Me)Cu(CN)Li_2$ (**41**), taken by Singer and Oehlschlager,[47] suggest that one major species is present in solution.[44] Upon addition of another equivalent of MeLi, however, the spectra become more complex, pointing to an equilibrium mixture of species. Reagent **41** is synthetically attractive in that it is highly selective for release of the silyl ligand in Michael additions, leaving MeCu(CN)Li behind which generates methane as the reagent-derived organic by-product, thereby simplifying workup (Eq. 7). The corresponding homocuprate oftentimes leads to oxidation ($PhMe_2SiSiMe_2Ph$, $PhMe_2SiOH$) and reduction ($PhMe_2SiH$) products which may require chromatographic separation,[47] as originally noted by Fleming.[22a–c]

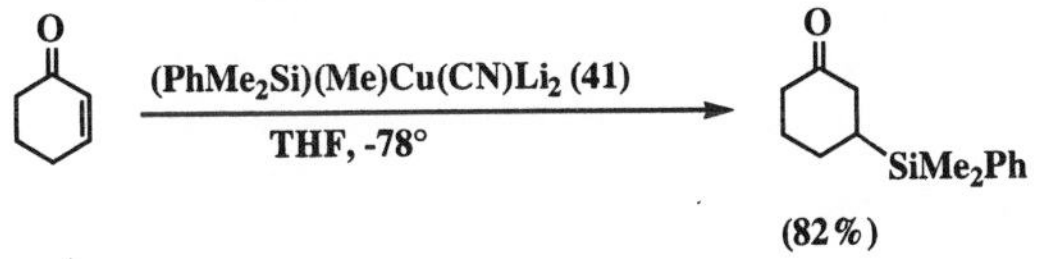

$$\text{cyclohexenone} \xrightarrow[\text{THF, -78°}]{(PhMe_2Si)(Me)Cu(CN)Li_2\ (\mathbf{41})} \text{3-}(SiMe_2Ph)\text{cyclohexanone}\ (82\%) \qquad (7)$$

Admixture of Me_3SnLi and CuCN, in the usual 2:1 ratio, led Sharma and Oehlschlager to observe signals in the ^{13}C NMR spectrum suggesting the presence of at least three species (Eq. 8).[44] The mixed system prepared from Me_3SnLi, MeLi, and CuCN in a 1:1:1 ratio, however, gave a clean ^{13}C NMR spectrum for the presence of $Me_3Sn(Me)Cu(CN)Li_2$. As with mixed silyl cuprate **41**, the tin analog reacts with enones to deliver the trimethylstannyl moiety in high yields (e.g., Eq. 9).

$$2Me_3SnLi + CuCN \xrightarrow[-50°]{THF} \boxed{(Me_3Sn)_2Cu(CN)Li_2 + (Me_3Sn)_3CuLi_2 + (Me_3Sn)_3Cu_2Li} \quad (8)$$

$$\text{cyclohexenone} \xrightarrow[THF, -78°]{(Me_3Sn)(Me)Cu(CN)Li_2} \text{3-}SnMe_3\text{-cyclohexanone (92\%)} \quad (9)$$

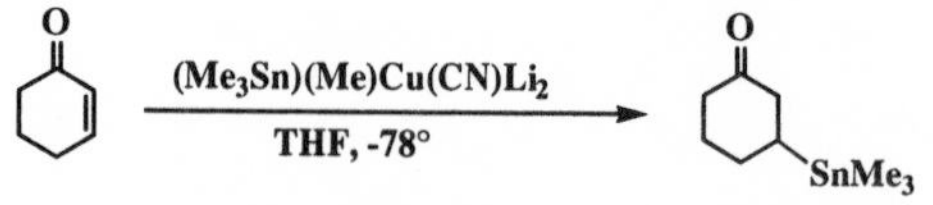

Ligand exchange between H.O. cuprates and carbon–tin bonds, in particular with vinylstannanes, have also proven to be facile, as originally discovered by Behling.[48] In light of our program on polyene constructions, it was suggested to Professor Jae In Lee, a visiting faculty scholar from Daksung Women's University in Korea, that it might be possible to use the 1,4-bis-tributylstannylbutadiene **42** as a lynchpin for further polyene construction. The project obviously called for the availability of the diene itself, and it was most fortunate that Arthur Ashe and co-workers had just recently described some dilithiations of this exact molecule.[49] Moreover, he generously provided us with experimental details on its preparation in gram quantities, the procedure for which starts with (*E*)-1,2-dichloroethylene (Scheme 22).

Exposure of butadiene **42** to $Me_2Cu(CN)Li_2$ leads to rapid transmetallation to **44** (R = Me) even at –78 °C. This is in contrast to the tributylstannylethylene analog **43**, which (as is true for vinylic tributylstannanes, in general) requires conditions on the order of room temperature over ca. 1 h (compare conditions in Scheme 23).[48] Addition of an enone at this stage leads to good yields of conjugate adducts bearing the trimethylstannylbutadiene moiety.[50] Two representative examples leading to **46** and **47** are shown in Schemes 24 and 25. Note that in Scheme 25, a mixed reagent **45** derived from MeLi (1 equiv), 2-lithiothiophene (2-ThLi, 1 equiv), and CuCN (1 equiv) can also be used to induce ligand exchange, albeit at higher temperatures.

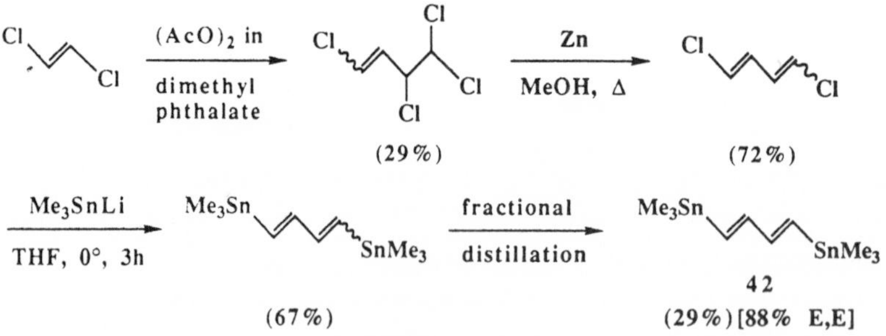

Scheme 22.

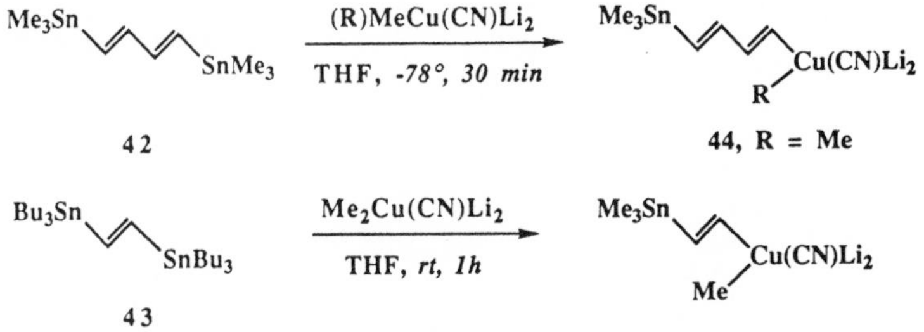

Me_3Sn ... $SnMe_3$ (42) — $(R)MeCu(CN)Li_2$, THF, -78°, 30 min → Me_3Sn ... $R(Cu(CN)Li_2)$ (44, R = Me)

Bu_3Sn ... $SnBu_3$ (43) — $Me_2Cu(CN)Li_2$, THF, rt, 1h → Me_3Sn ... $Me(Cu(CN)Li_2)$

Scheme 23.

42 — $Me_2Cu(CN)Li_2$, THF, -78°, 30 min → 44 → 46 (85%) ($SnMe_3$)

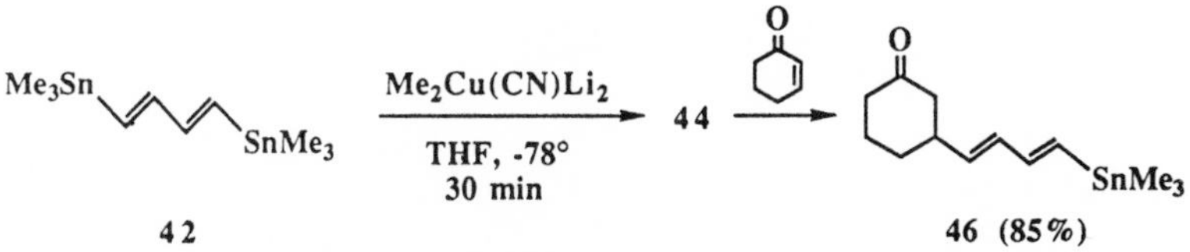

Scheme 24.

Assuming that the intermediate from these 1,4-additions is the corresponding enolate **48** (Figure 6), it seemed reasonable that a second equivalent of a H.O. cuprate might operate at the remaining trimethylstannyl site. Initial attempts to effect this one-pot double coupling met overall with marginal success, clearly resulting from problems at the second transmetallation stage. The culprit turned out

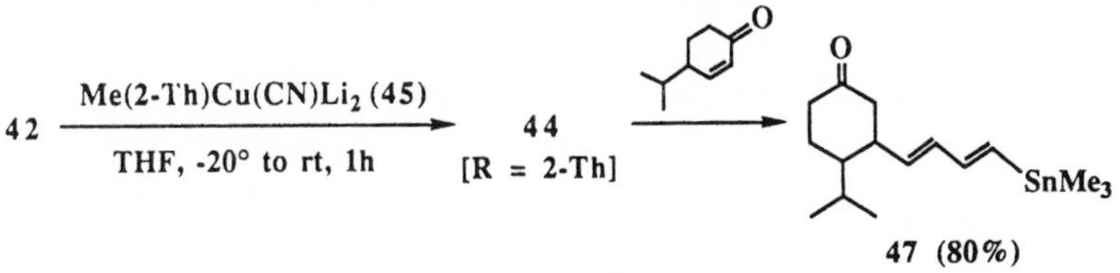

42 — $Me(2\text{-}Th)Cu(CN)Li_2$ (45), THF, -20° to rt, 1h → 44 [R = 2-Th] → 47 (80%) ($SnMe_3$)

Scheme 25.

O^-M^+

$SnMe_3$

48

site of 2nd transmetalation

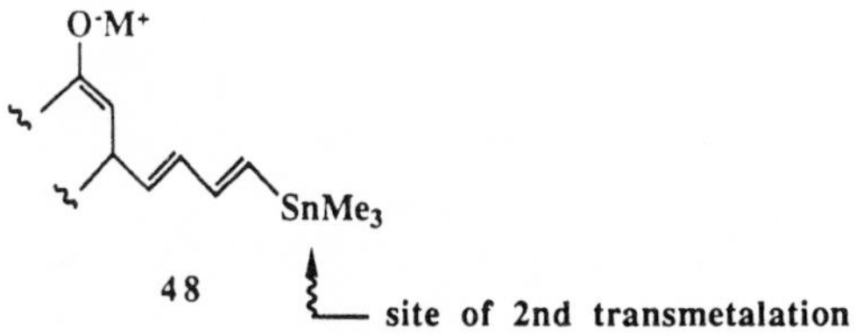

Figure 6.

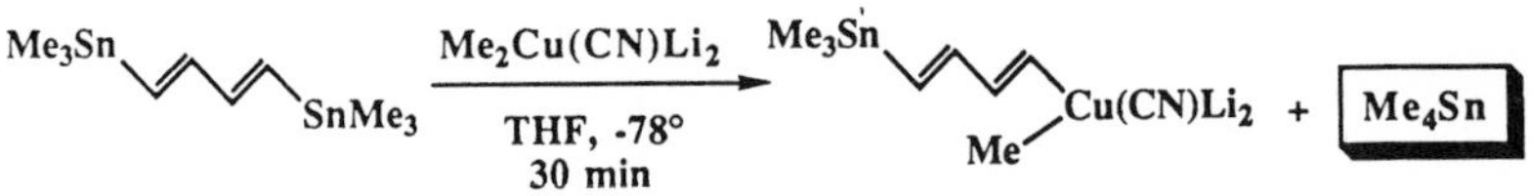

Me_3Sn $SnMe_3$ 42 → $Me_2Cu(CN)Li_2$, THF, -78°, 30 min → Me_3Sn $Cu(CN)Li_2$ Me 44, R = Me + Me_4Sn

Scheme 26.

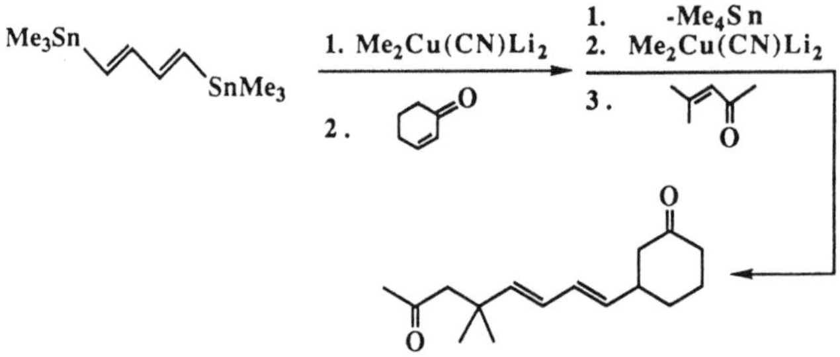

Me_3Sn $SnMe_3$ → 1. $Me_2Cu(CN)Li_2$ 2. (cyclohexenone) → 1. $-Me_4Sn$ 2. $Me_2Cu(CN)Li_2$ 3. (enone) → (74%) [97% E,E]

Scheme 27.

to be the Me_4Sn by-product present in solution which arises from the first transmetallation on **42** (Scheme 26). Thus, the Me_4Sn was competing with the butadienyltrimethylstannane portion of **48** for the newly introduced $Me_2Cu(CN)Li_2$, and while formally a degenerate process, considerable methyl transfer to the subsequently introduced enone was noted. The solution to the problem turned out to be trivial, and takes advantage of the volatility of Me_4Sn (bp 74–75 °C). Thus, by simply placing the reacting solution containing enolate **48** under vacuum (at <0.1 mm Hg) at temperatures between –15 and 0 °C, the volume can be reduced by ca. 50–75%. Reintroduction of fresh THF is followed by the second equivalent of cuprate, which effects transmetallation between 0° and room temperature over the course of one hour. Recooling to –78 °C followed by addition of the second enone leads to the desired Michael additions in good overall yields after workup and purification, typified by the example below (Scheme 27).[50]

Although success had come our way in terms of these single-flask, double-conjugate additions, the desired goal of realizing the corresponding *alkylations* has

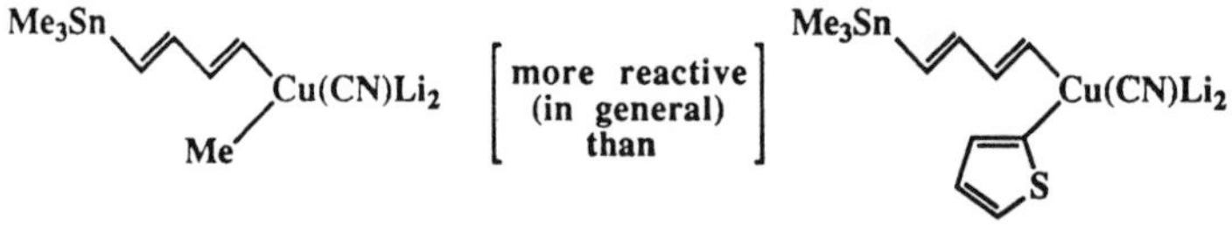

Me_3Sn $Cu(CN)Li_2$ Me — 44, R = Me [more reactive (in general) than] Me_3Sn $Cu(CN)Li_2$ S — 44, R = 2-Th

Figure 7.

been far more elusive. Alkylations of mixed alkyl vinyl cuprates akin to **44** (R = alkyl) are *not* useful in this regard, since unlike Michael additions where the vinylic ligand is selectively transferred to the substrate,[10,51] it is the alkyl residue which is preferentially released by the cuprate.[52] In many cases, the use of mixed cuprates such as $Me(2\text{-}Th)Cu(CN)Li_2$ (**45**) solves this problem.[53] Fortunately for us, we had found some years ago that the thienyl moiety tends to be a good non-transferable or "dummy" ligand for H.O. cuprates, the original observation to this effect having been made for Gilman L.O. cuprates by Nilsson and Ullenius years earlier.[54] However, the transmetallation chemistry of **45** need not follow that of $Me_2Cu(CN)Li_2$, and the relative reactivity of the mixed thienyl butadienyl cuprate **44** (R = 2-Th) is surely lower than that of **44** (R = Me) (Figure 7). Thus, all of our numerous attempts to date using these mixed reagents, together with vinyl triflates as reaction partners, have not been synthetically fruitful. Alternatives, however, have come along as our program on organozirconium chemistry continues to evolve (*vide infra*).

IV. LIGAND EXCHANGE BETWEEN VINYLALANES AND CU (I) REAGENTS

One of the fundamental reactions of organoaluminum reagents involves the addition of an aluminum–hydrogen or aluminum–carbon bond across an acetylene.[55] The resulting vinylalane, or its alanate derivative, can then be utilized in selected synthetic transformations. More often, however, these intermediates are quenched with halogen, from which chemistry of other metals, in particular, copper, is relied upon to form critical carbon–carbon bonds. In order to bypass these intermediate steps (i.e., halogenation, lithium–halogen exchange, cuprate generation; cf. Scheme 28, path A),[56] a graduate student in our group, Duy Nguyen, was asked to investigate the conversion of vinylalanes to vinylcuprates using $R_2Cu(CN)Li_2$ (path B).

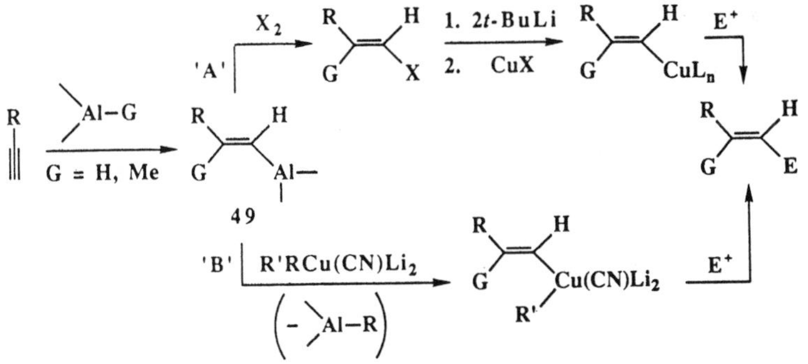

Scheme 28.

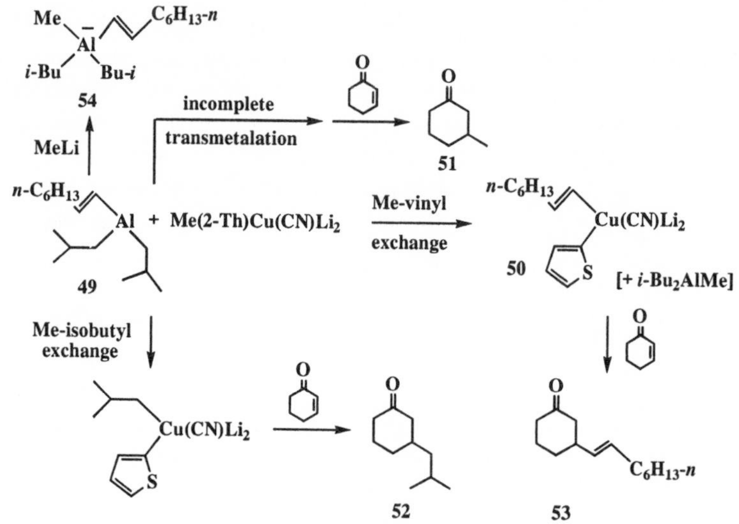

Scheme 29.

We chose to hydroaluminate 1-octyne to **49** as the point of departure, and Me(2-Th)Cu(CN)Li_2 as the cuprate with which to examine the transmetallation to **50** (Scheme 29). In brief, we were never successful in realizing a clean ligand exchange, as introduction of cyclohexenone to the vinylalane/H.O. cuprate mix afforded multiple products of 1,4 ligand transfer (i.e., to **51**, **52**, and **53**, from methyl, isobutyl, and 1-octenyl transfer, respectively), as well as considerable amounts of octene (ca. 30–50%). Incomplete transmetallation between the alane and cuprate could readily explain the presence of 3-methylcyclohexanone, but the appreciable buildup of the product from isobutyl delivery, **52** (ca. 20–40% of the mix relative to the desired product **53**), was most unexpected. Normally, according to George Zweifel with whom we consulted on these findings, one could expect to see ca. 2–3% competitive release of an isobutyl group on aluminum relative to most other ligands, especially when one is vinylic. We attempted to load the dice by converting **49** to the methylated alanate **54**, to which the L.O. cyanocuprate "in a bottle" [(2-Th)Cu(CN)Li][57] was added. In this way, the ratio of **53** to **52** was improved to 8:1, although the absolute yield of **53** was still <30%, most of the material being accounted for in the formation of 1-octene. Variations in temperature, RLi (e.g., *n*-BuLi, LiOMe, LiBr), and cuprate [Me_2Cu(CN)Li_2] only made matters worse. And yet, control experiments showed that both the vinylalane as well as the derived alanate **54** are completely ineffective at producing either **52** or **53** (by capillary GC analysis) at these low temperatures (–78 to 0 °C), notwithstanding prior art having demonstrated that 1,4-additions from alanates have on occasion been utilized (e.g., in prostaglandin syntheses) under ambient temperature conditions.[58]

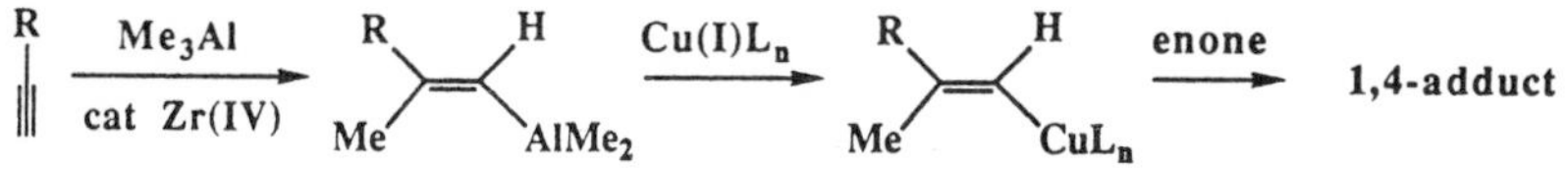

R≡ $\xrightarrow[\text{cat Zr(IV)}]{Me_3Al}$ (R)(Me)C=CH($AlMe_2$) $\xrightarrow{Cu(I)L_n}$ (R)(Me)C=CH(CuL_n) $\xrightarrow{\text{enone}}$ 1,4-adduct

Scheme 30.

We concluded, therefore, that the stoichiometric use of $R_2Cu(CN)Li_2$ was not going to afford a clean transmetallation on DIBAL-acetylene addition products. The solution, we felt, might come (in principle) from use of an unlikely reagent, the dimethyl analog of DIBAL (i.e., Me_2AlH, since methyl ligand exchange is degenerate and transfer of this group from copper to carbon is slower than that of other alkyl cuprates (e.g., that containing an isobutyl ligand). Assuming this hydride would add across an acetylene, the resulting dimethylvinyl alane might readily exchange the vinyl ligand on aluminum for a methyl group on copper, thereby arriving at the desired cuprate **50**, plus Me_3Al rather than i-Bu_2AlMe. But because there were just too many questions surrounding this "plan B" approach (especially with regard to Me_2AlH), we decided to check the concept by carrying out a Negishi carbometallation using Me_3Al catalyzed by zirconocene dichloride (Cp_2ZrCl_2), which affords a vinylic dimethylalane directly and in high yields. Since we observed that vinylalanes do not react with enones at sub-ambient temperatures, it occurred to us that formation of an aluminum enolate as driving force might facilitate the use of only catalytic amounts of a cuprate, in principle, provided that the transmetallation is facile (Scheme 30). Moreover, the presence of only catalytic Cu(I) might actually help to drive the ligand exchange toward the cuprate if CuCN is used, given the strength of the Cu–CN bond. Although it took us a while, Stuart Dimock was able to devise a procedure where, in fact, only 10 mol % CuCN is needed for effecting this carbometallation–transmetallation-conjugate addition scenario (Scheme 31). By simply adding the vinylalane **55** to a THF/Et_2O solution of CuCN·2LiCl at 0 °C in the presence of an enone, the desired adduct **56** is realized in good isolated yields.[60]

Competitive with the desired vinyl ligand transfer, depending upon substrate, some 1,4 transfer of a methyl group does take place due to the presence of excess Me_3Al, as prescribed in the Negishi carbometallation recipe.[59] This could be minimized, however, if the CH_2Cl_2 solution containing the initially formed vinyl-

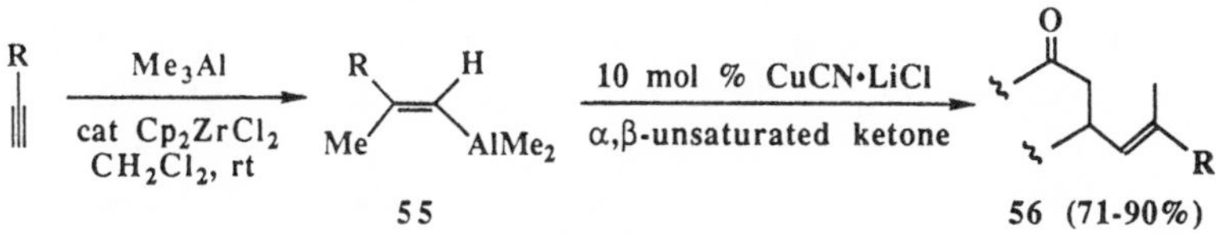

R≡ $\xrightarrow[\text{cat } Cp_2ZrCl_2,\ CH_2Cl_2,\ \text{rt}]{Me_3Al}$ (R)(Me)C=CH($AlMe_2$) **55** $\xrightarrow[\alpha,\beta\text{-unsaturated ketone}]{\text{10 mol \% CuCN·LiCl}}$ **56** (71-90%)

Scheme 31.

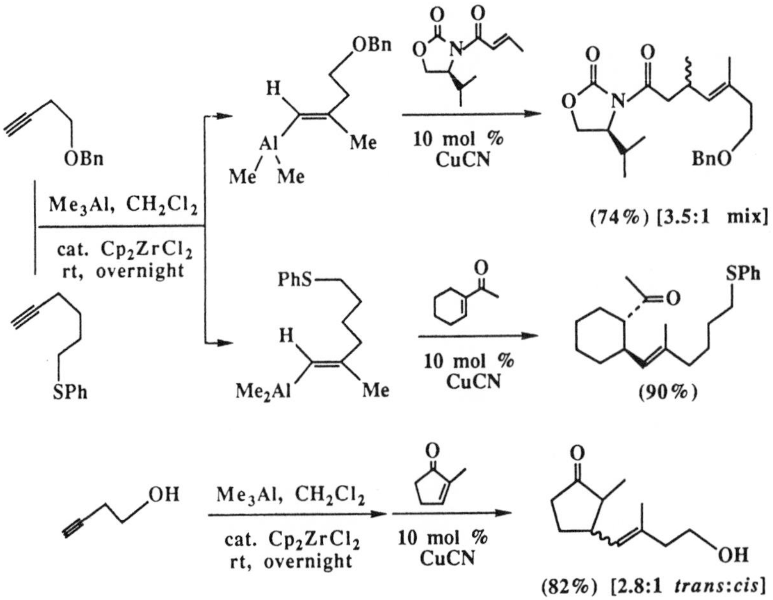

Scheme 32.

alane is placed under a vacuum pump to remove residual Me_3Al prior to its addition to the Cu(I)/enone solution. This modified procedure can usually account for yield improvements on the order of 15–25% when methyl transfer becomes problematic. Some representative examples follow (Scheme 32).

Scheme 32, last entry.[60] A solution of zirconocene dichloride (84.0 mg, 0.287 mmol) in 2.25 mL CH_2Cl_2 was treated with 1.75 mL (3.0 mmol) of a 2.0 M solution of trimethylaluminum in hexane at 0 °C followed by the careful dropwise addition of 3-butyn-1-ol (80.6 mg, 1.15 mmol) dissolved in 1.0 mL of CH_2Cl_2. The resulting solution was allowed to warm to room temperature over 12 h with stirring. The resulting yellow/orange solution was taken up into a gas-tight 5 mL motor-driven syringe together with a 0.5 mL CH_2Cl_2 rinse. A 1.0 M solution of CuCN·2LiCl was prepared by dissolving an admixture of CuCN (85.0 mg, 0.949 mmol) and LiCl (81.7 mg, 1.93 mmol) which had been previously flame-dried *in vacuo* in 0.95 mL of dry THF at room temperature with stirring. An initial aliquot from the syringe was added to 4.0 mL of dry Et_2O in a separate flame-dried round bottom flask under Ar at 0 °C with stirring giving rise to a pale yellow color. CuCN·LiCl solution (0.10 mL, 0.10 mmol) was added dropwise to the ethereal solution followed by the dropwise addition of a solution of 2-methylcyclopent-2-en-1-one (95.9

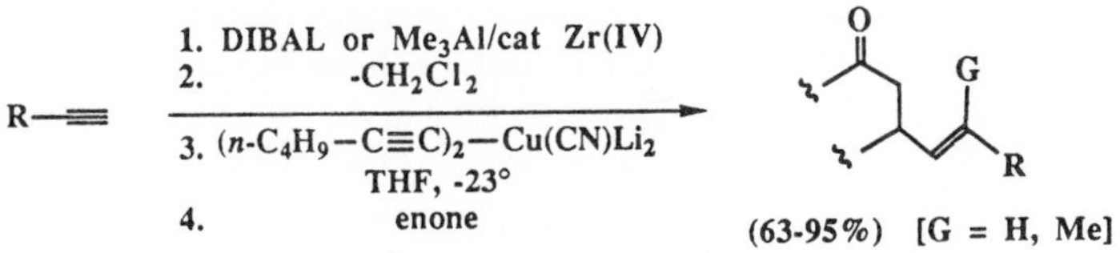

1. DIBAL or Me_3Al/cat Zr(IV)
2. -CH_2Cl_2
3. $(n\text{-}C_4H_9-C\equiv C)_2-Cu(CN)Li_2$ THF, -23°
4. enone

(63-95%) [G = H, Me]

Scheme 33.

mg, 0.998 mmol) dissolved in 1.0 mL of dry Et_2O. Slow dropwise addition of the remainder of the vinylalane solution was then conducted at 0 °C over 1 h and the resulting suspension stirred for an additional 1.5 h, at which point the reaction was complete by TLC (9:1 CH_2Cl_2/EtOAc). The resulting suspension was then poured into a biphasic mixture of 20 mL of 1.0 M aqueous tartaric acid and 20 mL Et_2O at 0 °C with vigorous stirring which was continued until all of the solids had dissolved. The mixture was separated and the aqueous phase washed with three 20 mL portions of Et_2O. The combined organic fractions were then washed with brine and dried over Na_2SO_4. Flash chromatography (silica gel, CH_2Cl_2/EtOAc, 17/3) yielded 148.8 mg (82%) of the desired keto alcohol as a 2.8:1 mixture of disastereomers by capillary GC analysis.

It was not long after Dimock had started developing this method that we read a communication to the editor in *J. Org. Chem.* by Ireland and Wipf[61] which told us how to successfully do the exact chemistry Nguyen had attempted years earlier (*vide supra*).[62] Thus, a H.O. cuprate could in fact be employed to effect vinylalane transmetallations in both cases of DIBAL and Me_3Al additions to acetylenes. And what type of H.O. cuprate does this? Apparently, the one derived from 1-hexynyllithium [i.e., $(n\text{-}C_4H_9C\equiv C)_2Cu(CN)Li_2$] leads to good overall yields of Michael adducts (Scheme 33). A representative example follows (Eq. 10).

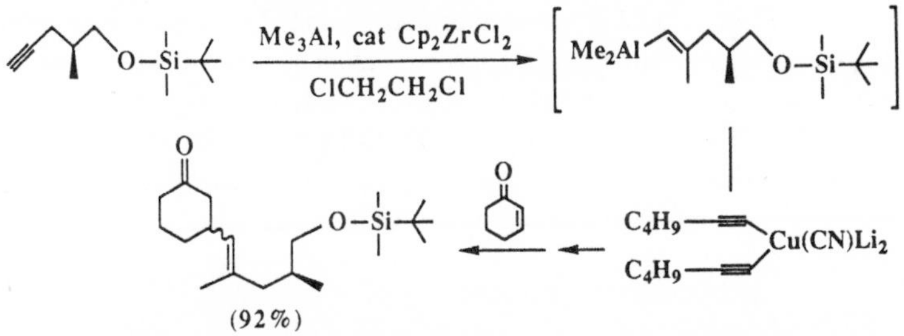

Me_3Al, cat Cp_2ZrCl_2 / $ClCH_2CH_2Cl$ → [Me_2Al ... O–Si] ; C_4H_9–≡ / C_4H_9–≡ Cu(CN)Li_2 ; cyclohexenone → (92%) (10)

General procedure for the carboalumination/transmetallation/conjugate addition of alkynes (Eq. 10):[61] 3-[(1*E*,4*S*)-5-[(*t*-Butyldimethylsilyl)oxy]-2,4-dimethyl-1-pentenyl]cyclohexanone. A suspension of Cp_2ZrCl_2 (28 mg, 0.10 mmol) in dry 1,2-dichloroethane (2 mL) was treated at 0 °C with a 2.0 M solution of trimethylaluminum in toluene (0.71 mL, 1.42 mmol), followed by addition of a solution of (4*R*)-5-(*t*-butyldimethylsilyl)oxy]-4-methyl-1-pentyne (100 mg, 0.47 mmol) in 1,2-dichloroethane (0.3 mL). The reaction mixture was stirred at room temperature for 3 h, the solvent was removed *in vacuo*, and dry Et_2O (2 mL) was added. The solution of the alkenylalane was added at –23 °C to a mixture of flame-dried CuCN (45 mg, 0.51 mmol) in THF (4 mL) and a 0.5 M solution of 1-hexynyllithium in THF/hexane (5:1) (2.2 mL, 1.1 mmol). The reaction mixture was stirred for 5 min at –23 °C, and a solution of 2-cyclohexenone (38 mg, 0.40 mmol) in THF (1 mL) was added dropwise. Stirring at –23 °C was continued for another 20 min. The mixture was quenched into a cold (0 °C) solution of saturated ammonium chloride/ammonium hydroxide (9:1) and extracted three times with Et_2O. The combined organic layers were dried ($MgSO_4$), filtered through silica gel, and chromatographed (EtOAc/hexane, 1:5) to yield 119 mg (92%) of the 1,4 adduct as a clear oil.

The Ireland and Wipf procedure[61] is rather remarkable for several reasons. Recall that we had tried extensively with alkyl [e.g., $Me_2Cu(CN)Li_2$] and mixed [e.g., $Me(2\text{-}Th)Cu(CN)Li_2$] reagents to induce transmetallation (*vide supra*). Why does the *acetylenic* cuprate do this so well? Moreover, further extension of the procedure to vinylic alanes bearing isobutyl groups on the metal seems likely, for although only two examples have been recorded to date, the overall yields are similar to those observed from the carbometallation/transmetallation sequence (Eq. 11).[63] Why is

$$n\text{-}C_6H_{13}\text{—}\equiv \xrightarrow[\text{2. } (n\text{-}C_6H_{13})_2Cu(CN)Li_2 \quad \text{3. cyclohexenone}]{\text{1. DIBAL}} \text{3-[(E)-CH=CH-}C_6H_{13}\text{-}n\text{]cyclohexanone} \quad (72\%) \qquad (11)$$

there not significant isobutyl transfer in these cases? Most astounding, however, is the realization that for exchange to occur with these particular ligands on these metals, the process is doubly contrathermodynamic. In other words, aluminum tends to give up ligands in the order: alkyne > alkene > alkane, presumably due to its Lewis acidity and the electronegativity of the various hybrid orbitals on carbon. Copper, on the other hand, shows just the opposite selectivity of ligand release:[64] alkyl > alkenyl >> alkynyl, where acetylenic groups are notoriously nontransferable (i.e., "dummy") ligands,[65] presumably due to $d\pi^*$ backbonding from copper into

Order of ligand release from Al: $sp > sp^2 > sp^3$

Order of ligand release from Cu: $sp^3 > sp^2 > sp$

the π network. And yet, here we have aluminum giving up an alkenyl for alkynyl group, and copper trading an acetylenic for an olefinic ligand. Quite frankly, I never would have asked Nguyen to attempt such a reaction. Clearly, there must be more (at least in this case) to this transmetallation game than just net stability associated

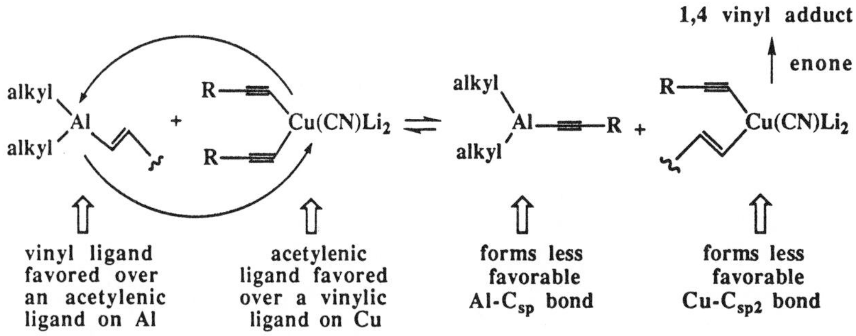

1,4 vinyl adduct

enone

vinyl ligand favored over an acetylenic ligand on Al

acetylenic ligand favored over a vinylic ligand on Cu

forms less favorable Al-C_{sp} bond

forms less favorable Cu-C_{sp2} bond

Scheme 34.

with the ligands on the metals. Recently, in a full account of their work,[63] Wipf recognizes these relationships (cf. Scheme 34) and suggests that the exchanges need not be thermodynamically controlled. Rather, these phenomena and the subsequent Michael additions which ensue are likely to be kinetically driven, where an initial transfer occurs between the H.O. reagent and the alane to produce an alanate and L.O. cyanocuprate. The alanate then releases its vinylic ligand to copper, arriving at the mixed H.O. species, with all events part of an equilibrium which is shifted by the trapping of the mixed vinyl cuprate by the enone.

V. COMMUNICATION BETWEEN ORGANOMETALLICS OF ZIRCONIUM AND COPPER

Back in 1987, while collaborating on the vinyl stannane/H.O. cuprate transmetallations initiated by Jim Behling at Searle,[48] I asked graduate student J.C. Barton to repeat some standard Schwartz-type hydrozirconation reactions[66] in anticipation of utilizing zirconocene intermediates as transmetallation partners with H.O. cuprates (Scheme 35). Unfortunately, we made the mistake of relying on an aged bottle of $Cp_2Zr(H)Cl$ (i.e., Schwartz' reagent)[67] and never achieved satisfactory results. My lack of experience with this reagent and naivete associated with its handling cost us not only some of Barton's time, but set the project back literally over a year. Nonetheless, while on a consulting trip to Searle soon thereafter, I discussed with Behling the prospects for preparing vinylic cyanocuprates directly from terminal

$$Cp_2Zr(H)Cl \equiv \text{Schwartz' reagent}$$

acetylenes via intermediates from hydrozirconation reactions. The incentives for realizing these goals were numerous, including as examples: (1) the need for vinylic halides or stannanes (which are patented intermediates for prostaglandin production)[68] as precursors to the corresponding organolithiums would be avoided; (2) hydrozirconation is far more regio- and stereoselective in its additions across 1-alkynes relative to hydrostannylations[69] on similar starting materials; and (3)

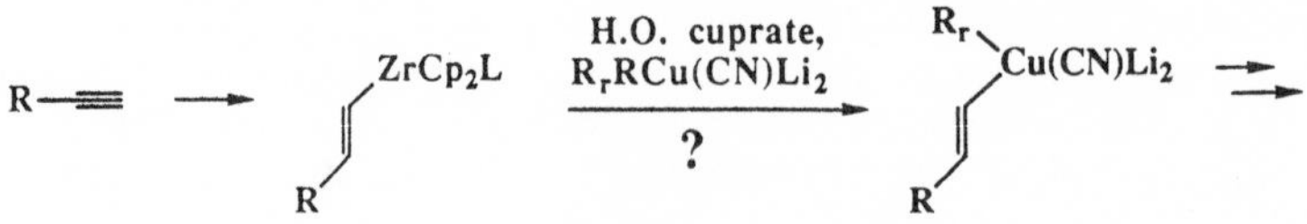

Scheme 35.

zirconium-containing by-products would be polar, essentially nontoxic, and easily lost during workup, while tetraalkylstannanes[30] are nonpolar, potentially quite toxic, and would require some form of physical separation (e.g., chromatography, distillation, etc.) to remove. This looked so good on paper that I could not help but recall a line once told to me in graduate school, tongue in cheek, by former mentor Harry Wasserman, who said concerning one idea: "There's so much theory in favor of it, let's just write it up!" Who in our group was the "right" person to bring this proposal to fruition?

The nod eventually went to Edmund Ellsworth, who within a span of two years in graduate school had already coauthored close to a dozen contributions. And so as his final project at UCSB before taking a postdoctoral position in the Corey group, I outlined to Edmund what looked to be the way in which the transmetallation would proceed. Of course, there were many tough questions which surfaced during this discussion, such as: (1) what is the impact of the chlorine group on vinyl zirconocenes in the presence of a H.O. cuprate?; (2) was $Me_2Cu(CN)Li_2$ the best H.O. cuprate to effect the transmetallation, as it is toward vinylstannanes?;[48] (3) how do the conditions of ligand exchange with vinylzirconocenes compare with the corresponding stannanes, which take ambient temperatures over time?;[48] and (4) is the resulting H.O. mixed cuprate going to be compatible with the zirconium (IV) Lewis acid also produced as a by-product?

Well, three months went by and after an extensive effort at varying many parameters such as solvent, additives, concentrations, etc., we had reached an impass. Although the transmetallation seemed to take place on **57** using $Me_2Cu(CN)Li_2$, after the enone has been added our best yields of desired 1,4-adduct were on the order of 50–60% (Scheme 36). We observed that starting material was

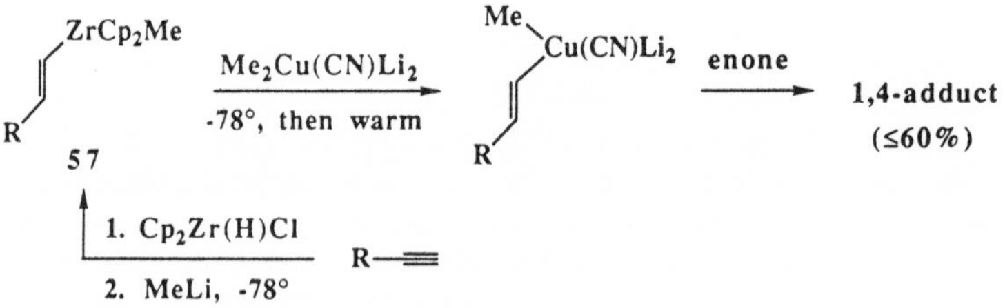

Scheme 36.

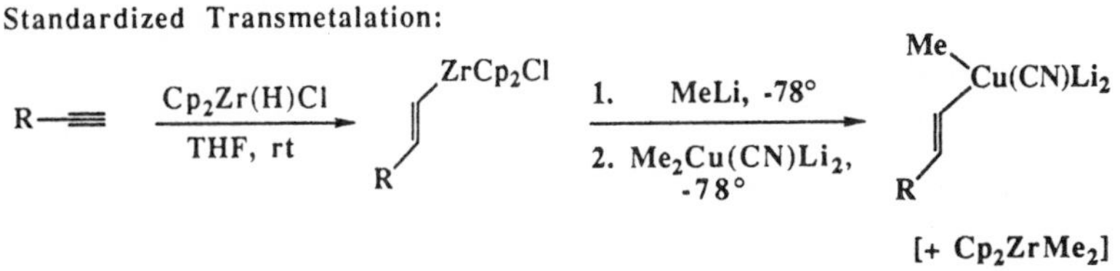

Standardized Transmetalation:

R—≡ $\xrightarrow[\text{THF, rt}]{Cp_2Zr(H)Cl}$ (R)CH=CH–$ZrCp_2Cl$ $\xrightarrow[\text{2. } Me_2Cu(CN)Li_2\text{, -78°}]{\text{1. MeLi, -78°}}$ (R)CH=CH–Cu(CN)Li_2(Me) [+ Cp_2ZrMe_2]

Scheme 37.

not fully consumed, and occasionally the product of 1,4-methyl transfer became dominant. It was time to regroup; to reexamine what exactly had been tried, and assess what we knew thus far. From Edmund's summary, there was one glaring omission: an experiment, indeed, originally thought to have been our very first trial in this series. That is, it was now obvious that what all of our experiments had in common was the warming as part of the transmetallation step, sometimes to –50 °C, and many times to even higher temperatures. Our best results emanated from the reactions where this portion of the experiment was conducted at colder temperatures. So I asked Edmund: "By the way, whatever happened in the original experiment calling for the transmetallation to occur all at –78 °C"? *It had never been done!* Another lesson had just been learned for both of us in communication skills! And so, in the very next trial, by maintaining the –78 °C bath and in literally less than five minutes time, the transmetallation from Zr to Cu was complete and extremely clean (Scheme 37). Isolated yields of conjugate addition products jumped into the 80's and 90's; no additives were needed, no further manipulations called for; it was that simple. I guess in retrospect, it always is.

And so our program in organozirconium chemistry was born. Edmund got yet another paper in *JACS*,[70] as well as a write-up in *ChemTracts* by Bruce Ganem[71] for his efforts.

A. Stoichiometric Transmetallations

This method for effecting transmetallations with H.O. cuprates is quite flexible in terms of how reagents are added to the intermediate vinyl zirconocenes, as long as the net amounts of MeLi and CuCN (i.e., 3MeLi + 1CuCN) remain the same. Thus, while several combinations exist for arriving at the 3MeLi + 1CuCN stoichiometry, including (1) MeLi + $Me_2Cu(CN)Li_2$, (2) 2MeLi + MeCu(CN)Li, and (3) 3MeLi + 1CuCN·2LiCl, there is little sensitivity to which of the three is chosen for use (Figure 8). The first, which calls for treatment of vinyl zirconocene **58** with MeLi at –78 °C, presumably forms the methylated product **59**. Subsequent addition of $Me_2Cu(CN)Li_2$ induces the transmetallation to cuprate **60** and the presumed by-product **61** (Scheme 38). Examples using this approach are illustrated in Scheme 39.

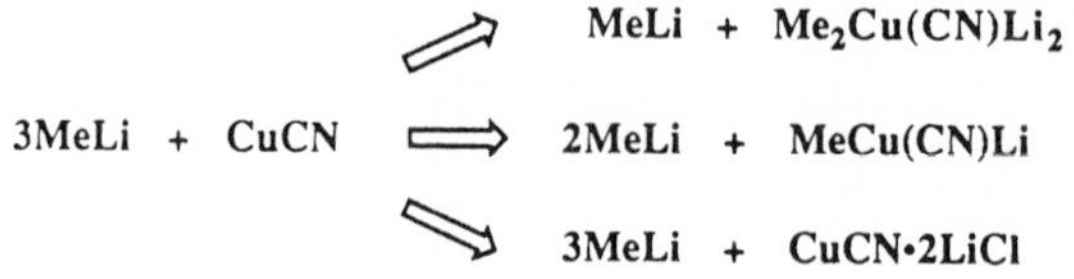

$3MeLi + CuCN \Rightarrow MeLi + Me_2Cu(CN)Li_2$

$3MeLi + CuCN \Rightarrow 2MeLi + MeCu(CN)Li$

$3MeLi + CuCN \Rightarrow 3MeLi + CuCN \cdot 2LiCl$

Figure 8.

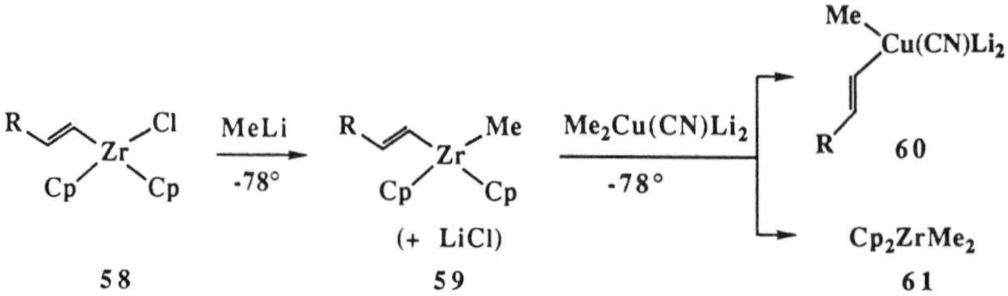

Scheme 38.

Scheme 39, top case.[70] A 10 mL round bottom flask equipped with a stir bar was charged with zirconocene chloride hydride (0.258 g, 1.00 mmol) and sealed with a septum. The flask was evacuated with a vacuum pump and purged with argon, the process being repeated 3 times. THF (2.0 mL) was injected and the mixture stirred to generate a white slurry which was treated with a THF solution (0.75 mL) of 1-ethynyl-1-trimethylsilyloxycyclohexane (0.22 mL, 1.00 mmol). The mixture was stirred for 30 min to yield a yellow solution which was cooled to –78 °C and treated via syringe with ethereal MeLi (0.71 mL, 1.0 mmol). Concurrently, $Me_2Cu(CN)Li_2$ was prepared using the following amounts of reagents: CuCN (0.90 g, 1.00 mmol), MeLi

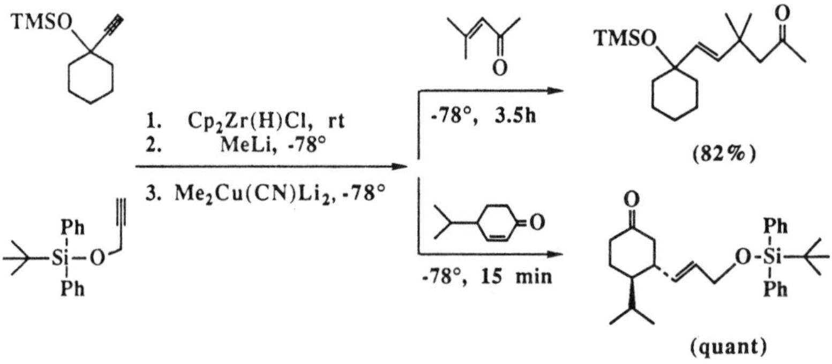

Scheme 39.

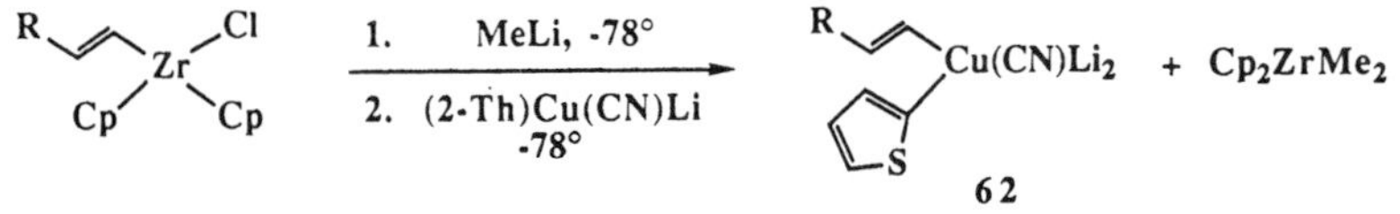

Scheme 40.

(1.43 mL, 2.0 mmol) in ether, and THF (1.0 mL). This solution was added via cannula to the vinylzirconocene. The mixture was stirred for 15 min at −78 °C to yield a bright yellow solution which was treated with mesityl oxide (0.048 mL, 0.50 mmol). The reaction was allowed to proceed for 3.5 h at −78 °C before being quenched with 15 mL of 10% NH_4OH in saturated aqueous NH_4Cl. The product was extracted with 3 × 30 mL of ether and dried over Na_2SO_4. The solution was then filtered through a pad of celite and the solvent removed *in vacuo*. The resulting residue was submitted to flash chromatography on silica gel (petroleum ether/ethyl acetate, 9/1) to yield 0.120 g (82%) of desired product as a colorless oil.

As for the second combination (cf. Figure 8), there is little reason to use this alternative; if a cyanocuprate is to be made in a separate flask from MeLi + CuCN, why not just use MeLi + $Me_2Cu(CN)Li_2$? The benefit to this scenario, however, comes in the form of an option to use the commercially available L.O. cyanocuprate (2-Th)Cu(CN)Li,[72] described by us years ago.[57] Thus, by adding 2MeLi and then (2-Th)Cu(CN)Li, both taken from stock solutions, to a vinylzirconocene, transmetallation occurs under the same mild conditions, only in this case to give rise to mixed cuprate **62** (Scheme 40). An example of this process, followed by the reaction of **62** (R = Ph) with isophorone is shown in Scheme 41.

The third scenario is, perhaps, the easiest to carry out, for it relies on only one transfer of an organometallic (i.e., MeLi). It is made possible by the ability of LiCl to solubilize CuCN in THF, as originally reported by Knochel,[73] and hence two

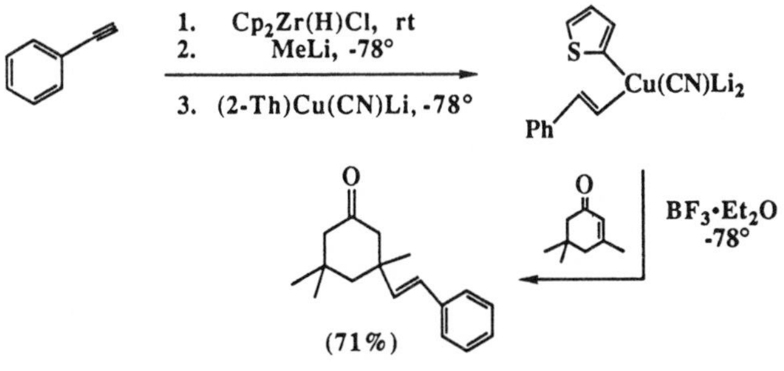

Scheme 41.

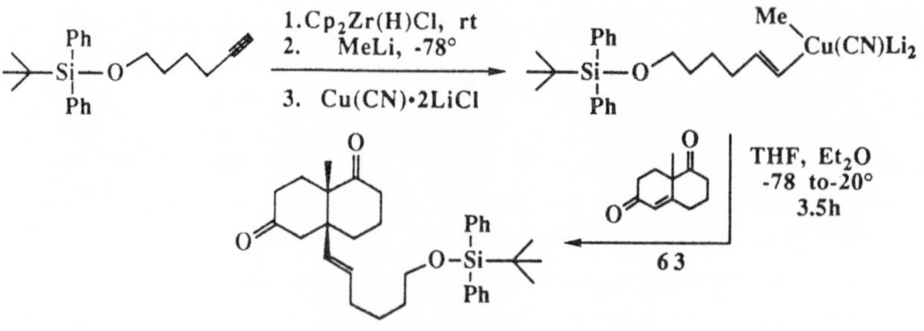

Scheme 42.

solutions are added to the preformed zirconocene prior to addition of the final solution containing an enone. Two examples of this strategy are illustrated in Schemes 42 and 43, which include the challenging cases of a keto octalone **63** and a fully functionalized prostaglandin precursor **64**.[70]

Before completing his study on these transmetallations and just prior to his departure for Harvard, Edmund discovered a new method for generating $Cp_2Zr(H)Cl$ *in situ* from the far less-expensive Cp_2ZrCl_2. Although all of his early work had used Schwartz' reagent of superb quality, generously supplied by Lee Kelly and Jeff Sullivan at Boulder Scientific in Colorado,[74] we thought it worth-

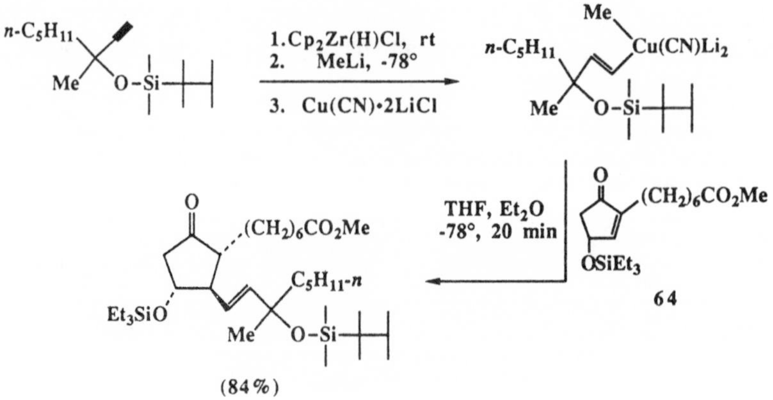

Scheme 43.

Table 4. *In Situ* Generation of $Cp_2Zr(H)Cl$ with Various Hydride Sources

$$Cp_2ZrCl_2 \xrightarrow{[H]} Cp_2Zr(H)Cl + \text{by-product from reducing agent}$$

[H]	*Reference*	*By-products(s)*
t-BuMgCl	Negishi[75]	magnesium salts
Red-Al	Schwartz[76]	aluminum salts
$LiAl(O\text{-}t\text{-}Bu)_3H$	Wailes[77]	aluminum salts
$LiEt_3BH$	Lipshutz[78]	Et_3B

while to find a reducing agent that would give $Cp_2Zr(H)Cl$ along with a by-product tolerable by both cuprates and other functionality (e.g., hydroxyl protecting groups). Previous work in this area[75–78] (as well as that subsequent to our own, most recently by Negishi),[75] had also provided alternatives to arrive at $Cp_2Zr(H)Cl$[79] while simultaneously generating aluminum or magnesium salts. Our choice of hydride source evolved toward $LiEt_3BH$[80] ("SuperHydride"™),[81] for it presumably leads to $Cp_2Zr(H)Cl$ and relatively innocuous Et_3B (Table 4). To establish the generality of this procedure and do a direct comparison with (1) existing *in situ* literature methods,[75–77] and (2) a fresh bottle of $Cp_2Zr(H)Cl$, the torch was passed to a new recruit, Robert Keil. Although Robert was an undergraduate at Berkeley, the "copper boys" in the group tried hard not to hold his background against him! Starting his graduate career on zirconium chemistry alone allowed him to eventually (*vide infra*) break into the cuprate game slowly, for while he had experience with organometallics as an undergraduate, the techniques learned there, as anywhere, of course, will vary according to the metals involved.

Robert's study on the $Cp_2ZrCl_2/LiEt_3BH$ mix was in fact highly successful, as we were able to establish the compatibility of this combination with tetrahydropyranyl (THP, see Table 5) and trimethylsilyl (TMS) alcohol derivatives, as well as its tolerance by an ester, sulfide, and imide.[78] The intermediate vinylic zircono-

Table 5. Compatibility on *In Situ* Generated $Cp_2Zr(H)Cl$ with a THP Ether

1. $Cp_2Zr(H)Cl$ (commercial or *in situ*)
2. H_2O

Source	*Yield (%)*
Boulder Scientific	77
$Cp_2ZrCl_2 + BiEt_3BH$	75
Cp_2ZrCl_2 + Red-Al	50
Cp_2ZrCl_2 + *t*-BuMgCl	<10

Table 6. Representative Comparison Reactions for Commercial Versus *In Situ* Prepared $Cp_2Zr(H)Cl$ with 1-Alkynes

$$Cp_2ZrCl_2 \xrightarrow[\text{2. substrate} \quad \text{3. } H_2O]{\text{1. } LiEt_3BH,\ THF,\ rt} \text{olefinic product}$$

Entry	*1-Alkyne*	*Product*		*Yield using commercial $Cp_2Zr(H)Cl$*
1	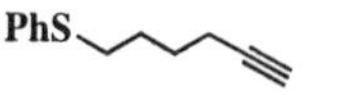 PhS (alkyne)	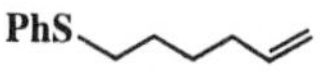 PhS (alkene) (88%)	vs.	(85%)
2	 OTMS (alkyne)	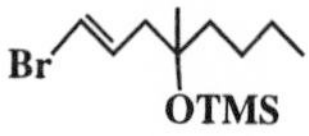 Br (alkene), OTMS (83%)	vs.	(93%)
3	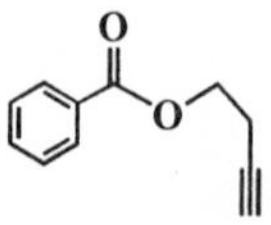 PhC(O)O (alkyne)	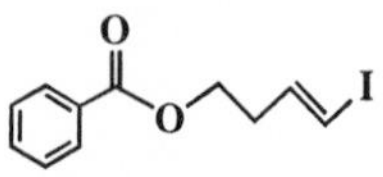 PhC(O)O (alkene), I (93%)	vs.	(86%)

Table 6, entry 2.[78] A flame-dried 10 mL flask with stir bar, cooled under argon, was charged with 163 mg (0.56 mmol) of Cp_2ZrCl_2 followed by 4 mL of THF. To this solution was added, dropwise over 2 min, 0.56 mL of a 1 M solution of $LiEt_3BH$ in THF ("SuperHydride", Aldrich). The solution was stirred, shielded from light, for 1 h at ambient temperature. After this time, 59 mg (0.28 mmol) of 4-methyl-4-trimethylsilyloxy-1-octyne (dried azeotropically with toluene at room temperature) was added and the mixture stirred for 10 min to provide a clear, yellow solution. Introduction of *N*-bromosuccinimide (recrystallized and dried azeotropically with toluene) led to a cloudy mixture which was stirred for 5 min at room temperature. The reaction was then poured into 20 mL of saturated aqueous $NaHCO_3$ and extracted with 10% EtOAc/hexanes (15 mL, 2x). The combined organic layers were washed with saturated aqueous NaCl, dried over Na_2SO_4, and filtered through a pad of Celite atop silica gel. Rotary evaporation *in vacuo* afforded a clear oil (68 mg, 83%), pure by VPC analysis.

cenes could readily be quenched, following the classical Schwartz protocol,[66,76] with electrophiles such as H_2O, D_2O, I_2, and NBS (Table 6).

With some experience in organozirconium chemistry under his belt, it was time for Robert to meld hydrozirconations with cuprate-based transmetallations—but with an important twist. That is, would it be possible to effect the transmetallation

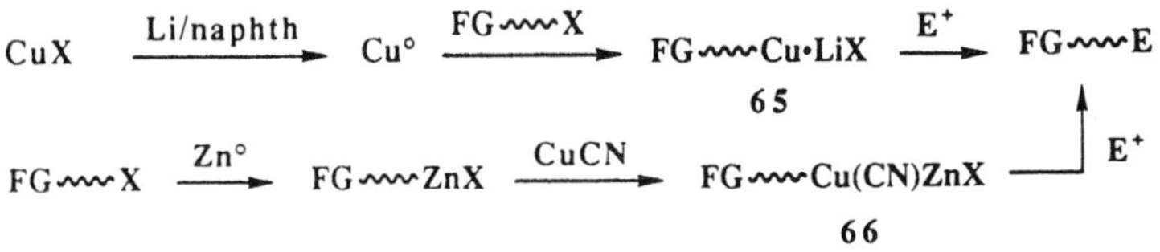

CuX $\xrightarrow{\text{Li/naphth}}$ Cu° $\xrightarrow{\text{FG\textasciitilde X}}$ FG~Cu·LiX (65) $\xrightarrow{E^+}$ FG~E

FG~X $\xrightarrow{\text{Zn°}}$ FG~ZnX $\xrightarrow{\text{CuCN}}$ FG~Cu(CN)ZnX (66) $\xrightarrow{E^+}$ FG~E

Scheme 44.

of a vinylic zirconocene *containing an electrophilic functional group* (e.g., an ester, nitrile, etc.,) at the other end of the molecule? This goal was set having been influenced by the extensive work from the Rieke[82] and Knochel[83] groups, which had shown that cuprates of varying types could be prepared containing a variety of electrophilic centers therein. The trade-off for this compatability manifests itself in the reactivity patterns of these reagents relative to standard lithio cuprates. This can be ascribed mainly to either the formally non "ate" character of the organocopper complexes prepared via Rieke copper (i.e., "RCu", **65**),[82,84] or the use of organozinc halides en route to L.O. zinc cyanocuprates, RCu(CN)ZnX, **66** (Scheme 44).[83]

We felt that our proposal had a reasonable chance for success, since the transmetallation of a funtionalized vinylic zirconocene (e.g., **67**) should take place rapidly under the same mild (i.e., –78 °C) conditions observed in the earlier Ellsworth study where electrophilic centers were not present.[70] However, the $64,000 question which loomed had to do with the next step in the handling of **67**; that is, its reaction with MeLi. Would it go cleanly to zirconium, or would the "FG", e.g., a carboalkoxy group, interfere (Scheme 45)? While pondering whether MeMgX or, perhaps, yet another source of "Me⁻" would be better choices than MeLi, I recalled the common expression "never overlook the obvious". In this situation, the "obvious" was the question: Is MeLi really necessary? Ellsworth had previously addressed this in his sequence, and found, in a single experiment, that the MeLi led to a yield that was between 5–10% higher than those transmetallations performed with $Me_2Cu(CN)Li_2$ alone. This determination, however, was done at a very early stage of that chemistry's development. It didn't take long for Robert to establish that we could simply leave it out! This was obviously good news, and with the Ellsworth procedure further streamlined to reflect Keil's recent observation, we

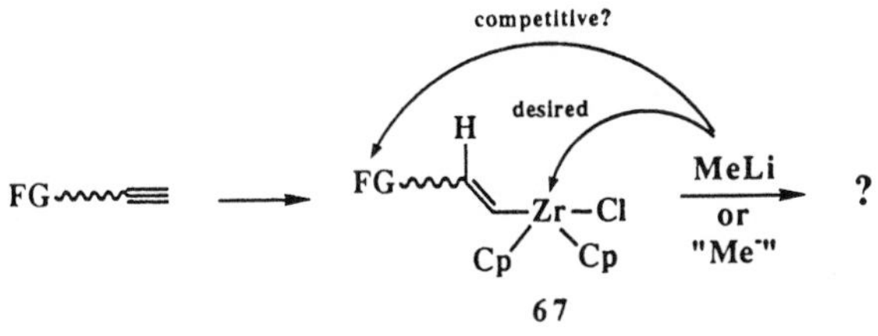

competitive?
desired
FG~≡ → FG~CH=CH–Zr(Cl)Cp₂ (67) $\xrightarrow[\text{"Me}^-\text{"}]{\text{MeLi or}}$?

Scheme 45.

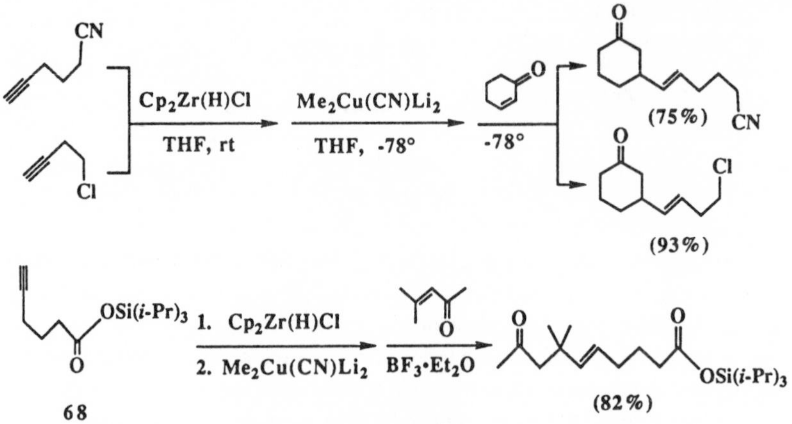

Scheme 46.

were able to deliver vinyl ligands to enones which contained ω-nitrile, chloro, and ester functionalities, typified by those examples in Scheme 46.[85] Noteworthy is the fact that these are true *lithio* cuprates, and their conjugate additions occur, therefore, under very mild conditions even in cases of β,β-disubstituted enones, as long as $BF_3{\cdot}OEt_2$ (1 equiv) is present.

Some limitations to this coupling do exist, however. Aldehydes and ketones are *not* tolerated within the l-alkyne, since $Cp_2Zr(H)Cl$ is well known to effect their reduction; e.g., acetone is routinely used as a quenching agent for analysis of activity.[79] The nature of the ω-ester which can be present is also not general, as both methyl and ethyl analogs of **68** do not afford the corresponding Michael adducts. Rather, only the products of hydrozirconation-protio quenching (i.e. the olefins) are formed. It may well be that the presence of the basic carbonyl oxygen is inhibiting transmetallation via a Lewis acid–Lewis base interaction, as in **69** or **70**. Although the triisopropylsilyl [TIPS] ester completely eliminates this presumed phenomenon, other silyl esters are only partially effective; e.g., *t*-butyldimethyl-silyl (TBDMS) esters afford ca. 25% conjugate adduct, the remainder being the olefin.

While the hydrozirconation/transmetallation/1,4-addition story was unfolding, we already were cognizant of the fact that related alkylations would not be successful using mixed alkyl/vinylic cuprates generated by this protocol. One of the more curious aspects of cuprate chemistry is the observation that reagents

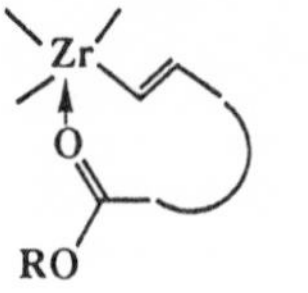

69 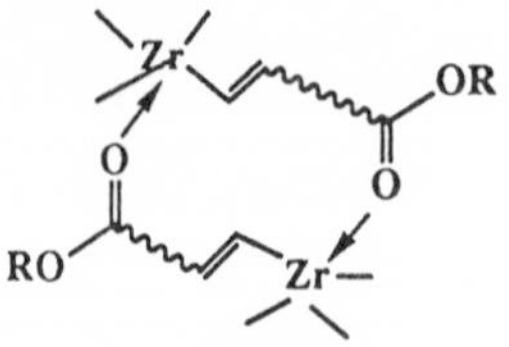 70

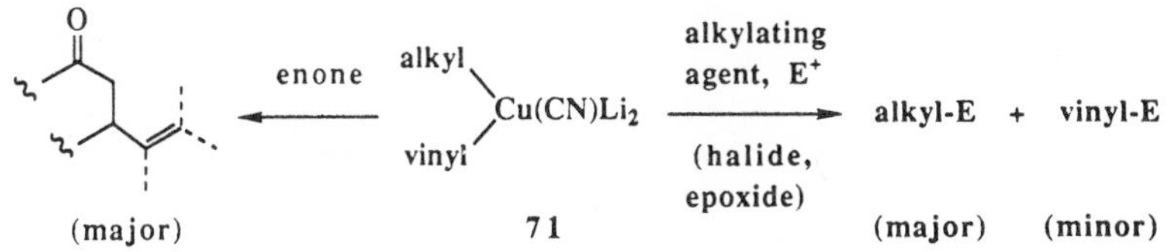

Scheme 47.

bearing both sp^3 and sp^2 ligands (i.e., as in **71**) behave quite differently when exposed to an α,β-unsaturated ketone versus an alkylating agent. The former substrate reacts with **71** very selectively (> 100:1) to accept the vinylic group, while the latter preferentially (ca. 5–10:1) forms a bond with the sp^3 ligand from copper (Scheme 47).[52,65] This is not unreasonable, since these are mechanistically distinct processes, with alkylations potentially proceeding either through radical intermediates[86a] or, as shown below, initial Cu(III) intermediates followed by reductive elimination, while 1,4-additions are electrophilically driven and likely to involve an initial dπ* complexation[86b] as in **72**, and then Cu(III)-β-adduct formation (**73**). Ultimately, upon reductive coupling, the enolate and a Cu(I) species is afforded (Scheme 48).

To enable the vinyl ligand to be selectively transferred from copper to the alkylating agent, it would be necessary to devise a procedure which allowed for arrival at a mixed H.O. cuprate containing another non-transferrable ligand, in addition to the nitrile group. So I asked Dr. Kaneyoshi Kato to investigate this possibility, knowing that he would work very hard to see the project to completion during his one year stint in our group. My confidence also derived from the fact that he had been graciously sent to our laboratory from Takeda Chemical Company in Osaka by their then Director of Chemistry, Dr. Shinji Terao, who just happened to be my laboratory partner during my first year in graduate school at Yale in Harry Wasserman's group. Needless to say, Kato did not disappoint me. He tested a variety of possibilities as to mode of cuprate generation from intermediate vinyl zircono-

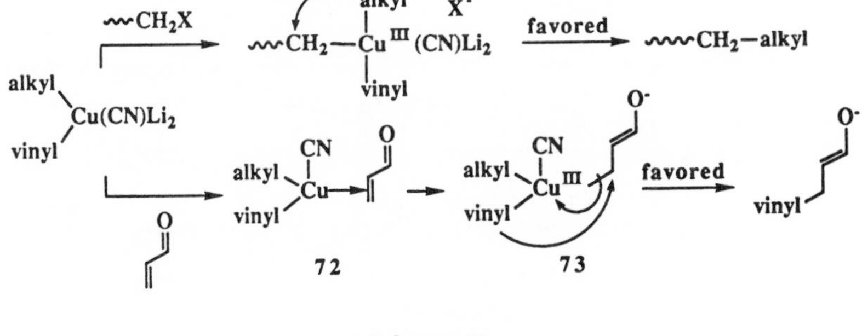

Scheme 48.

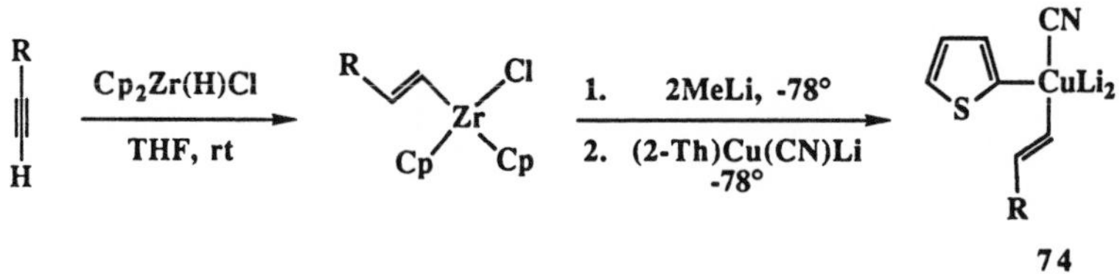

Scheme 49.

cenes, and in the end, found that the most effective called for the 2MeLi/(2-Th)Cu(CN)Li scenario thereby giving the mixed reagent **74** (Scheme 49). Couplings with **74**, all performed in this 1-pot operation, were straightforward with alkylating agents which include epoxides, activated (allylic, benzylic) halides, and a vinyl triflate, typified by the examples below (Scheme 50).[87]

Scheme 50, last example.[87] A 10 mL 2-necked round-bottom flask equipped with a stir bar was charged with zirconocene chloride hydride (126 mg, 0.48 mmol). The flask was evacuated and purged with argon, the process being repeated 3 times. THF (3 mL) was injected followed by the addition of 1-dimethyl-*t*-butylsilyloxy-3-butyne (95 mg, 0.51 mmol). The mixture was stirred for 15 min to yield a clear yellow solution which was cooled to –78 °C and treated with ethereal MeLi (0.95 mL, 1.04 mmol). Concurrently, thiophene (46 mL, 48 mg, 0.57 mmol) was placed in a round-bottom flask with a stir bar. The flask was evacuated and purged with argon as above and THF (1 mL) was introduced via syringe. The solution was cooled to –20 °C

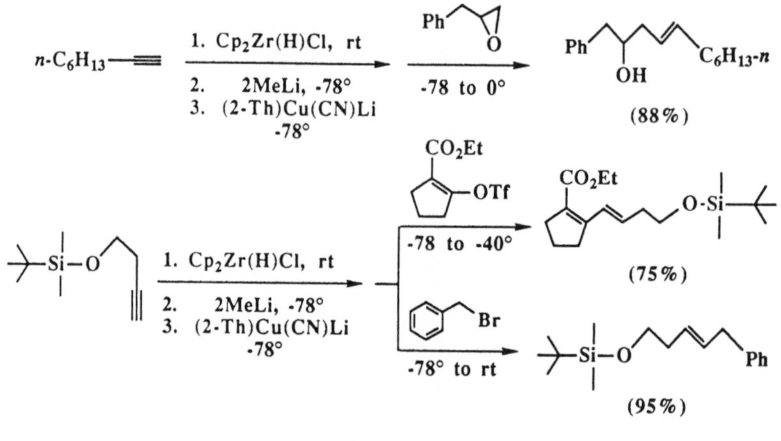

Scheme 50.

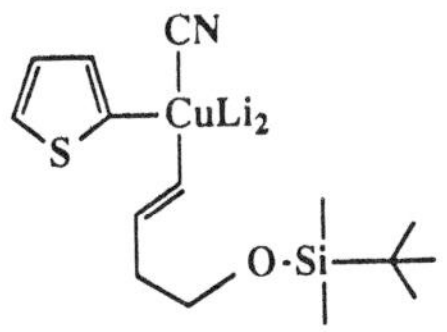

75

to which was added *n*-BuLi (0.25 mL, 0.50 mmol). The solution was then stirred for 1 h at –20 °C and transferred via cannula to a suspension of CuCN (45 mg, 0.51 mmol) in THF (1 mL) which was precooled to –78 °C. The bath was subsequently removed and the suspension warmed to room temperature to obtain a clear orange solution which was cooled again to –78 °C and transferred via cannula to the solution containing the vinyl zirconocene. After being stirred for 30 min at –78 °C, the resulting solution was treated with benzyl bromide (40 mg, 0.23 mmol) and then kept for 1 h at –78 °C followed by warming to room temperature and stirring at this temperature for an additional 3 h. Quenching was carried out using 10% conc NH_4OH in saturated aqueous NH_4Cl, and was followed by extraction with ether. The extracts were washed with water, dried, and concentrated *in vacuo*. Column chromatography with silica gel (10% EtOAc in hexane) gave 1-dimethyl-*t*-butylsilyloxy-6-phenyl-3-pentene (59 mg, 95%).

Although Kato could dispense with this portion of the project rather quickly, what demanded much of his time was the unexpected problems associated with the reactions of cuprates **74** (e.g., **75**) with simple primary, *unactivated* halides. We reasoned that the explanation for the essentially 0% yield observed with such unexceptional cases as 1-bromooctane lies in the incompatability of cuprates **74** with the zirconium(IV) Lewis acidic by-product of the transmetallation at the higher temperatures (usually ca. 0 °C–rt) required for displacements with these less reactive mixed cuprates (cf. Figure 7). In other words, it was our tendency to focus on the *copper* portion of the transmetallation event, normally not having to worry about the fate of zirconium. Well, that "ignorance is bliss" approach was no longer valid, for it was time to recognize the fully balanced equation, where Cp_2ZrMe_2 is

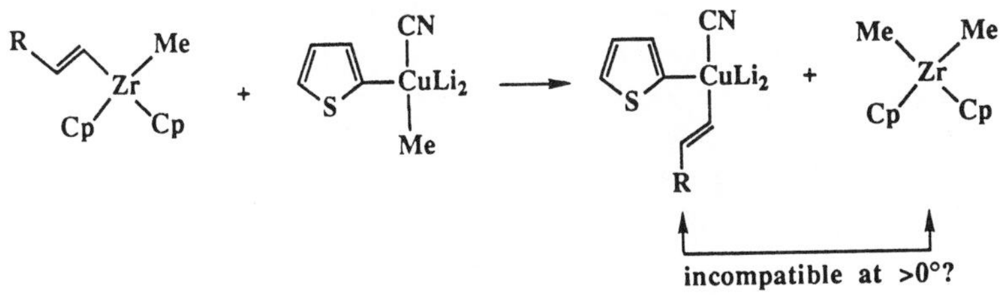

incompatible at >0°?

Scheme 51.

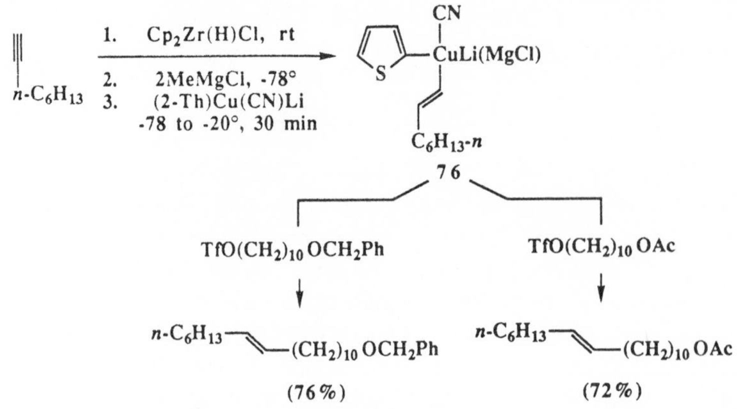

n-C_6H_{13} (1-octyne)
1. $Cp_2Zr(H)Cl$, rt
2. 2MeMgCl, -78°
3. (2-Th)Cu(CN)Li
-78 to -20°, 30 min

(2-thienyl)Cu(CN)Li(MgCl)(CH=CH-C_6H_{13}-n)
76

$TfO(CH_2)_{10}OCH_2Ph$ → n-C_6H_{13}-CH=CH-$(CH_2)_{10}OCH_2Ph$ (76%)

$TfO(CH_2)_{10}OAc$ → n-C_6H_{13}-CH=CH-$(CH_2)_{10}OAc$ (72%)

Scheme 52.

not likely to be an innocent bystander as reaction temperatures rise during these alkylations (vs., e.g., 1,4-additions which proceed at –78 °C; Scheme 51).

Eventually, Kato discovered that by utilizing the more reactive primary triflates as electrophiles and replacing MeLi with MeMgCl in the transmetallation sequence, good yields of alkylated products could be realized (Scheme 52). Here, the presumed cuprate is the mixed metal reagent **76**, known from our earlier work to be capable of displacing primary leaving groups.[88]

During Kato's search for conditions to maximize vinyl zirconocene to vinyl cuprate transmetallation for subsequent alkylations, his labmate, a talented undergraduate by the name of Paul Fatheree, was looking into the possibility that an imidazole ring lithiated at nitrogen could serve as a new "dummy" ligand in cyanocuprate reactions. By treating sublimed imidazole with n-BuLi in THF and then adding this solution to CuCN or CuCN·2LiCl,[73] a pale green solution is produced at room temperature, suggesting the presence of (Imid)Cu(CN)Li, **77** (Scheme 53). When cooled to –78 °C, however, homogeneity is lost only to be regained upon slight warming following introduction of an equivalent of an organolithium, the stoichiometry now implying formation of R(Imid)Cu(CN)Li_2, **78**. Low temperature ^{1}H NMR analysis of these mixed H.O. reagents led to a surprising observation regarding their composition. Thus, at –78 °C, it appears that **78** is in fact only one component of an equilibrium established with the lower order cyanocuprate **80** and free lithioimidazole **79** (Scheme 53).[89] Fortunately, the occurrence of **79** does not interfere in the Michael addition chemistry of **78**, although we have noted occasional competitive N-alkylation with unactivated alkylating reagents.

When the benzyl ether of glycidol was then added to **78** (R = n-Bu) and warmed from –78 to 0 °C over 4 h, an 82% yield of the ring opened product **81** was isolated

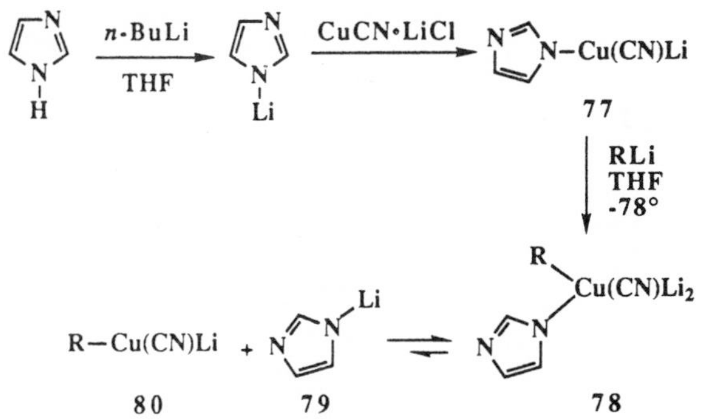

Scheme 53.

(Scheme 54). By contrast, the identical reaction *sans* the lithioimidazole (i.e., using *n*-BuCu(CN)Li) was quite sluggish, returning only 22% of **81** along with 55% of recovered epoxide.[89]

Bill Hagen, a graduate student also in the group whose thesis was being built around the functionalization and reactions of the imidazole ring,[90] was the logical person with whom Fatheree collaborated on this chemistry. To further share the load, a new recruit, Kirk Stevens, was added to the team to do a few more representative examples, and with Hagen, to determine if the analogous pyrrolo ligand (from *N*-lithiopyrrole) could function in a related capacity. Stevens' study on these cuprates (e.g., **82**) showed, as with imidazolo reagents **78**, that better yields are to be expected from the H.O. rather than L.O. species. In this case, their 1,4-additions to 4-isopropylcyclohexenone under otherwise identical conditions demonstrated the value of the pyrrolo ligands' presence (Scheme 55).[89]

Unlike the lithiated imidazole + CuCN gemisch, water-white solutions of L.O. pyrrolo cuprate **83** remain homogeneous, even at –78 °C. Conversion to the H.O. cuprate **84** imparts a slight yellow coloration. Both L.O. cuprates **77** and **83** are stable at refrigerator temperatures (ca. 4 °C) as 0.2 M solutions in THF for at least a week (Scheme 56).[89]

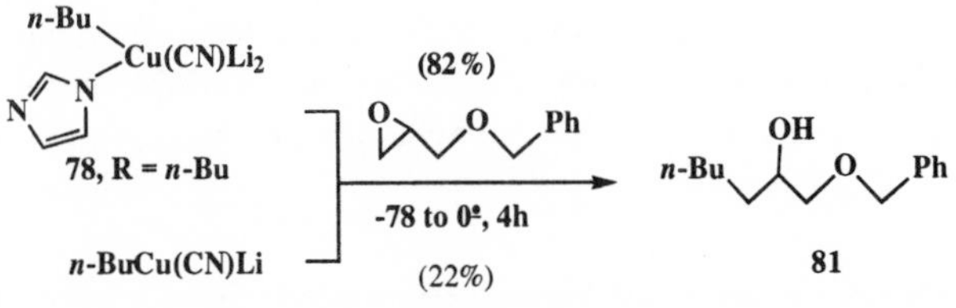

Scheme 54.

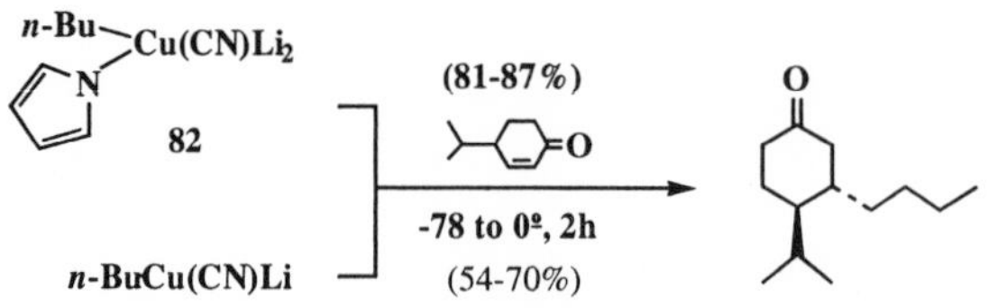

Scheme 55.

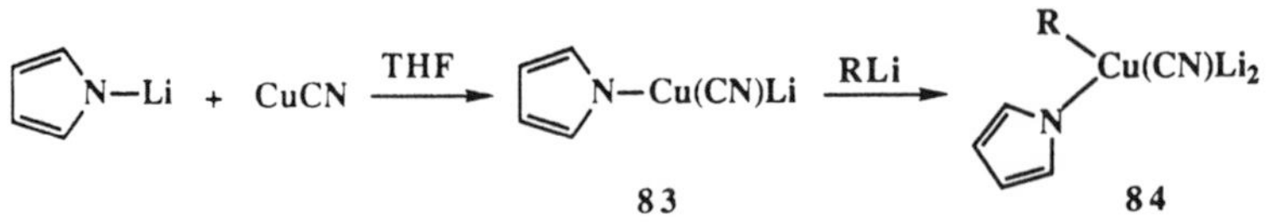

Scheme 56.

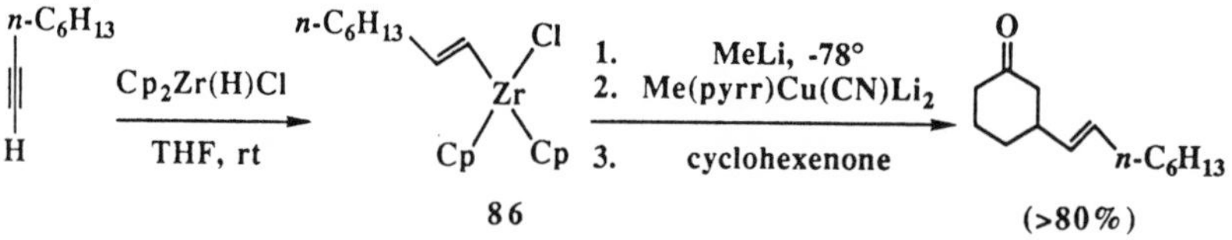

Scheme 57.

In assessing the previously discussed Ellsworth and Kato contributions, which show that Me(2-Th)Cu(CN)Li_2 (**45**) is effective at transmetallations of vinyl zirconocenes, Hagen questioned the propensity of Me(Pyrr)Cu(CN)Li_2 (**85**) to induce ligand exchange under similar mild conditions. He therefore treated zirconocene **86** with MeLi at –78 °C and then cuprate **85**, followed by the enone. The results from this quick test indeed suggested that the pyrrolo group could, likewise, serve as the equivalent of the 2-thienyl moiety for these transmetallations (Scheme 57). The overall benefits therefore associated with the imidazole- and pyrrole-derived mixed H.O. cyanocuprates lie in the avoidance of sulfur chemistry and the stability of *N*-lithioimidazole (it can be stored at room temperature for months and even weighed out quickly in air).[89] Potentially most significant, from a practical standpoint, is the lack of any homocoupling product (i.e., 2,2′-dithiophene) so often seen after quenching reactions involving thienyl cuprates.

B. Catalytic Processes

At the time the details of the vinyl zirconocene-H.O. cuprate transmetallations were being worked out, Siegmann was developing the biaryl couplings (*vide supra*)[11] directly across the hall from Ellsworth. The newly introduced zirconium chemistry to our group jogged Siegmann's memory of his former labmate at the ETH in Zurich, Dr. Roman Lehmann, who had received his degree on the basis of extensive studies involving *catalyzed* alkylations and conjugate addition reactions of alkyl and vinyl zirconocenes![91] I can vividly recall Cony coming into my office, Lehmann's thesis in hand, and telling me: "Bruce, you had better have a look at this." He didn't need to say more, for here was a tome of beautiful, unpublished methodology from Venanzi's laboratory, "buried" in this opus. By the time I eventually got through the German, relying quite heavily on assistance by Siegmann, it seemed that there should be a way to bring this methodology to light. I thought that we might be able to expand on the Lehmann study, using allylic phosphates[92] as coupling partners. Our collaborative study with Eiichi Nakamura and Masayuki Arai on copper-catalyzed allylations of alkyl titanium reagents[93] taught us to expect allylic phosphates to react in a highly S_N2' fashion (Scheme 58). I presented this scenario to Keil, and we agreed (with Venanzi's blessing)[94] that upon completion of his study on functionalized H.O. lithio cuprates (*vide supra*), he would take on this assignment. In the interim, a manuscript (submitted to *J. Org. Chem.*) appeared in my mail box for review from Clay Heathcock, written by a then new faculty member at the University of Pittsburgh, who just happened to be the same co-author with Ireland on the paper dealing with vinyl alane/H.O. cuprate transmetallations (*vide supra*), Prof. Peter Wipf. The subject of the paper was: copper-catalyzed 1,4-additions of alkyl zirconocenes! I felt obliged to alert Heathcock to the Venanzi work, with the upshot being the publication of the Wipf contribution on Michael additions.[95] Eventually, our joint efforts with Venanzi and Lehmann led to a manuscript on the corresponding allylic alkylations.[96]

The simplicity of these one-pot processes is striking; the hydrozirconation of an olefin occurs in <1 h at 25–40 °C, after which is added the Cu(I) salt (10 mol %) and then the enone[95] or alkylating agent.[96] The Michael additions occur, not surprisingly, much more rapidly than the alkylations, especially if run at 40 °C (Scheme 59). Any one of several sources of Cu(I) is acceptable (e.g., CuBr·DMS, CuBr, CuI, CuCN, etc.), although such is not the case with the $Cp_2Zr(H)Cl$ required.[95] Commercially prepared material seems to be the best choice, although *in situ* derived reagent using the Cp_2ZrCl_2/LiAl(O-*t*-Bu)$_3$H combination[77] is also "active". Oddly, $Cp_2Zr(H)Cl$ prepared from the Keil procedure (i.e., Cp_2ZrCl_2 +

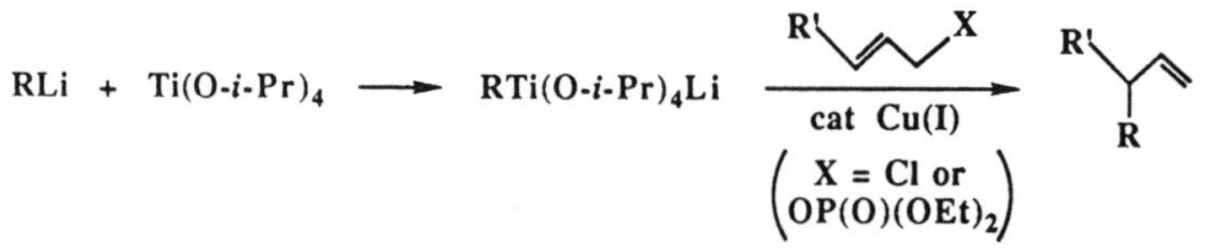

$$\text{RLi} + \text{Ti(O-}i\text{-Pr)}_4 \longrightarrow \text{RTi(O-}i\text{-Pr)}_4\text{Li} \xrightarrow[\text{cat Cu(I)} \; (\text{X = Cl or OP(O)(OEt)}_2)]{\text{R}'\text{CH=CHCH}_2\text{X}} \text{R}'\text{CH(R)CH=CH}_2$$

Scheme 58.

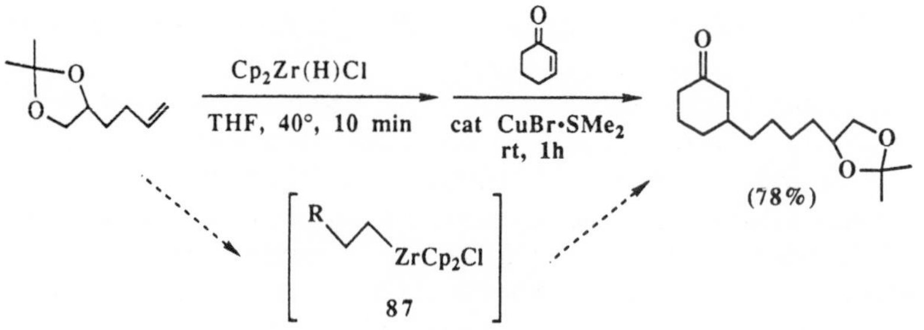

Scheme 59.

$LiEt_3BH$, *vide supra*) did *not* lead to a 1,4-adduct, although the control reaction using commercial $Cp_2Zr(H)Cl$ in the presence of by-product Et_3B was unaffected. Likewise, generation of Schwartz' reagent from Cp_2ZrCl_2 + *t*-BuMgCl gave rise to an alkyl zirconocene;[75] however, subsequent 1,4-addition to cyclohexenone was *not* observed.

3-n-Hexylcyclohexan-1-one:[95] A solution of 200 mg (2.38 mmol) of 1-hexene in 5 mL of THF was treated at room temperature with 674 mg (2.61 mmol, 1.1 equiv) of zirconocene chloride hydride and stirred at 40 °C for 10 min. After the mixture was cooled to room temperature, 228 mg (2.38 mmol) of 2-cyclohexenone and 50 mg (0.24 mmol, 0.10 equiv) of copper(I) bromide–dimethyl sulfide complex were added. The reaction mixture was stirred at 40 °C for 10 min, quenched with 25 mL of wet Et_2O, and extracted twice with a saturated aqueous solution of $NaHCO_3$. The organic layer was dried (Na_2SO_4), filtered through silica gel, and concentrated. The residue was purified by chromatography on silica gel (10% AcOEt in hexanes) to afford 344 mg (79%) of the product as a colorless oil.

The allylations of intermediates **87** using 10 mol % CuCN were no less intriguing.[96] Ratios for S_N2':S_N2 products in unsymmetrical cases are consistently high, and yields of chain-elongated materials isolated ranged from the low 80's to nearly quantitative. The nature of the leaving group, being either chloride, bromide, or phosphate, did not seem to influence the outcome, as the following examples show (Scheme 60).

Perhaps most curious is the comparison Lehmann made of the $RZrCp_2Cl$/cat. CuCN system with other copper reagents toward S_N2' attack on allylic halides.[91,96] Thus, while the Yamamoto reagent ("RCu·LiI + BF_3")[97,98] gives product ratios almost as selectively as the Zr/Cu mix, the Gilman cuprate can favor either pathway depending upon substrate (Table 7).

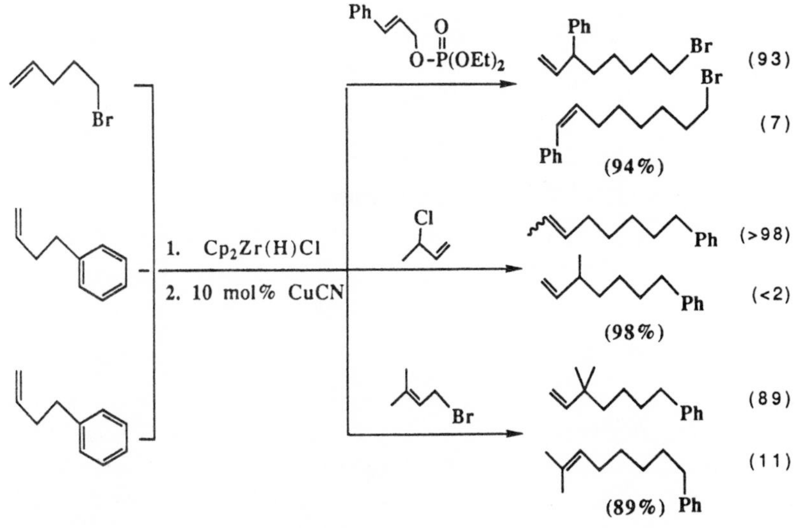

1. $Cp_2Zr(H)Cl$
2. 10 mol% CuCN

Ph, O, O-P(OEt)$_2$ — Br (93), Br (7), Ph (94%)

Cl — Ph (>98), Ph (<2) (98%)

Br — Ph (89), (11) (89%) Ph

Scheme 60.

Table 7. Comparison of Copper Reagents Toward S_N2'/S_N2 Reactions

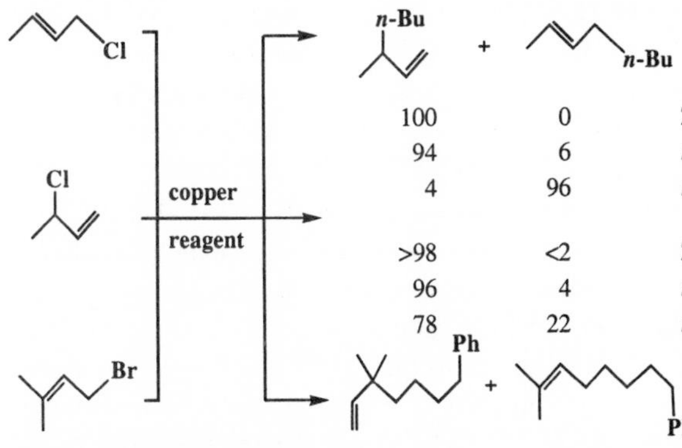

Educt	*Products (%)*		*Copper Reagent*
Cl (copper reagent)	n-Bu (SN2')	n-Bu (SN2)	
	100	0	Zr/Cu*
	94	6	n-BuCu·BF$_3$
	4	96	n-Bu$_2$CuLi
Cl	>98	<2	Zr/Cu*
	96	4	n-BuCu·BF$_3$
	78	22	n-Bu$_2$CuLi
Br	Ph (SN2')	Ph (SN2)	
	89	11	Zr/Cu*
	90	10	n-BuCu·BF$_3$
	0	100	n-Bu$_2$CuLi

*Zr/Cu = Ph(CH$_2$)$_4$ZrCp$_2$Cl, cat CuCN

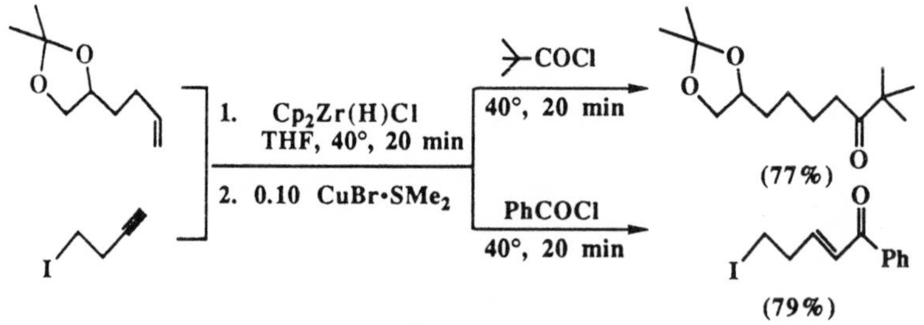

1. $Cp_2Zr(H)Cl$ THF, 40°, 20 min
2. 0.10 $CuBr \cdot SMe_2$

COCl, 40°, 20 min (77%)

PhCOCl, 40°, 20 min (79%)

Scheme 61.

Surprisingly, there are several electrophiles that do not participate in these copper-catalyzed couplings, some of which are equally as activated as are allylic halides and phosphates. Those that were attempted that do *not* react (in our hands) include benzylic halides, allylic acetates, and mono-substituted epoxides. Acid halides, however, do appear to react with alkyl and vinyl zirconocenes, as recently shown by Wipf,[99] to afford ketones and enones in moderate to good yields. These acylations proceed readily at 40 °C, and the electrophile may contain other functional groups such as nitrile, ester, protected alcohols, and even a primary iodide (Scheme 61).

VI. FUTURE DIRECTIONS AND PROSPECTS

Having written several reviews on organocopper chemistry,[3,65,100] I might have thought at this point that it would be somewhat straightforward to see what lies ahead in this field. Each time I consider making such projections, a flashback to 1980 reminds me that, as a second year assistant professor, I once said to Robert Wilhelm, my first Ph.D. student (now at Syntex) who initially developed H.O. cyanocuprates,[101] that we would "take our one *JACS* communication on these reagents and then move on to other things." Right! What does seem almost obvious though, to us at least, is that the continued development of transmetallation chemistry mediated by a Cu(I) species, whether a salt (e.g., CuCN, CuBr·DMS, etc.) or cuprate ($R_2Cu(CN)Li_2$), will surely run its course. It is also quite reasonable to assume, given the extraordinary wealth of organometallic chemistry at ones fingertips, that many other metals beyond Sn, Si, Al, and Zr are "ripe" for participation in these synthetically valuable ligand exchange processes. Comasseto and Barriel have added tellurium to the list,[102] since vinyl tellurides **88** undergo transmetallations with $R_2Cu(CN)Li_2$ in THF at ambient temperatures and subsequently deliver the vinylic ligand to enones in a Michael sense (e.g., see Scheme 62). Very recently, a joint effort by the Curran and Wipf groups has led to a novel tandem radical cyclization/transmetallation sequence initiated by samarium iodide, although in this case CuI·$P(OEt)_3$ proved to be the Cu(I) source of choice rather

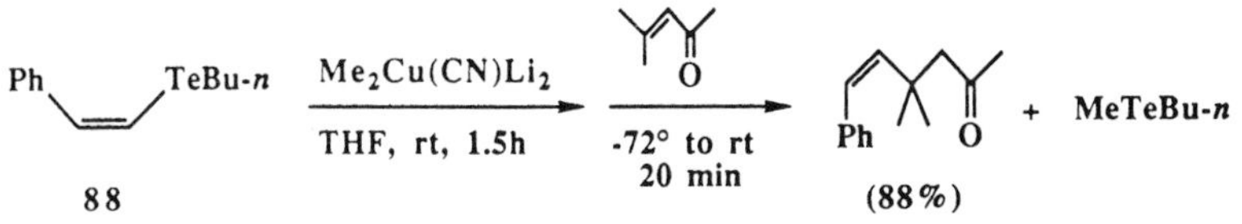

Ph TeBu-n; $Me_2Cu(CN)Li_2$, THF, rt, 1.5h; -72° to rt, 20 min; 88; Ph, O (88%); + MeTeBu-n

Scheme 62.

than CuCN or a cyanocuprate (Scheme 63).[103] Zirconium, however, seems today at least, to be a "hot" metal in organocopper chemistry. Of course, its use in tandem with Cu(I) reagents is just one particular aspect of this metals' rich and popular synthetic chemistry, noteworthy contributions on which can be found from the Negishi,[104] Buchwald,[105] and Livinghouse[106] groups, to name a few. Surely at first glance it is tempting to consider, as examples, the numerous cyclizations of enynes[104] which proceed through cyclic vinylic zirconocenes, and their related heteroatom-containing systems[105,106] as viable candidates for further manipulation via a Cu(I)-induced transmetallation.

It should be mentioned that all of the Cu(I)-initiated transmetallation chemistry to date has focused on only two types of situations: (1) a single metal within the organic framework, as in **89** or **90**, and (2) an ethylene or butadiene system containing the Bu_3Sn residue at the 1,2- and 1,4-sites, respectively, in an E-disposition (cf. **91** and **92**).

But there are several other configurations possible (e.g., see Figure 9) for multimetal relationships which could prove especially attractive for making multiple C–C bonds, potentially in a one-pot operation, as was achieved with **92**, $M_1 = M_2 = Me_3Sn$ (*vide supra*). These include: (1) formation of a differentiated sp^2 1,1-dimetallo species **94**, akin to that used in the elegant studies by Normant and Knochel (i.e., **93**),[107] and (2) a mono-metalated species **95** with functionality (FG) susceptible to insertion by another metal, as represented in Scheme 64.

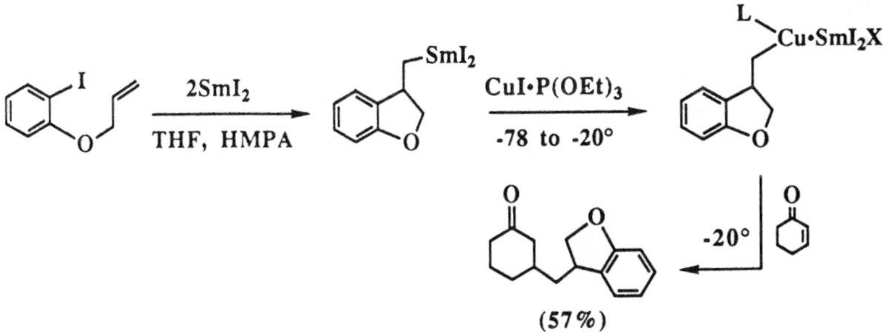

I, O; $2SmI_2$, THF, HMPA; SmI_2, O; $CuI{\cdot}P(OEt)_3$, -78 to -20°; L, $Cu{\cdot}SmI_2X$, O; -20°, O; O, O (57%)

Scheme 63.

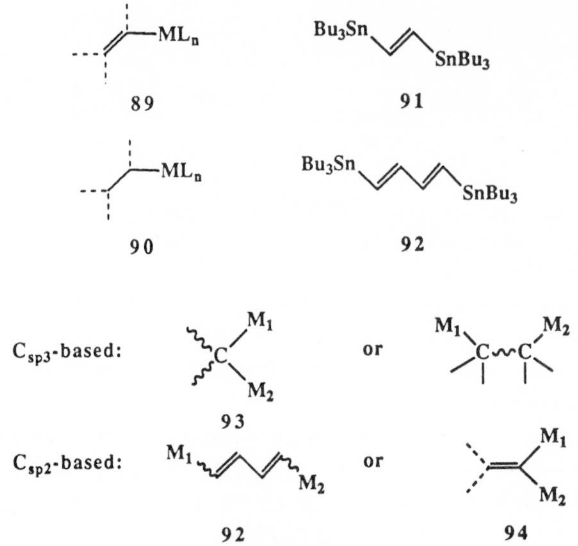

Figure 9.

The origin of our ongoing work concerning the sp^2-based diorganometallic **94** actually grew out of quite a different line of thought. Part of J.C. Barton's Ph.D. research in our group required his preparation of a specific (*Z*)-vinylstannane, **96**, in chiral, nonracemic form. After many unsuccessful attempts using, e.g., chiral hydride reductions of an acetylenic ketone,[108] we settled on lactic acid as the precursor source of chirality. All of our routes, however, proceeded through an acetylenic stannane, since the alternatives for (*Z*)-vinylstannane formation were not amenable to arriving at **96** in that they called for the C_{sp^2}-chiral C_{sp^3} bond to be made (Scheme 65).[109] On the other hand, the intermediacy of an acetylenic stannane **98** greatly simplifies the analysis by allowing for tin to be inserted at a later stage (e.g., using the single-flask sequence **97**→**98**, Scheme 66).[110] This was hardly a stroke of genius, however, since Lindlar reduction to **99** is not only unprecedented, but as one expert in the field, Terry Mitchell, in his review, "Transition-Metal Catalysis in Organotin Chemistry", put it:[111] "...transition-metal-catalyzed hydrogenation of vinyl- and alkynyltins does not occur,...". Only recently, in fact, has a vinylstannane been hydrogenolyzed by Lautens using (dppb)Rh(nbd)BF_4 [dppb = 1,4-bis(diphenylphosphino)butane; nbd = norbornadiene] under rather severe condi-

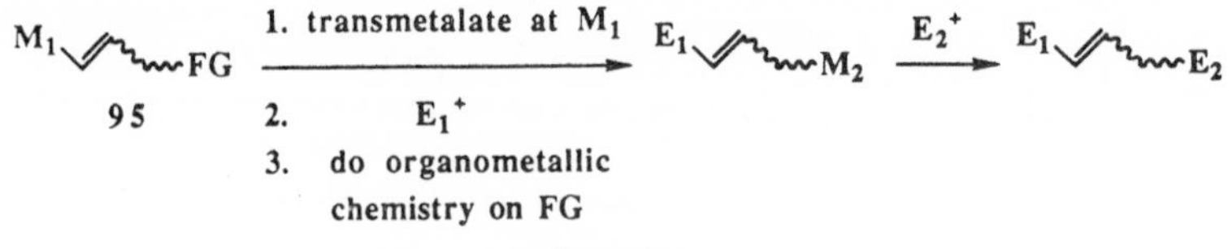

Scheme 64.

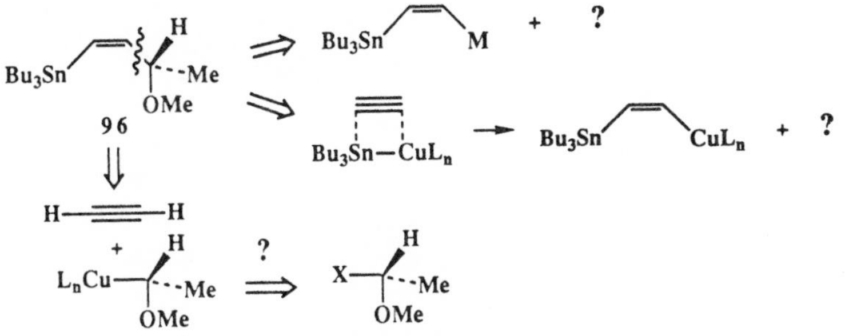

Scheme 65.

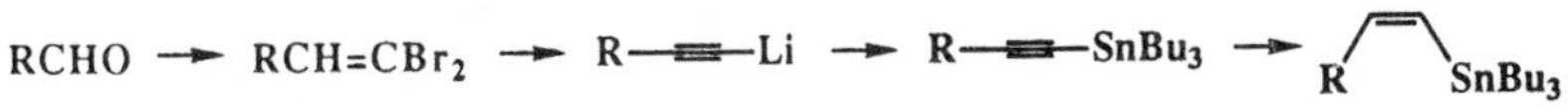

$RCHO \rightarrow RCH{=}CBr_2 \rightarrow R{-}{\equiv}{-}Li \rightarrow R{-}{\equiv}{-}SnBu_3 \rightarrow$ (Z)-$RCH{=}CHSnBu_3$

97 98 99

Scheme 66.

tions of pressure (ca. 1500 psi) over 24–36 hours.[112] Our "ace in the hole" here was the knowledge that hydrozirconation of trialkylstannyl alkynes takes place rapidly under "normal" conditions (in THF at rt), and in very high yields (Eq. 12), an

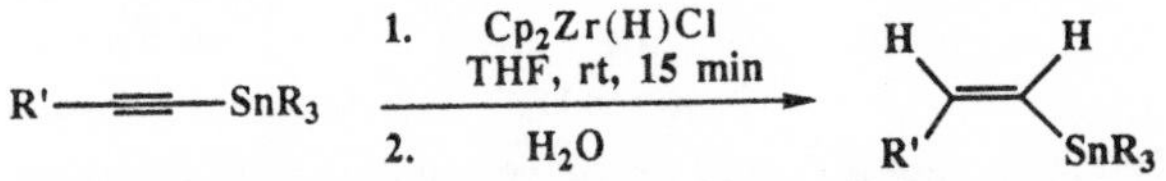

$$R'{-}{\equiv}{-}SnR_3 \xrightarrow[\text{2. } H_2O]{\text{1. } Cp_2Zr(H)Cl,\ \text{THF, rt, 15 min}} (Z)\text{-}R'CH{=}CHSnR_3 \qquad (12)$$

observation made originally by Keil as part of his study on the *in situ* generation of Schwartz' reagent (*vide supra*). By merely working up the initially formed adduct, J.C. obtained his (Z)-vinylstannane essentially quantitively. With J.C. feeling the pinch of severe time constraints on him (i.e., he was already months late for his job), Keil went back to these hydrozirconations and established the methods' generality as a facile entry to (Z)-vinylstannanes,[113] some typical examples being illustrated in Scheme 67.

Scheme 67, last case.[113] A dry 10-mL round-bottom flask purged with and maintained under a blanket of argon, was charged with 40 mg (0.15 mmol) of $Cp_2Zr(H)Cl$. To this was added 3 mL of dry THF and 58 mg (0.13 mmol) of 1-tributylstannylbutyn-4-ol benzyl ether. The suspension was stirred for 15 min during which the mixture became a clear yellow solution. It was then diluted with 5 mL of pentane and after 10 min of additional stirring, the supernatant was filtered through a short plug of silica gel and concentrated *in vacuo* to afford 53 mg (90%) of (Z)-tributylstannylbuten-4-ol benzyl ether.

A more exciting aspect to this work, however, was recognition of the regiochemical issues surrounding the hydrozirconation step. To which position of the acetylene

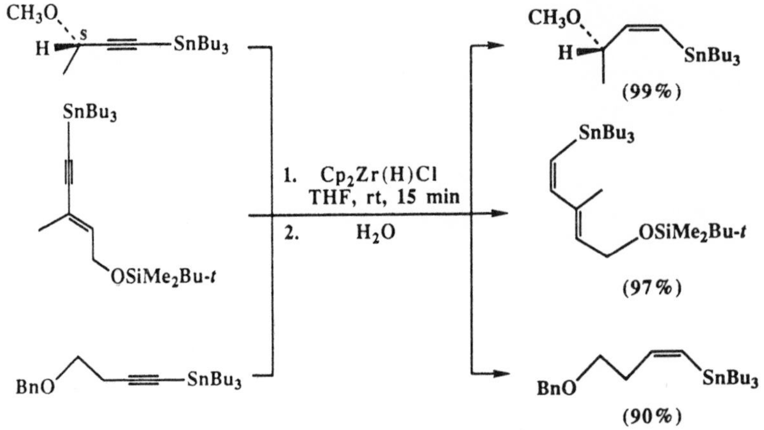

Scheme 67.

does the $ZrCp_2Cl$ go? From literature reports on an analogous silylacetylene,[114] it seemed likely that path **a** (Scheme 68) would be preferred, as opposed to path **b**, thereby leading to 1,1-dimetallo intermediate **100**. Such a species presents some very interesting possibilities, given the known differences between vinylstannanes[48] and vinylzirconocenes[70,78] toward H.O. cyanocuprate transmetallations. In other words, **100** (Scheme 69) could be viewed first as a vinyl zirconocene, prone to transmetallation with $R_2Cu(CN)Li_2$ at –78 °C. Once the resulting cuprate has made the desired C–C bond, it might then be possible, in a one-pot operation, to take advantage of the remaining vinylstannane toward a similar transmetallation (at room temperature) so as to construct a second C–C. This might be achieved by simply adding MeLi (1 equiv) after the initial coupling with E_1 (i.e., at the stage of **101**), which can either combine with **102** to reform the H.O. cuprate **104** and then

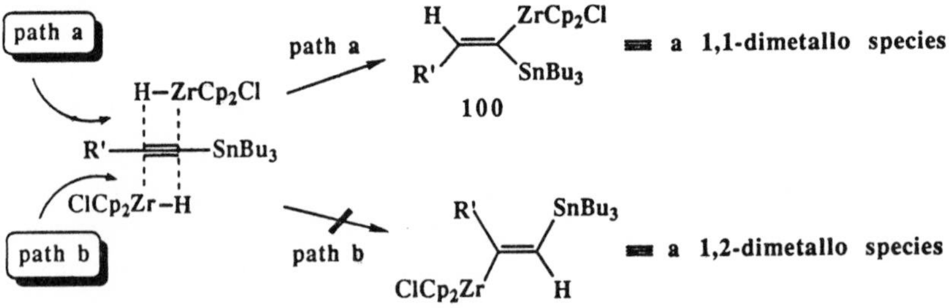

Scheme 68.

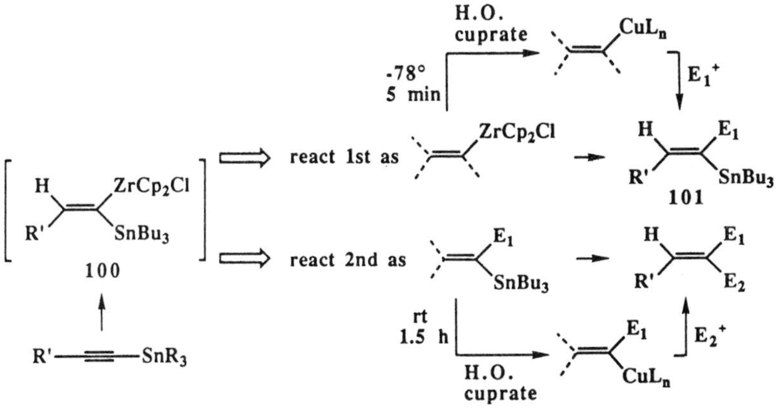

H.O.
cuprate
-78°
5 min
CuL_n
E_1^+
H
$ZrCp_2Cl$
R'
$SnBu_3$
100
react 1st as
$ZrCp_2Cl$
H
E_1
R'
$SnBu_3$
101
react 2nd as
E_1
$SnBu_3$
H
E_1
R'
E_2
R'—≡—SnR_3
rt
1.5 h
H.O.
cuprate
E_1
CuL_n
E_2^+

Scheme 69.

H
$ZrCp_2Cl$
R'
$SnBu_3$
100
H.O.
cuprate
L
CN
H
$CuLi_2$
R'
$SnBu_3$
E_1^+
H
E_1
R'
$SnBu_3$ + L-Cu(CN)Li
101
102
MeLi
H
E_1
R'
Li
103
L-Cu(CN)Li
102
L
$Cu(CN)Li_2$
Me
104
H
E_1
R'
$SnBu_3$
101
H
E_1
R'
$Cu(CN)Li_2$
L
E_2^+
H
E_1
R'
E_2
107
E_2^+
H
E_1
R'
$Cu(CN)Li_2$
L
105
same cuprate
106

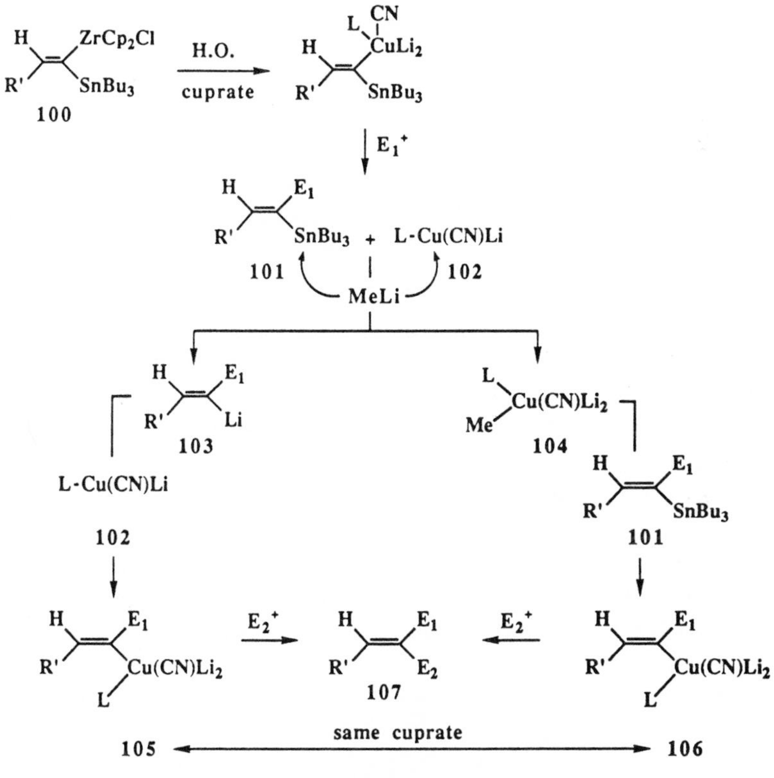

Scheme 70.

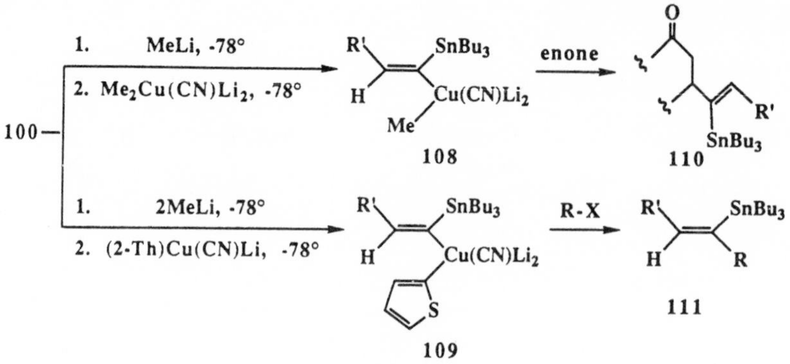

Scheme 71.

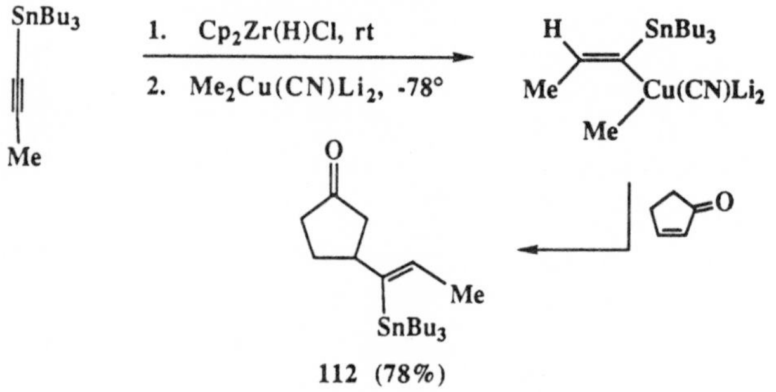

Scheme 72.

transmetallate the remaining vinylstannane **101**, or undergo Li/Sn exchange to the vinyllithium **103** and then recombine with **102** to the H.O. cuprate **105** (Scheme 70). Either pathway leads to the same species (i.e., **105** = **106**), which ultimately reacts with another electrophile (E_2^+) making the second C–C bond (**107**). Moreover, the presence of the vinylstannane moiety in **101** could alternatively be viewed as a Stille coupling partner.[40]

Thus far, Keil has established the selective transmetallation of the vinyl zirconocene portion of **100**, and depending upon the nature of the electrophile, converted **100** to either mixed cuprate **108** or **109** (Scheme 71).[115] Representative examples of the former conversion, ultimately leading to keto vinylstannanes **110** (e.g., stannylated 1,4-dienes **112** and **113**) are illustrated in Schemes 72 and 73. We are currently working on the single-flask, double transmetallation concept outlined above, as well as several modifications involving tandem cuprate/palladium couplings of these dimetallo intermediates.

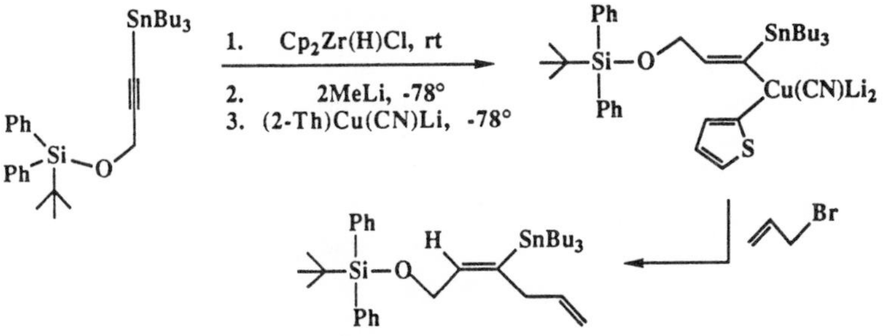

Scheme 73.

VII. CONCLUDING REMARKS

This review highlights, in a personal way, many of the events surrounding the chemistry described by our group in the literature over the past few years, updating a prior account.[3] The new methodology based on "higher order" cyanocuprate technology being developed in numerous laboratories throughout the world is but a small yet highly valued subdivision of the field, indicative of the vitality of organocopper chemistry as a whole. Whether reagents designated as "$R_2Cu(CN)Li_2$" are rightfully characterized as cyanide-bound species or otherwise, while an interesting question in its own right, bears no impact whatsoever on their synthetic utility. Indeed, higher order cyanocuprates, unlike Gilman reagents (i.e., R_2CuLi) have a unique reactivity profile that enables them to undergo efficient ligand exchange processes with other organometallic intermediates. This property alone not only streamlines some traditional modes of reagent generation, but greatly extends their scope in that highly basic species (i.e., organolithiums and Grignards) need not serve as cuprate precursors. Thus, "transmetallations" between Cu(I) and transition metal complexes may be a key buzzword in organocopper chemistry for the 1990s.

ACKNOWLEDGMENTS

It is a pleasure for me to acknowledge the insightful and skilled contributions by the talented students at the undergraduate, graduate and postdoctoral levels, whose names appear in the text. The interactions with them, their predecessors who have built our program in copper chemistry, as well as those currently within the group who are the basis of tomorrow's findings, are for me what makes this chemistry so enjoyable. The financial support for our programs extended to us by the National Institutes of Health, National Science Foundation, and the Petroleum Research Fund, administered by the American Chemical Society, is also warmly acknowledged.

ADDENDUM

With this manuscript securely in the editor's hands, I felt that it would probably be a while until the story I had told concerning the "Bertz Affair" would require updating. Well, not quite! Less than a month after submission, I visited Michigan as an invited seminar speaker. During a leisurely drive from Detroit's Metro Airport to Ann Arbor, Masato Koreeda broke the news to me that my schedule included a session with Jim Penner-Hahn, who in collaboration with Paul Knochel, has utilized EXAFS spectroscopy to investigate the nature of the environment around copper for the species generated from the 2*n*-BuLi + CuCN combination. The point to their experiments, I was told, was to show via this technique that CuCN, BuCu(CN)Li, and $Bu_2Cu(CN)Li_2$ all contain a copper-bound cyanide ligand. Unfortunately for me, it didn't work out as planned. In fact, the data have led these workers to disclose that while a lower order cyanocuprate clearly possesses a cyanide group on copper, the species formed from 2RLi + CuCN is also *dicoordinate* (i.e., with two butyl but no cyano ligands on copper). Hence, their *JACS* manuscript in press, a copy of which was graciously left in my possession, concludes that the more accurate representation is Bertz' expression "$R_2CuLi{\cdot}LiCN$"!

Could the nature of the EXAFS experiment be misleading here? Is it possible, or perhaps even probable, that the EXAFS data cannot take into account the likelihood of a bridging cyanide, making the environment around the Cu(I) atom nonlinear? Depending upon the angular array of ligands around the metal, the effectiveness of the EXAFS technique can, in fact, vary.

Although one could take issue with the authors' conclusion that their EXAFS data are "unambiguous", the study is very intellectually provocative in that the $R_2CuLi{\cdot}LiCN$ formulation *cannot* be an accurate description of the nature and location of the cyanide ion (i.e., it is unequivocally not part of free, intact LiCN). After all, as shown herein (cf. Figure 2), when LiCN is added to Me_2CuLi, complete loss of this salt occurs (by both IR and NMR experiments) to give the same species arrived at by adding 2MeLi to CuCN. Thus, the intriguing question arises: if the CN ligand is not on copper, then where is it?

REFERENCES AND NOTES

1. Bertz, S.H. *J. Am. Chem. Soc.* **1990**, *112*, 4031.
2. Gilman, H.; Jones, R.G.; Woods, L.A. *J. Org. Chem.* **1952**, *17*, 1630.
3. Lipshutz, B.H. *Synlett.* **1990**, *3*, 119.
4. Lipshutz, B.H.; Sharma, S.; Ellsworth, E.L. *J. Am. Chem. Soc.* **1990**, *112*, 4032.
5. Bertz, S.H.; Gibson, C P.; Dabbagh, G. *Tetrahedron Lett.* **1987**, *28*, 4251.
6. Bertz, S.H. *J. Am. Chem. Soc.* **1991**, *113*, 5470.
7. Posner, G.H. *Org. React.* **1975**, *22*, 253 (see Table IIc therein).
8. Bringmann, G.; Walter, R.; Weirich, R. *Agnew. Chem. Int. Ed. Engl.* **1990**, *29*, 977; Kaufmann, T. *Agnew. Chem. Int. Ed. Engl.* **1974**, *13*, 291; Camus, A.; Marsich, N., *J. Organomet. Chem.* **1972**, *46* 385.
9. Bertz, S.H.; Gibson, C.P. *J. Am. Chem. Soc.* **1986**, *108*, 8286.
10. Lipshutz, B.H.; Wilhelm, R.S.; Kozlowski, J.A. *J. Org. Chem.* **1984**, *49*, 3938.

11. Lipshutz, B.H.; Siegmann, K.; Garcia, E. *J. Am. Chem. Soc.* **1991**, *113*, 8161.
12. Lipshutz, B.H.; Siegmann, K.; Garcia, E. *Tetrahedron* **1992**, *48*, 2579.
13. Lipshutz, B.H.; Siegmann, K.; Garcia, E.; Kayser, F. *J. Am. Chem. Soc.* **1993**, *115*, 9276.
14. Watanabe, T.; Miyaura, N.; Suzuki, A. *Synlett* **1992**, 207.
15. Inoue, S.; Takaya, H.; Tani, K.; Otsuka, S.; Sato, T.; Noyori, R. *J. Am. Chem. Soc.* **1990**, *112*, 4897; Noyori, R.; Tomino, I.; Tanimoto, Y. *ibid.* **1979**, *101*, 3129.
16. For representative studies, see (a) Lipshutz, B.H.; Kozlowski, J.A.; Wilhelm, R.S. *J. Org. Chem.* **1984**, *49*, 3943 [^{1}H NMR data on $R_2Cu(CN)Li_2$]; (b) Lipshutz, B.H.; Ellsworth, E.L.; Siahaan, T.J. *J. Am. Chem. Soc.* **1988**, *110*, 4834 [^{1}H NMR data on $R_2Cu(CN)Li_2 + BF_3{\cdot}Et_2O$]; (c) Lipshutz, B.H.; Ellsworth, E.L.; Siahaan, T.J.; Shirasi, A. *Tetrahedron Lett.* **1988**, *29*, 6677 [effects of Me_3SiCl on $R_2Cu(CN)Li_2$].
17. Bergdahl, M.; Lindstedt, E-L.; Nilsson, M.; Olsson, T. *Tetrahedron* **1988**, *44*, 2055; Bergdahl, M.; Lindstedt, E-L.; Nilsson, M.; Olsson, T. *ibid.* **1989**, *45*, 535; Bergdahl, M.; Lindstedt, E-L.; Olsson, T. *J. Organomet. Chem.* **1989**, *365*, C11; Lindstedt, E-L.; Nilssop, M.; Olsson, T. *ibid.* **1987**, *334*, 255; Hallnemo, G.; Olsson, T.; Ullenius, C. *ibid.* **1984**, *265*, C22; **1985**, *282*, 133; Hallnemo, G.; Ullenius, C. *Tetrahedron* **1983**, *39*, 1621; Christenson, B.; Olsson, T.; Ullenius, C. *ibid.* **1989**, *45*, 523; Christenson, B.; Hallnemo, G.; Ullenius, C. *ibid.* **1991**, *47* 4739; *Tetrahedron Lett.* **1986**, *27*, 395; Christenson, B.; Hallnemo, G.; Ullenius, C. *Chem. Ser.* **1987**, *27*, 511.
18. Lipshutz, B.H.; Kayser, F.; Siegmann, K. *Tetrahedron Lett.* **1993**, *34*, 6689. In principle, at least, if LiCN is introduced to these L.O. cuprates under "kinetic" conditions, ratios similar to those observed from oxidations of H.O. cuprates should result.
19. Lipshutz, B.H.; Reuter, D.C.; Ellsworth, E.L. *J. Org. Chem.* **1989**, *54*, 4975.
20. Oehlschlager, A.C.; Hutzinger, M.W.; Aksela, R.; Sharma, S.; Singh, S.M. *Tetrahedron Lett.* **1990**, *31*, 165.
21. Lipshutz, B.H.; Ellsworth, E.L.; Dimock, S.H.; Reuter, D.C. *Tetrahedron Lett.* **1989**, *30*, 2065.
22. (a) Fleming, I.; Waterson, D.J. *J. Chem. Soc., Perkin Trans. I* **1984**, 1809; (b) Fleming, I.; Newton, T.W. *ibid.* **1984**, 1805; (c) Fleming, I.; Newton, T.W.; Roessler, F. *ibid.* **1981**, 2527; (d) see also Sharma, S.; Oehlschlager, A.C. *Tetrahedron* **1991**, *47*, 1179.
23. (a) For a recent use and detailed procedure relying on the method of Still,[23b] see Cunico, R.F. *J. Org. Chem.* **1990**, *55*, 4474; (b) Still, W.C. *ibid.* **1976**, *41*, 3063.
24. Corriu, R.J.P.; Huynh, V.; Iqbal, J.; Moreau, J.J.E. *J. Organomet. Chem.* **1984**, *276*, C61.
25. Capella, L.; Degl'Innocenti, A.; Reginato, G.; Ricci, A.; Taddei, M. *J. Org. Chem.* **1989**, *54*, 1473.
26. Hudrlik, P.F.; Wasugh, M.A.; Hudrlik, A.M. *J. Organomet. Chem.* **1984**, *271*, 69.
27. (a) Amberg, W.; Seebach, D. *Chem. Ber.* **1990**, *123*, 2439; (b) see also, *ibid.* **1990**, *123*, 2429 and 2413 in the series.
28. Lipshutz, B.H.; Reuter, D.C. *Tetrahedron Lett.* **1989**,*30*, 4617.
29. Still, W.C. *J. Am. Chem. Soc.* **1977**, *99*, 4836; Tamborski, P.C.; Ford, F.E.; Soloski, E.J. *J. Org. Chem.* **1963**, *28*, 237.
30. Pereyre, M.; Quintards J-P.; Rahm, A. In: *Tin in Organic Synthesis*, Butterworths, London, 1987.
31. (a) Lipshutz, B.H.; Sharma, S:, Reuter, D.C. *Tetrahedron Lett.* **1990**, *32*, 7253; (b) U.S. Patent No. 4,282,165, issued August 4, 1981.
32. Chenard, B.L.; Van Zyl, C.M. *J. Org. Chem.* **1986**, *51*, 3561. We are indebted to Dr. Chenard for the extremely helpful advice and suggestions offered in the early stages of our work here.
33. Piers, E.; Tillyer, R.D. *J. Org. Chem.* **1988**, *53*, 2065.
34. (a) Barbero, A.; Cuadrado, P.; Fleming, I.; Gonzalez, A.M.; Pulido, F.J. *J. Chem. Soc. Chem. Commun.* **1992**, 351; (b) Degl'Innocenti, A.; Stucchi, E.; Capperucci, A.; Mordini, A.; Reginato, G.; Ricci, A. *Synlett* **1992**, 332.
35. Aksela, R.; Oehlschlager, A.C. *Tetrahedron* **1991**, *47*, 1163.
36. Capella, L.; Degl' Innocenti, A.; Mordini, A.; Reginato, G.; Ricci, A.; Seconi, G. *Synthesis* **1991**, 1201.

37. Barbero, A.; Cuadrado P.; Fleming, I.; Gonzalez, A.M.; Pulido, F.J. *J. Chem. Soc., Chem. Commun.* **1990**, 1030.
38. Marek, I.; Alexakis, A.; Normant, J.-F. *Tetrahedron Lett.* **1991**, *32*, 6337.
39. Beauder, I.; Parrain, J-L.; Quintard, J.-P. *Tetrahedron Lett.* **1991**, *32*, 6333.
40. Labadie, J.W.; Stille, J.K. *J. Am. Chem. Soc.* **1983**, *105*, 6129.
41. Kende, A.S.; DeVita, R.J. *Tetrahedron Lett.* **1990**, *31*, 307; Connell, R.D.; Helquist, P.; Akermark, B. *J. Org. Chem.* **1989**, *54*, 3359; Kann, N.; Rein, T.; Akermark, B.; Helquist, P. *ibid.* **1990**, *55*, 5312; Connell, R.D.; Rein, T.; Akermark, B.; Helquist, P. *ibid.* **1988**, *53*, 3845; Nikaido, M.; Aslanian, R.; Scavo, F.; Helquist, P.; Akermark, B.; Backvall, J-E. *ibid.* **1984**, *49*, 4738; Rein, T.; Akermark, B.; Helquist, P. *Acta. Chem. Scand.* **1988**, *B42*, 569.
42. Bury, R.W. In: *Agricultural Uses of Antibiotics*; Moats, W.A., Ed.: ACS Symposium Series 320, American Chemical Society; Washington, DC, 1986, pp. 61–72.
43. Sharma, S.; Oehlschlager, A.C. *J. Org. Chem.* **1989**, *54*, 5383.
44. Sharma, S.; Oehlschlager, A.C. *J. Org. Chem.* **1991**, *56*, 770.
45. Ashby, E.C.; Watkins, J.J. *J. Am. Chem. Soc.* **1977**, *99*, 5312; *J. Chem. Soc., Chem. Commun.* **1976**, 784.
46. Bertz, S.H.; Dabbagh, G. *J. Am. Chem. Soc.* **1988**, *110*, 3668.
47. Singer, R.D.; Oehlschlager, A.C. *J. Org. Chem.* **1991**, *56*, 3510.
48. Behling, J.R.; Babiak, K.A.; Ng, J.S.; Campbell, A.L.; Moretti, R.; Koerner, M.; Lipshutz, B.H. *J. Am. Chem. Soc.* **1988**, *110*, 2641.
49. Ashe, A.; Mahmoud, S. *Organometallics* **1988**, *7*, 1878.
50. Lipshutz, B.H.; Lee, J.I. *Tetrahedron Lett.* **1991**, *32*, 7211.
51. Posner, G.H.; Sterling, J.J.; Whitten, C.E.; Lentz, C.M.; Brunelle, D.J. *J. Am. Chem. Soc.* **1975**, *97*, 107.
52. Lipshutz, B.H.; Wilhelm, R.S.; Kozlowski, J.A. *J. Org. Chem.* **1984**, *49*, 3928.
53. Lipshutz, B.H.; Kozlowski, J.A.; Parker, P.A.; Nguyen, S.L.; McCarthy, K.E. *J. Organomet. Chem.* **1985**, *285*, 437.
54. Malmberg, H.; Nilsson, M.; Ullenius, C. *Tetrahedron Lett.* **1982**, *23*, 3823.
55. Zwiefel, G.; Miller, J.A. *Org. React.* **1984**, *32*, 375.
56. For example, see Baker, R.; Castro, J.L. *J. Chem. Soc., Chem Commun.* **1989**, 378.
57. Lipshutz, B.H.; Koerner, M.; Parker, D.A. *Tetrahedron Lett.* **1987**, *28*, 945.
58. Floyd, M.B.; Schaub, R.E.; Weiss, M.J. *Prostaglandins* **1975**, *10*, 289; Pappo, R.; Collins, P.W. *Tetrahedron Lett.* **1972**, 2627.
59. Negishi, E. *Pure Appl. Chem.* **1981**, *53*, 2333; *Chem. Scripta* **1989**, *29*, 457.
60. Lipshutz, B.H.; Dimock, S.H. *J. Org. Chem.* **1991**, *56*, 5761.
61. Ireland, R.E.; Wipf, P. *J. Org. Chem.* **1990**, *55*, 1425.
62. Nguyen, D.T. M.S. Thesis, UCSB, 1989.
63. Wipf, P.; Smitrovich, J.H.; Moon, G.W. *J. Org. Chem.* **1992**, *57*, 3178.
64. The alkyl/alkenyl order of release is reversed, however, for conjugate additions; the alkynyl ligand remains attached to copper in both situations.
65. Lipshutz, B.H.; Sengupta, S. *Org. React.* **1992**, *41*, 135.
66. Schwartz, J.; Labinger, J.A. *Angew. Chem. Int. Ed. Engl.* **1976**, *15*, 333.
67. Referred to as such by the Aldrich Chemical Company, catalog no. 22,367-0, listed as zirconocene chloride hydride.
68. For representative examples, see [stannanes] *Chem. Abs.* 115(21): 231689r; 113(19): 171773; 102(25): 220650b; 91(17): 140417v; [iodides] *ibid.*, 115(21): 231689r, 95(5): 42475a; 86(17): 120876n; 109(17): 149196v.
69. Odic, Y.; Pereyre, M. *J. Organomet. Chem.* **1973**, *55*, 273; Negishi, E. In: *Organometallics in Organic Synthesis*; Wiley, Vol. 1, 1980; Kikukowa, K.; Umekawa, H.; Wada, F.; Matsuda, T. *Chem. Lett.* **1988**, 881.

70. Lipshutz, B.H.; Ellsworth, E.L. *J. Am. Chem. Soc.* **1990**, *112*, 7440; Babiak, K.A., et al. *J. Am. Chem. Soc.* **1990**, *112*, 7441.
71. Ganem, B. *Chemtracts Organic Chemistry* **1991**, *4*, 44.
72. Sold by Aldrich as lithium 2-thienylcyanocuprate, catalog no. 32,417-5.
73. Knochel, P.; Yeh, M.C.P.; Berk, S.C.; Talbert, J. *J. Org. Chem.* **1988**, *53*, 2390.
74. We are indebted to Boulder Scientific for supplying us with quantities of $Cp_2Zr(H)Cl$ and Cp_2ZrCl_2.
75. Swanson, D.R.; Nguyen, T.; Noda, Y.; Negishi, E. *J. Org. Chem.* **1991**, *56*, 2590.
76. Carr, D.B.; Schwartz, J. *J. Am. Chem. Soc.* **1979**, *101*, 3521.
77. Wailes, P.C.; Weigold, H. *Inorg. Syn.* **1979**, XIX, 223.
78. Lipshutz, B.H.; Keil, R.; Ellsworth, E.L. *Tetrahedron Lett.* **1990**, *31*, 7257.
79. For a recent preparation of this solid reagent, see Buchwald, S.L.; LaMaire, S.J.; Nielsen, R.B.; Watson, R.T.; King, S.M. *Tetrahedron Lett.* **1987**, *28*, 3895.
80. Brown, H.C.; Kim, S.C.; Krishnamurthy, S. *J. Org. Chem.* **1980**, *45*, 1.
81. Sold by Aldrich as Super-Hydride, catalog no 17,972-8.
82. Stack, D.E.; Dawson, B.T.; Rieke, R.D. *J. Am. Chem. Soc.* **1991**, *113*, 4672, and references therein.
83. Rao, S.A.; Knochel, P. *J. Am. Chem. Soc.* **1991**, *113*, 5735, and references therein.
84. Ebert, G.W.; Klein, W.R. *J. Org. Chem.* **1991**, *56*, 4744, and references therein.
85. Lipshutz, B.H.; Keil, R. *J. Am. Chem. Soc.* **1992**, *114*, 7919.
86. (a) Betz, S.H.; Dabbagh, G.; Mujsce, A.M. *J. Am. Chem. Soc.* **1991**, *113*, 631; Lipshutz, B.H.; Wilhelm, R.S. *ibid.* **1982**, *104*, 4696; Ashby, E.C.; Coleman, D. *J. Org. Chem.* **1987**, *52*, 4554; (b) Hallnemo, G.; Olsson, T.; Ullenius, C. *J. Organomet. Chem.* **1985**, *282*, 133; Corey, E.J.; Boaz, N.W. *Tetrahedron Lett.* **1985**, *26*, 6015.
87. Lipshutz, B.H.; Kato, K. *Tetrahedron Lett.* **1991**, *32*, 5647.
88. Lipshutz, B.H.; Nguyen, S.; Parker, D.A.; McCarthy, K.E.; Barton, J.C.; Whitney, S.; Kotsuki, H. *Tetrahedron* **1986**, *42*, 2873.
89. Lipshutz, B.H.; Fatheree, P.; Hagen, W.; Stevens, K.L. *Tetrahedron Lett.* **1992**, *33*, 1041.
90. See, for example, Lipshutz, B.H.; Hagen, W. *Tetrahedron Lett.* **1992**, *33*, 5865.
91. Dissertation ETH Nr. 8507, Zurich, 1988.
92. Yanagisawa, A.; Nomura, N.; Noritake, Y.; Yamamoto, H. *Synthesis* **1991**, 1130; Yanagisawa, A.; Noritake, Y.; Nomura, N.; Yamamoto, H. *Synlett* **1991**, 251.
93. Arai, M.; Nakamura, E.; Lipshutz, B.H. *J. Org. Chem.* **1991**, *56*, 5489; Arai, M.; Lipshutz, B.H.; Nakamura, E. *Tetrahedron* **1992**, *48*, 5709.
94. Venanzi, L.M., personal communication.
95. Wipf, P.; Smitrovich, J.H. *J. Org. Chem.* **1991**, *56*, 6494.
96. Venanzi, L.M.; Lehmann, R.; Keil, R.; Lipshutz, B.H. *Tetrahedron Lett.* **1992**, *33*, 5757.
97. Yamamoto, Y.; Maruyama, K. *J. Am. Chem. Soc.* **1978**, *100*, 3240; Yamamoto, Y.; Yamamoto, S.; Yatagai, H.; Ishihara, Y.; Maruyama, K. *J. Org. Chem.* **1982**, *47*, 119; Yamamoto, Y. *Angew. Chem. Int. Ed. Engl.* **1986**, *25*, 947.
98. Lipshutz, B.H.; Ellsworth, E.L.; Dimock, S.H. *J. Am. Chem. Soc.* **1990**, *112*, 5869.
99. Wipf, P.; Yu, W. *Synlett* **1992**, 718; see also Wipf, P. *Synthesis* **1993**, 537.
100. Lipshutz, B.H.; Wilhelm, R.S.; Kozlowski, J.A. *Tetrahedron* **1984**, *40*, 5005; Lipshutz, B.H. *Synthesis* **1987**, 325; Lipshutz, B.H. In: *Comprehensive Organic Synthesis*; Trost, B.M., Ed.; Pergamon, 1992, Vol. 1, pp. 107–138; Lipshutz, B.H. In: *Metals in Organic Synthesis: A Manual*, Schlosser, M., Ed.; Wiley, 1995, pp. 283–382.
101. Lipshutz, B.H.; Wilhelm, R.S.; Floyd, D.M. *J. Am. Chem. Soc.* **1981**, *103*, 7672.
102. Comasseto, J.V.; Berriel, J.N. *Syn. Comm.* **1990**, *20*, 1681.
103. Totleben, M.J.; Curran, D.P.; Wipf, P. *J. Org. Chem.* **1992**, *57*, 1740.
104. Swanson, D.R.; Negishi, E. *Organometallics* **1991**, *10*, 825, and references therein.
105. Tidwell, J.H.; Senn. D.R.; Buchwald, S.L. *J. Am. Chem. Soc.* **1991**, *113*, 4685; Buchwold, S.L.; Nielsen, R.B. *Chem. Rev.* **1988**, *88*, 1044.

106. Jensen, M.; Livinghouse, T. *J. Am. Chem. Soc.* **1989**, *111*, 4495.
107. Knochel, P.; Normant, J.F. *Tetrahedron Lett.* **1986**, *27*, 1039, 1043, 4427, 4431, 5727; Tucker, C.E.; Knochel, P. *J. Am. Chem. Soc.* **1991**, *113*, 9888 and references therein; see also, Mitchell, T.; Amamria, A. *J. Organomet. Chem.* **1983**, *252*, 47.
108. Midland, M.M.; McLaughlin, J.J. *J. Org. Chem.* **1984**, *49*, 1316.
109. Corey, E.J.; Eckrich, T. *Tetrahedron Lett.* 1984, *25*, 2415, 2419; Marino, J.P.; Emonds, M.V.M.; Stengel, P.J.; Oliveira, A.R.M.; Simonelli, F.; Ferreira, J.T.B. *ibid.* **1992**, *33*, 49; Zhang, H.X.; Guibe, F.; Balavoine, G. *J. Org. Chem.* **1990**, *55*, 1857; Piers, E.; Tillyer, R.D. *J. Chem. Soc. Perkin Trans. I*, **1989**, 2124.
110. Corey, E.J.; Fuchs, P.L. *Tetrahedron Lett.* **1972**, 3769.
111. Mitchell, T.N. *J. Organomet. Chem.* **1986**, *304*, 1.
112. Lautens, M.; Zhang, C.H.; Crudden, C.M. *Angew. Chem. Int. Ed. Engl.* **1992**, *31*, 232. Using diimide, however, a vinylstannane can be reduced; cf. Rahm, A.; Grimeau, J.; Petraud, M.; Barbe, B. *J. Organomet. Chem.* **1985**, *286*, 297.
113. Lipshutz, B.H.; Keil, R.; Barton, J.C. *Tetrahedron Lett.* **1992**, *33*, 5861.
114. Erker, G.; Zwettler, R.; Kruger, C.; Noe, R.; Werner, S. *J. Am. Chem. Soc.* **1990**, *113*, 9620, and references therein.
115. Lipshutz, B.H.; Keil, R. *Inorg. Chim. Acta* **1994**, *220*, 41.

THE EVOLUTION OF A COMMERCIALLY FEASIBLE PROSTAGLANDIN SYNTHESIS

James R. Behling, Paul W. Collins, and John S. Ng

I. INTRODUCTION

The involvement of Searle with metal mediated conjugate addition chemistry began in the late 1960s and was interwoven with the goal of identifying therapeutically useful prostaglandins. The conjugate addition approach to prostaglandin structures (Scheme 1) was independently conceived and researched at Searle[1] and other

Advances in Metal-Organic Chemistry
Volume 4, pages 65–87.

ISBN: 1-55938-709-2

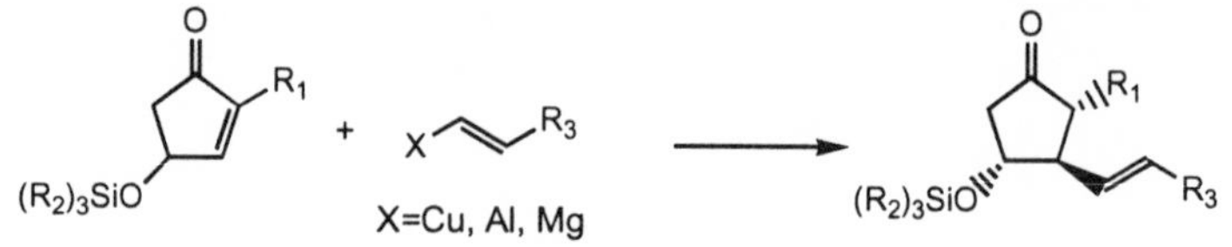

(R2)3SiO + X R3 X=Cu, Al, Mg → (R2)3SiO R1 R3

Scheme 1.

laboratories.[2–4] Today, it is the preferred manufacturing process for several synthetic prostaglandin drugs.[5] The discovery of misoprostol[6] at Searle and its subsequent development and commercialization as an antiulcer agent (Cytotec®) prompted a concerted effort to improve on the conjugate addition theme and adapt it to a process acceptable for manufacturing. In addition, extensive research was undertaken to find efficient methods for the production of the single bioactive stereoisomer of misoprostol.[7]

II. EARLY PROSTAGLANDIN SYNTHESIS

One of the first synthetic approaches in the prostaglandin field at Searle involved 1,2 Grignard addition of the acetylenic side chain **2** to enol ether derivatives of the substituted cyclopentanedione **1** (Scheme 2) to generate PGB-like structures **3**.[8,9] These compounds displayed weak or no biological activity. The cause for this weak activity was postulated to be the conformational restriction imposed by the $\Delta^{8,12}$ olefin of **3** and several methods were investigated to selectively reduce this double bond in the presence of $\Delta^{13,14}$ unsaturation. The repeated failures experienced in this approach to obtain PGE-type structures eventually led to the concept of conjugate addition as a means of solving the problem.

The first task in researching the conjugate addition strategy was preparation of the requisite cyclopentenone **5** (Scheme 3). Compound **4** which was available by a laborious multistep procedure[9,10] was a logical precursor to **5**. After investigation of numerous aluminum- and boron-based reducing reagents, sodium dihydro-bis-(2-methoxyethoxy)aluminum hydride (Red-Al®) was found to selectively reduce

OR COOH X O 1 X=H, OH R= Me BrMg H THPO 2 → O COOH X OH 3

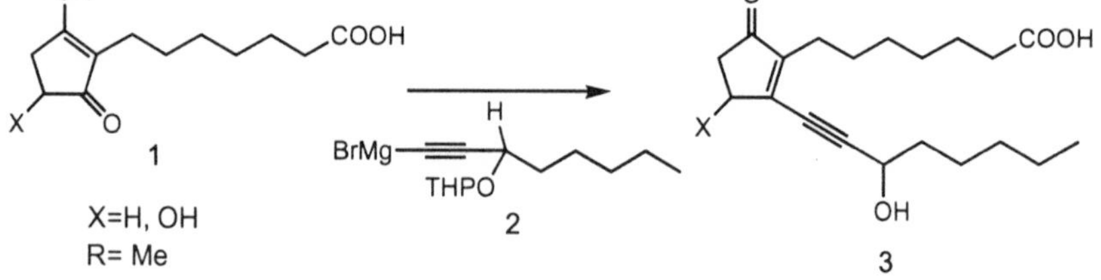

Scheme 2.

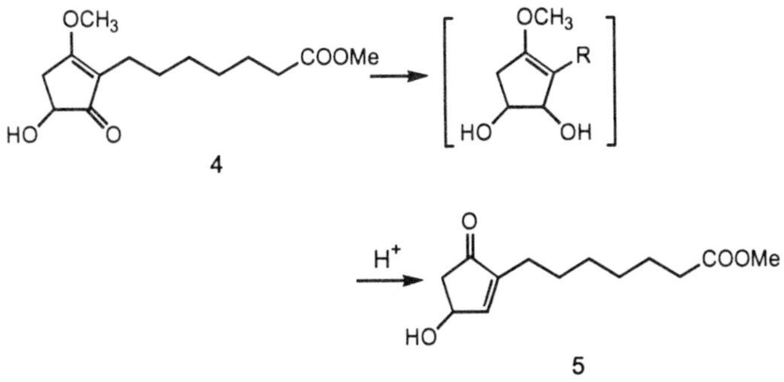

Scheme 3.

the ketone of **4** at –60 °C without significant ester attack and give, after acidic workup, the desired enone **5** in acceptable yield.[1,10]

The initial conjugate addition research involved reaction of **5** and its THP ether **6** (Scheme 4) with simple organometallic derivatives of the unfunctionalized prostaglandin omega side chain.[1] Both trialkynylaluminum and alkenyl copper reagents, **7** and **8**, were examined. The aluminum reagents were easily prepared by generation of the alkynyl lithium with *n*-BuLi followed by treatment with $AlCl_3$. In contrast, the alkenyl copper reagents were accessed through an elaborate cascade of organometallic species[1] (Scheme 5). Reaction of 1-octyne with catechol borane[11] followed by $HgCl_2$ generated the crystalline (*E*)-vinyl mercuric chloride **11** which was treated with magnesium metal to provide the vinyl Grignard **12**. Finally, a stoichiometric quantity of CuI was added at –50 °C to produce the desired alkenyl copper species **8**. This simple copper reagent, unlike cuprates, was quite unstable

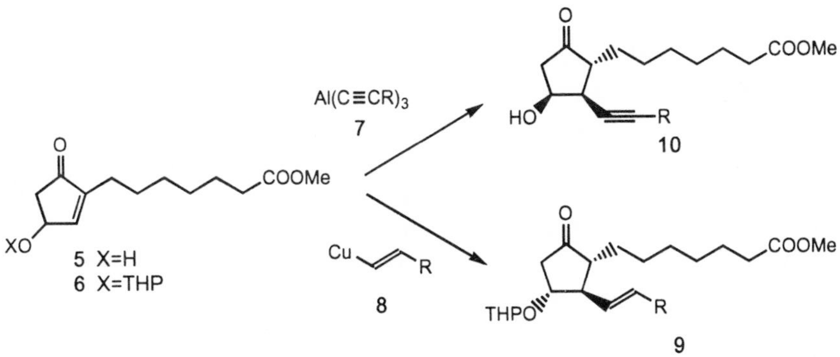

Scheme 4.

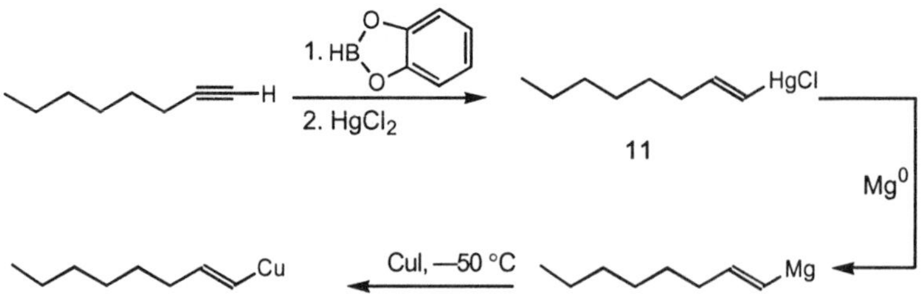

Scheme 5.

and rapidly decomposed at temperatures above –30 °C. This cascade of transformations became necessary when attempts to directly convert the corresponding (*E*)-vinyl halides, accessible by treatment of the vinyl boron intermediate with Br_2 or I_2, to vinyl lithium or magnesium species by reacting with the corresponding metals resulted in halide elimination and resultant alkyne formation.[12] The facile lithium/halogen exchange reaction was reported[13] subsequent to this work.

The aluminum reagent **7** readily added to **5** (see Scheme 4) in a conjugate fashion but was unreactive toward the protected compound **6**. Furthermore, only *cis* addition occurred with **5** to give the 11-epimer **10**, suggesting that the free hydroxy was directing the addition. In contrast, the copper reagent **8** added smoothly to **6** to provide exclusively **9** having the desired prostaglandin stereochemistry. The stereospecific nature of this organocopper reaction was a pivotal finding and is a major advantage of the conjugate addition strategy. Interestingly, a mixture of ring geometries was produced when **5** was the substrate for **8**.

The final hurdle was the preparation of fully functionalized prostaglandins with a hydroxy group appropriately positioned in the ω-chain. At that juncture, the only available acetylene derivatization techniques were the familiar catechol borane hydroboration procedure and hydroalumination with diisobutylaluminum hydride (DIBAL-H),[14] each followed by iodination to give (*E*)-vinyl iodides. Neither of these routes was very satisfactory with protected 1-alkynols because of poor and inconsistent yields, loss of protecting groups, and a laborious workup of intermediate vinylboronic acids in the case of catechol borane. Nevertheless, misoprostol (**14**, Scheme 6) was first synthesized in 1973 using these methodologies.[10,15] The cuprate reagent **13** was prepared by treatment of the (*E*)-vinyl iodide with *n*-BuLi at –60 °C followed by the addition of copper 1-pentyne solubilized with hexamethylphosphorous triamide.[16]

In 1975 Corey reported[17] that tributyltin hydride added to acetylenes at elevated temperature and in the presence of a free radical initiator (α,α′-azoisobutyronitrile) to provide (*E*)-vinylstannanes which, in turn, could be converted to either vinyl iodides or to vinyllithiums. At the time, we were interested in introducing unsatu-

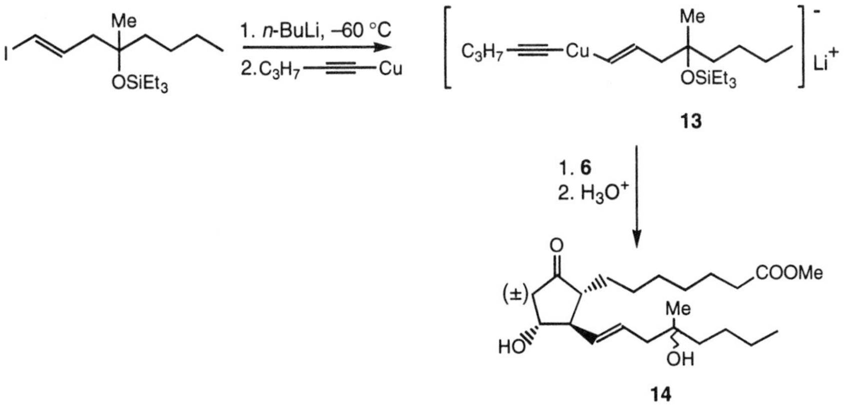

Scheme 6.

ration in the $\Delta^{17,18}$ position of the prostaglandin molecule and thought this methodology might offer an advantage over existing approaches. Under the reported conditions, a mixture of products was obtained in which the internal unsaturated position of the side chain precursors **15** had been attacked as well (Scheme 7).

In an attempt to find more selective reaction conditions, the use of UV irradiation was examined as a free radical promoter. Indeed, when a mixture of **15** and tri-*n*-butyltin hydride was placed in a pyrex glass flask without solvent, and irradiated at 0 °C with a 275-W sunlamp, selective hydrostannation of the terminal acetylene occurred to give exclusively the desired vinylstannanes as an 85/15 (*E*/*Z*) mixture.[18] The same *E*/*Z* ratio was encountered with the misoprostol side chain regardless of reaction time or conditions. This result was consistent with previous reports[19] that a steady-state ratio occurs with homopropargylic alcohols. In contrast, products from hydrostannation of propargylic alcohols, which are precursors to 15-hydroxy prostaglandins, will completely convert to (*E*)-vinylstannanes from the initially formed (*Z*)-isomers after several hours of irradiation.[18]

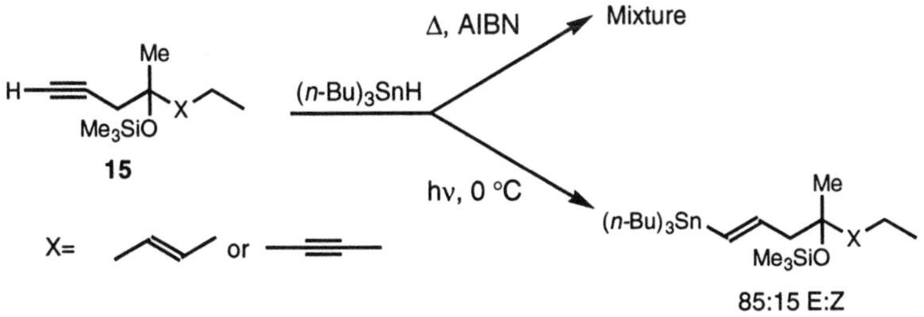

Scheme 7.

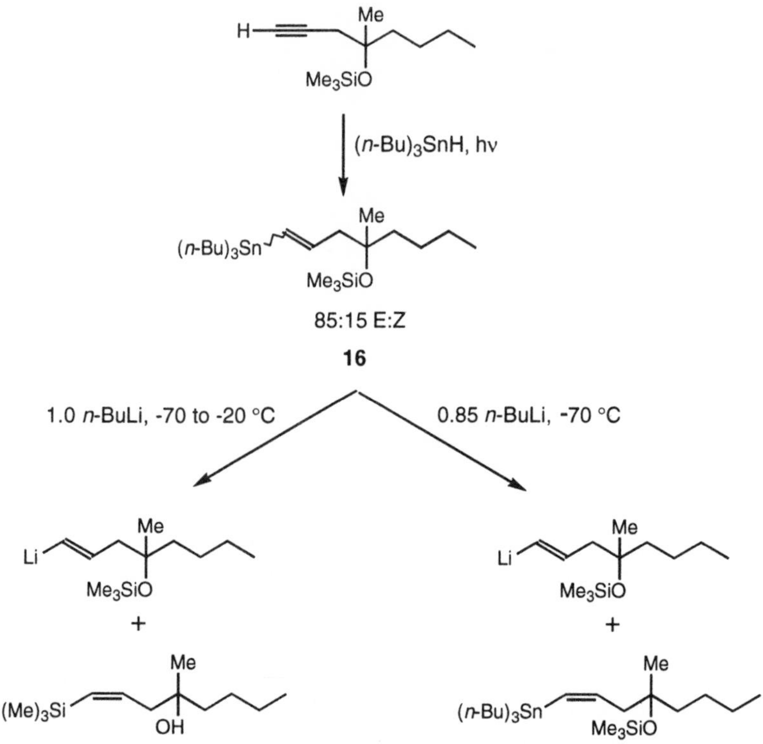

Scheme 8.

Interestingly, it was observed that the 85/15 (*E*/*Z*) ratio was not reflected in the misoprostol product obtained via the cuprate reagent generated by direct conversion of the vinylstannane mixture, yet it was preserved when the stannane mixture was converted to the corresponding vinyl iodides.[18] It was subsequently found that the rate of exchange of the (*Z*)-vinylstannane with *n*-BuLi was appreciably less than that of the (*E*)-isomer at the normal reaction temperature of –50 °C while both vinyl iodides exchanged rapidly at –60 °C. Thus, the production of (13*Z*)-misoprostol (which was separable by chromatography) could be virtually excluded by limiting the amount of *n*-BuLi to 0.85 equivalents during the tin–lithium exchange reaction and by maintaining the temperature of the reaction at –70 °C (Scheme 8).

We also discovered that the (*Z*)-vinyllithium species at a higher temperature (–20 °C) selectively underwent a silyl group migration to form an inert hydroxy vinylsilane (Scheme 8).[20] The incorporation of this hydrostannation and tin–lithium exchange technology into the synthesis of misoprostol greatly improved the conjugate addition strategy and allowed the first large scale (10–100g) synthesis of

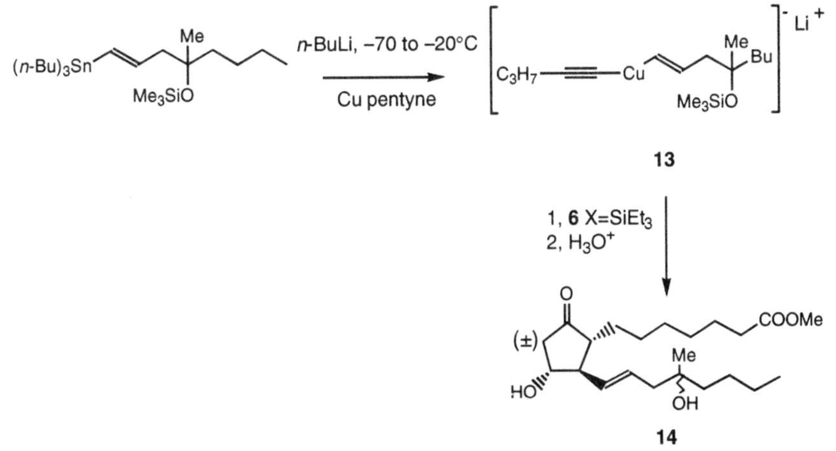

Scheme 9.

misoprostol **14** (Scheme 9). Additionally, the substitution of a triethylsilyl protecting group for the traditionally used THP in **6** also improved the process.

III. THE DEVELOPMENT CHALLENGES

This hydrostannation/pentynyl cuprate protocol provided technology that could be scaled to the kilogram level. However, several important aspects of this process required improvement before it could be considered optimal from a production standpoint. Copper(I) pentyne is not an item of commerce, and due to the potential decomposition properties of metal acetylides, presented storage and handling liabilities when prepared and used on a large scale. Although hydrostannation technology provided a convenient synthon for the misoprostol ω-chain, its efficiency suffered from the presence of 15% of the undesired (*Z*)-isomer. Additionally, the tetra-*n*-butyltin by-product of tin–lithium exchange reduced the efficiency of chromatographic purification of the product.

The conjugate addition chemistry as outlined above (Scheme 9) required that a solubilized copper pentyne solution be prepared in a separate reactor prior to addition to a vinyllithium solution in THF. This solution was unstable to a small concentration of oxygen and required a rigorously controlled atmosphere during its preparation and handling. Finally, in order to assure complete conversion of enone **6** (X = $SiEt_3$) to misoprostol **14** on a kilogram scale, three equivalents of vinyl cuprate **13** were required. This further reduced the efficiency of the process.

IV. VINYLSTANNANE–TRANSMETALLATION: CONJUGATE ADDITION IMPROVEMENTS

The first task associated with the definition of optimal conditions for the synthesis of misoprostol was to identify a cuprate reagent that was derived from a commercially available copper salt and that displayed suitable reactivity. We found that the mixed dilithio-methyl vinyl cyanocuprate reagent formed by reacting CuCN with one equivalent of MeLi and one equivalent of the vinyl–lithium species derived from **16** reacted cleanly with enone **6** to provide, after hydrolysis, misoprostol in 93% isolated yield (Scheme 10). This "higher order" cuprate reagent similar to the one first disclosed by Lipshutz[21] could be used in a 1.15 molar excess thereby reducing the amount of **16** needed for the conjugate addition. It did, however, require the use of multiple low-temperature reactors since a solution of MeCuCNLi **17** in THF required preparation at –40 °C prior to mixing with vinyllithium **18** in THF at –60 °C.

Although the exact nature of these "higher order" cuprate reagents remained unresolved,[22–23] it was clear that ligand transfer phenomena were operative. The reagent derived upon mixing an equivalent of $R_2CuCNLi_2$ with $R'_2CuCNLi_2$ (R = Me, R′ = vinyl) was identical in reactivity with the reagent derived from simple mixing of stoichiometric quantities of CuCN with MeLi and vinyllithium. Both modes of mixing resulted in a reagent that selectively transferred the vinyl group in a conjugate sense.[24] Moreover, another report by Lipshutz[25] indicated that the addition of *n*-BuLi to a solution of $Me_2CuCNLi_2$ resulted in the formation of MeLi in an NMR experiment and further solidified the notion that these cuprate reagents preferred to exist in a "mixed" state.

With these observations in mind, we postulated (Scheme 11) that if there were an equilibrium associated with a dilithio-dialkyl cyano cuprate **20**, the alkyllithium component **21** would be available to participate in a transmetallation reaction. If the transmetallation reaction were to provide a component of vinyllithium **22**, ligand mixing would occur to continually supply a new mixed vinyl-alkyl cuprate

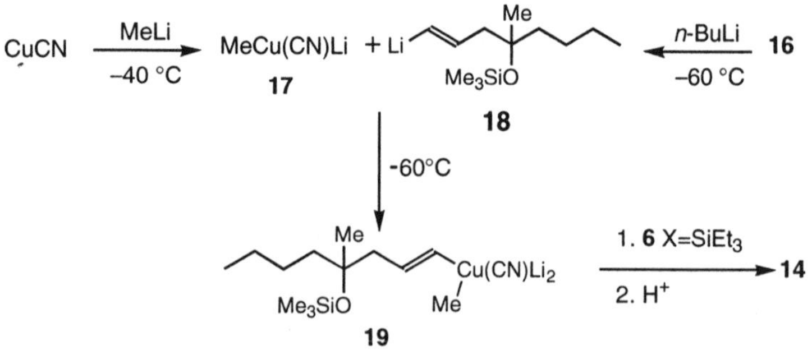

Scheme 10.

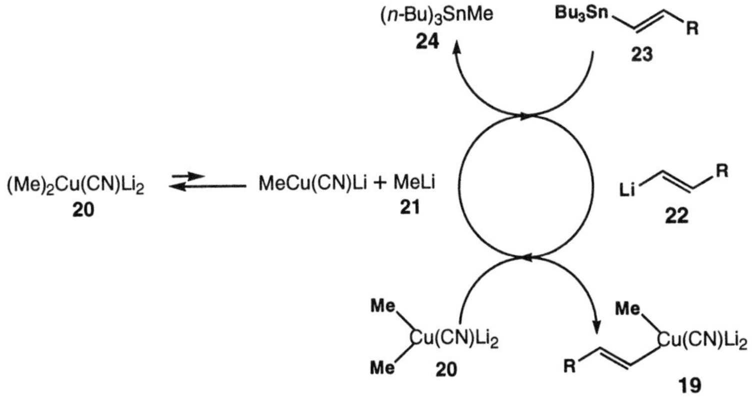

Scheme 11.

19 and regenerate the alkyllithium **21**. If the transmetallation partner were a trialkyl-vinylstannane **23**, the process would be driven to completion by the irreversible formation of tetraalkyl tin **24** (Scheme 11). Despite the report[25] that no such equilibrium existed, the hypothesis was tested, and the results were gratifying.[26]

Indeed, when vinylstannane **16** was treated with an equivalent of $Me_2CuCNLi_2$ at room temperature in THF, followed by the addition of enone **6** (X = $SiEt_3$) at –60 °C, protected misoprostol was produced (Scheme 12). These conditions were found to be suitable for the synthesis of a variety of prostaglandin analogs (Scheme 13) as well as for the introduction of other vinyl side chains to α,β unsaturated ketones (Table 1).

Misoprostol production. Copper cyanide (1.21 g, 13.5 mmol, flame dried under argon) in THF (15 mL) was treated with methyllithium (20.6 mL, 1.44 M in diethyl ether, 29.7 mmol) at 0 °C. The cooling bath was removed, and the vinylstannane **16** (7.65 g, 15.2 mmol) in THF (15 mL) was added. After 1.5 h at ambient temperature, the mixture was cooled to –64 °C (dry ice-isopropanol), and the enone **6** (3.2 g, 9.63 mmol) in THF (15 mL) was

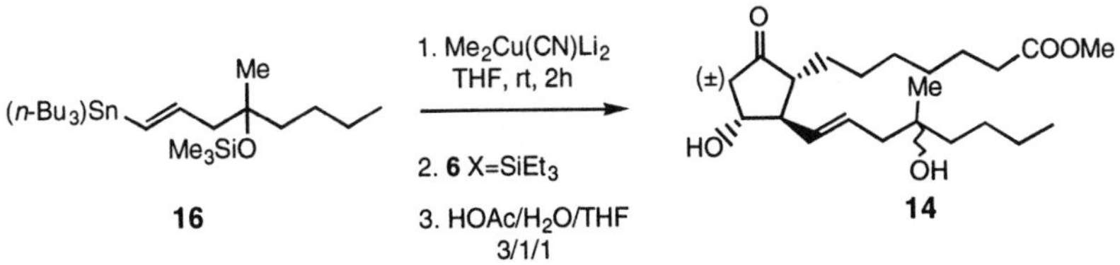

Scheme 12.

Table 1.

Enone	*Stannane*	*Product (%) Yield*
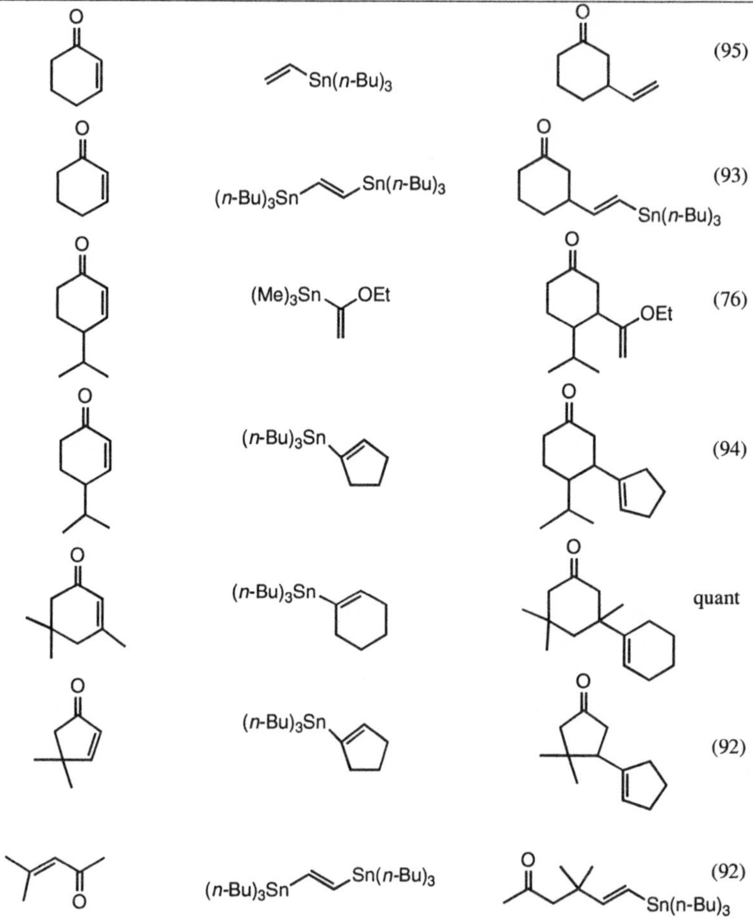 O	$Sn(n\text{-}Bu)_3$	O (95)
O	$(n\text{-}Bu)_3Sn$ $Sn(n\text{-}Bu)_3$	O $Sn(n\text{-}Bu)_3$ (93)
O	$(Me)_3Sn$ OEt	O OEt (76)
O	$(n\text{-}Bu)_3Sn$	O (94)
O	$(n\text{-}Bu)_3Sn$	O quant
O	$(n\text{-}Bu)_3Sn$	O (92)
O	$(n\text{-}Bu)_3Sn$ $Sn(n\text{-}Bu)_3$	O $Sn(n\text{-}Bu)_3$ (92)

added rapidly via a cannula. The temperature rose to –35 °C. After 3 min, the mixture was quenched into a 9:1 saturated ammonium chloride/ammonium hydroxide solution. Ethyl acetate extraction followed by solvent removal provided 11 g of residue which was solvolyzed (3:1:1, acetic acid:THF:water, 100 mL) and chromatographed (silica gel, ethyl acetate/hexane eluant) to provide the product, misoprostol **14** (3.15 g, 8.2 mmol, 91%).

This *in-situ* protocol for conjugate addition was applied successfully to the synthesis of misoprostol on a kilogram scale.[27] It permitted the entire cuprate

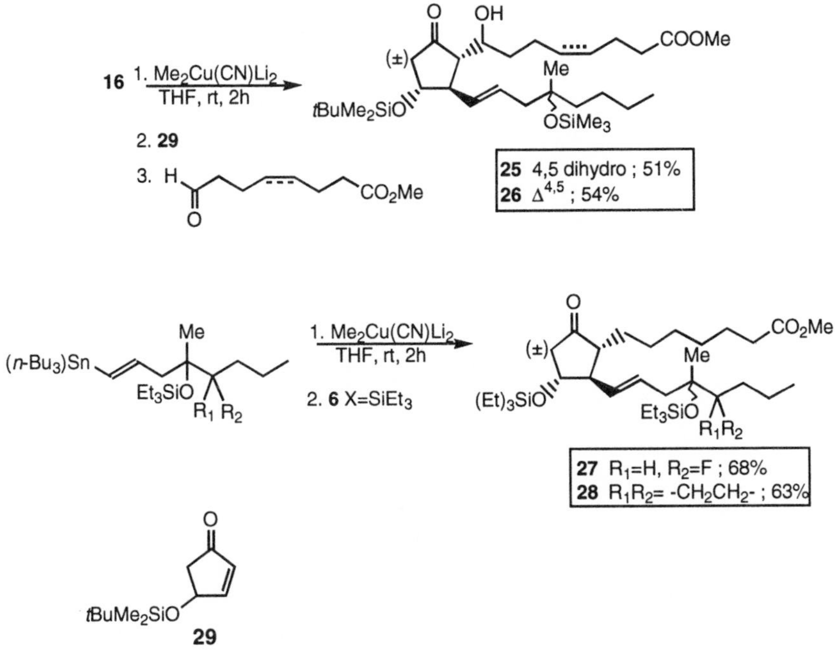

16 1. $Me_2Cu(CN)Li_2$, THF, rt, 2h; 2. **29**; 3. $H(C=O)\ldots CO_2Me$

25 4,5 dihydro ; 51%
26 $\Delta^{4,5}$; 54%

1. $Me_2Cu(CN)Li_2$, THF, rt, 2h; 2. **6** X=$SiEt_3$

27 R_1=H, R_2=F ; 68%
28 R_1R_2= $-CH_2CH_2-$; 63%

29

Scheme 13.

formation/conjugate addition protocol to be effectively carried out in one reactor, and eliminated the requirement for the separate synthesis of a copper salt containing a non-transferrable ligand. Unfortunately, the stereochemical profile of the product was unchanged with respect to olefin geometry at the $\Delta^{13,14}$ position. Even under these room temperature transmetallation conditions, an 85/15 mixture of olefin geometries was obtained. This result prompted us to search for a practical stereoselective synthesis of (*E*)-β-hydroxy-vinylstannanes. We developed a method for the synthesis of pure (*E*)-vinylstannane **16** which incorporated the regiospecific opening of a terminal epoxide **32** (see Scheme 15) with dilithio-2-thienyl-(2-(*E*)-tributylstannylethenyl)cyanocuprate.[28]

(*E*)-Bis(tributylstannyl)ethylene **31** was prepared by the hydrostannation of ethynyltributylstannane **30** (Scheme 14).[29] The product was isolated by high vacuum distillation on a wiped film evaporator to provide multikilogram quantities. The ethynyltributylstannane was prepared by treating lithium acetylide (prepared *in-situ* from acetylene and *n*-BuLi) with an equivalent of tri-*n*-butyltin chloride.[30]

The epoxide **32** was obtained by the addition of dimethylsulfoxonium methylide[31,32] to 2-hexanone. Despite the popularity of this methylide addition to an aldehyde or ketone to provide an epoxide,[33] the combination of sodium hydride and dimethylsulfoxide is unacceptably hazardous on a large scale.[34,35] The use of

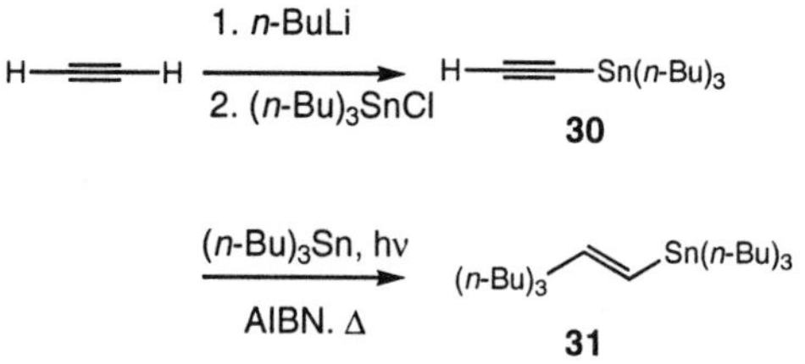

1. n-BuLi
H—≡—H ⟶ H—≡—Sn(n-Bu)$_3$
2. (n-Bu)$_3$SnCl
30

(n-Bu)$_3$Sn, hν ⟶ (n-Bu)$_3$Sn-CH=CH-Sn(n-Bu)$_3$
AIBN. Δ
31

Scheme 14.

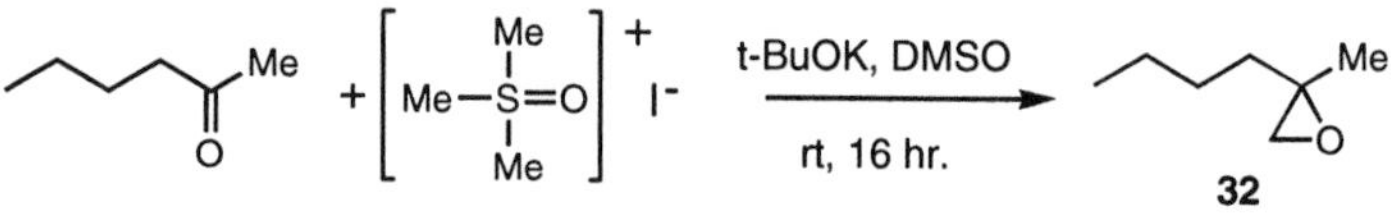

+ [Me—S(Me)(Me)=O]$^+$ I$^-$ ⟶ t-BuOK, DMSO; rt, 16 hr. 32

Scheme 15.

potassium-*t*-butoxide in place of sodium hydride (Scheme 15) provided a synthesis of epoxide **32** that was safely carried out on a multi-kilogram scale[36] in acceptable yield.

The cuprate **35** was prepared in the *in-situ* mode from (*E*)-bis(tributylstannyl)ethylene and dilithio methyl (2-thienyl) cyano cuprate **34** (Scheme 16).[28] The ensuing reaction of **35** with terminal epoxide **32** provided, after distillative isolation, the required (*E*)-vinylstannane **16**.[28,37] When the vinylstannane prepared using this protocol was used in the preparation of misoprostol, the prostaglandin was obtained

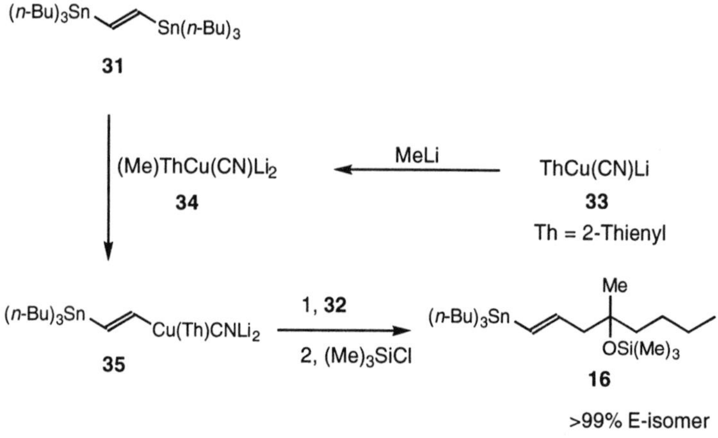

(n-Bu)$_3$Sn-CH=CH-Sn(n-Bu)$_3$
31

(Me)ThCu(CN)Li$_2$ ⟵ MeLi — ThCu(CN)Li
34 33
Th = 2-Thienyl

(n-Bu)$_3$Sn-CH=CH-Cu(Th)CNLi$_2$
35

1, 32
2, (Me)$_3$SiCl

(n-Bu)$_3$Sn ... Me, OSi(Me)$_3$
16
>99% E-isomer

Scheme 16.

Table 2.

Substrate	*Product*	*(Yield %)*
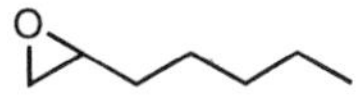 O	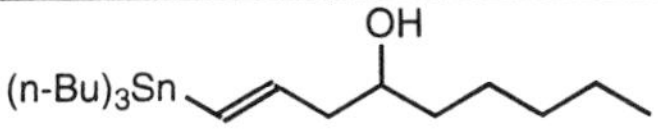 OH, $(n\text{-}Bu)_3Sn$	(69)
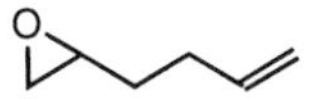 O	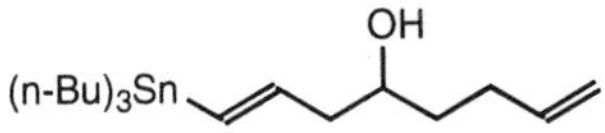 OH, $(n\text{-}Bu)_3Sn$	(58)
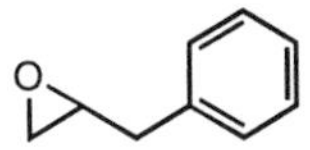 O	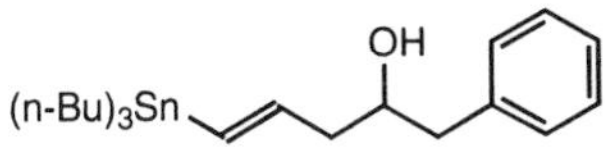 OH, $(n\text{-}Bu)_3Sn$	(68)
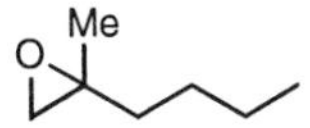 Me, O	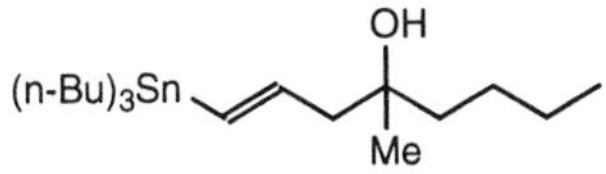 OH, $(n\text{-}Bu)_3Sn$, Me	(74)
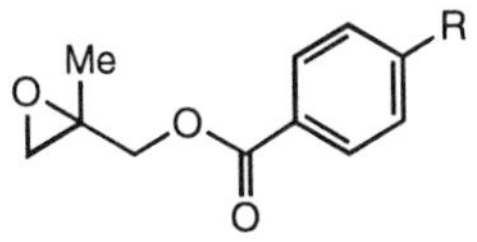 Me, O, O, O, R	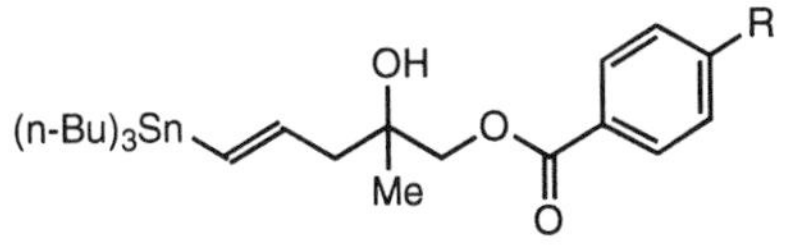 OH, $(n\text{-}Bu)_3Sn$, Me, O, O, R	R = –H, (59) R = –OMe, (61) R = –Cl, (66)
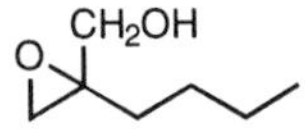 CH_2OH, O	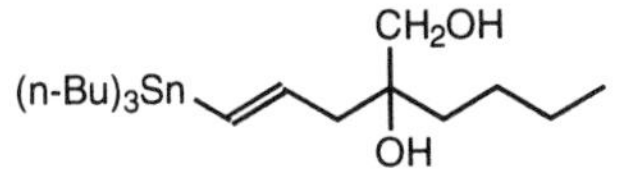 CH_2OH, $(n\text{-}Bu)_3Sn$, OH	(50)

in a 91% yield.[26,27] The product was exclusively the desired $\Delta^{13,14}$ (*E*)-isomer. This procedure was greatly simplified when the stable[38] lithio-(2-thienyl)-cyano cuprate **33** became available from Aldrich Chemical Company as a 0.25 M solution in THF. This methodology has been used for the opening of a variety of terminal epoxides[28] (Table 2).

B-Hydroxy-(E)-vinyl stannane preparation. Lithium 2-thienylcyanocuprate can be purchased from Aldrich Chemical Company (0.25 M in THF, catalog number 32,417-5) or alternatively can be prepared as a 0.5 M solution in THF as follows: To triply distilled thiophene (5.25 g, 62.5 mmol) in cold (–78 °C) anhydrous THF (24.7 mL, distilled from sodium benzophenone ketyl) was added via a syringe a solution of *n*-butyllithium (26.5 mL, 62.5 mmol, 2.4 M in hexane) at such a rate that the internal temperature did not exceed –20 °C. The resulting solution was stirred for 30 min, cooled to –60 °C and added to a cold (–60 °C) slurry of CuCN (5.59 g, 62.5 mmol, flame dried under argon) in anhydrous THF (64.7 mL). The resulting solution was allowed to come to ambient temperature and then stored at –5 °C under an atmosphere of argon. This solution was used from time to time.

The following is a typical procedure for the preparation of β-hydroxy-(*E*)-vinyl stannanes via epoxide opening with dilithio 2-thienyl(2-(*E*)-tributylstannylethenyl) cyanocuprate. To a cooled (–10 °C) solution of lithium 2-thienylcyanocuprate (2.0 mL, 0.5 M in THF, 1.0 mmol) was added methyllithium (0.76 mL, 1.45 M in diethyl ether, 1.10 mmol). The cooling bath was removed and to the resulting homogeneous solution was added (*E*)-bis (tributylstannyl)ethylene (0.53 mL, 606 mg, 1.0 mmol) via a syringe. The resulting solution was allowed to warm to room temperature for 30 min. The dark red solution of cuprate was then cooled (–78 °C) and an epoxide (0.88 mmol) was added via syringe. The reaction mixture was stirred at this temperature for 1 h and then warmed to room temperature for 1 h. It was then quenched into a vigorously stirred solution of saturated ammonium chloride/ammonium hydroxide (9/1). After stirring for 30 min the dark blue solution was extracted with ethyl acetate, the layers were separated, and the aqueous layer was reextracted with additional ethyl acetate. The organic layers were combined, washed with saturated aqueous sodium chloride solution, dried (Na_2SO_4), and concentrated to an oil. The oil was purified by medium pressure chromatography on silica gel (pretreated with triethylamine) using 95/5 hexane/ethyl acetate as the eluant to provide the desired product in which the olefin geometry was pure *E* (determined by ^{1}H NMR).

A similar three-component *one-pot* approach to prostaglandins has been demonstrated whereby (*E*)-bis(tributylstannyl)ethylene functioned as a vinylogous "linch pin"[39] (Scheme 17). Although the isolated yields of prostaglandin analogs using this extremely convergent approach are only in the 40–60% range (Table 3), the simplicity of the method makes it a viable approach.

Scheme 17.

Table 3.

E	*Product*	*Yield (%)*
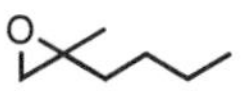 O	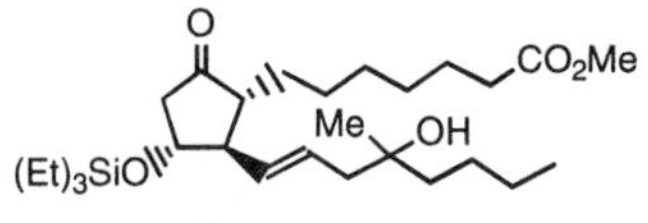 O, CO_2Me, Me, OH, $(Et)_3SiO$	(40)
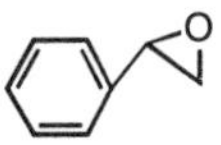 O	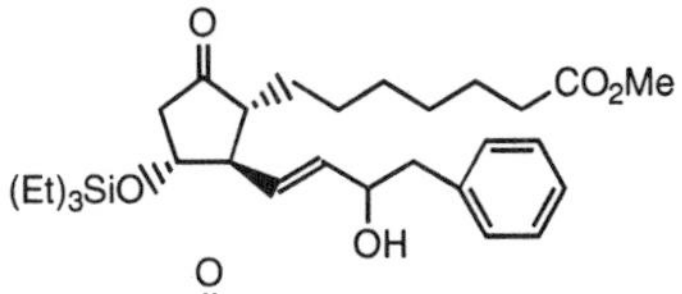 O, CO_2Me, $(Et)_3SiO$, OH	(60)
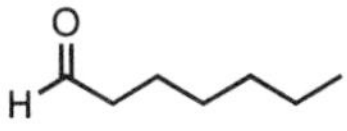 O, H	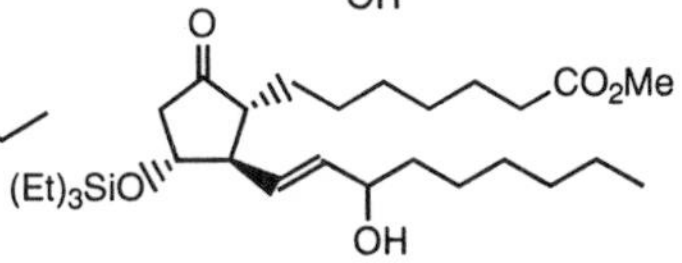 O, CO_2Me, $(Et)_3SiO$, OH	(40)

The direct transmetallation from tin to copper at room temperature offered a commercially feasible synthesis of misoprostol and other prostaglandin analogs. However, because the high molecular weight tin by-product of transmetallation reduced the efficiency of product purification, we sought to further improve our conjugate addition sequence by eliminating tin from the process.

It had been known for some time that hydrozirconation of a terminal acetylene proceeded via *cis* addition to afford 100% (*E*)-olefins.[40–42] Moreover, it was shown that (*E*)-vinylzirconates could be converted to cuprate reagents useful for the synthesis of prostaglandins, including misoprostol, via iodination followed by lithiation[43] (Scheme 18). Significantly, this hydrozirconation reaction was found to proceed rapidly in THF[43] rather than a solvent that required removal prior to treatment with an alkyllithium. Although this protocol provided prostaglandins in the misoprostol family having exclusively the (*E*)-olefin geometry in the lower side

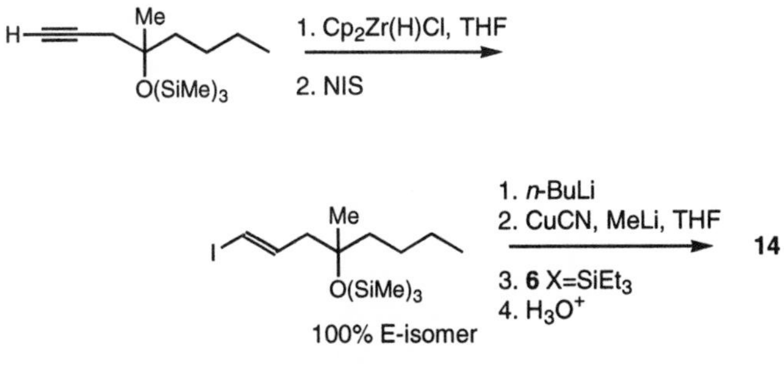

H, Me, $O(SiMe)_3$

1. $Cp_2Zr(H)Cl$, THF
2. NIS

I, Me, $O(SiMe)_3$

100% E-isomer

1. *n*-BuLi
2. CuCN, MeLi, THF
3. **6** X=$SiEt_3$
4. H_3O^+

14

Scheme 18

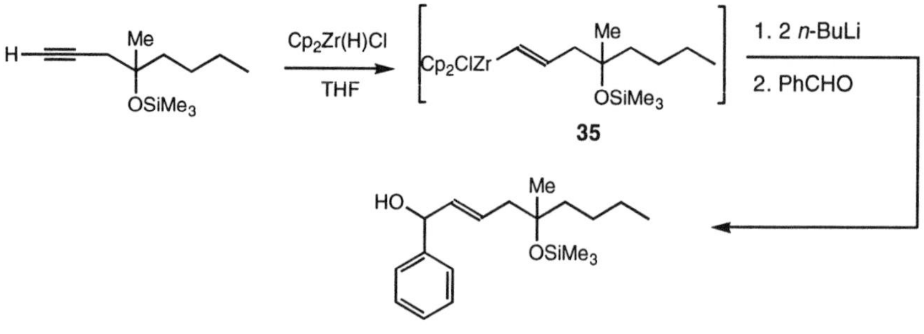

Scheme 19.

chain, the requirement for two pots was a digression in efficiency. Therefore, we developed a one-pot approach that employed a second generation of *in-situ* cuprate technology.[44]

At the onset of our studies, the direct transfer of a vinyl ligand directly from zirconium to a Michael acceptor using copper triflate,[45] or nickel salts,[46] had been known for some time. Neither of these methods for the conjugate delivery of a vinyl ligand has enjoyed widespread use in carbon–carbon bond construction. Furthermore, it was not clear that direct lithiation of a vinyl zirconate was possible.[47,48] We found, however, that lithiation occurred to the extent of at least 20% when the (*E*)-vinylzirconate **35** was treated with *n*-BuLi (Scheme 19).

We postulated that this vinyl anion component of transmetallation could participate in the same kind of ligand mixing reaction when exposed to a dilithio-dialkyl cyano cuprate that we observed previously with vinylstannanes[26] to provide a mixed alkyl–vinyl cuprate suitable for the conjugate delivery of the vinyl group (Scheme 20). This strategy proved to be extremely efficient for the synthesis of prostaglandins (Table 4) in the misoprostol family.[44,49] It has also been shown to be generally applicable to the conjugate addition of vinyl groups using a variety of preparative protocols.[50] Similar methodology has been reported for carboalumi-nated terminal acetylenes using stoichiometric,[51] and catalytic[52] Cu(I) salts.

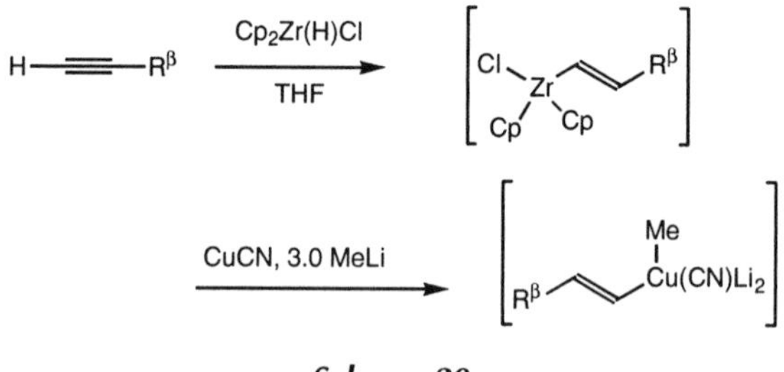

Scheme 20.

Table 4.

	Enone	*Alkyne*	*Product*	*Yield (%)*
A		(±) HC≡C, CH_3, $OSiMe_3$	O, (±), $(CH_2)_6COOMe$, CH_3, $OSiMe_3$, Et_3SiO	(73)
B		(±) HC≡C, CH_3, $OSiMe_3$	O, (±), COOMe, CH_3, $OSiMe_3$, Et_3SiO	(71)
B		(±) HC≡C, CH_3, $OSiMe_3$	O, (±), COOMe, CH_3, $OSiMe_3$, Et_3SiO	(86)
B		(±) HC≡C, CH_3, OPh, $OSiMe_3$	O, (±), COOMe, CH_3, OPh, $OSiMe_3$, Et_3SiO	(55)
B		(±) HC≡C, CH_3, $OSiMe_3$	O, (±), COOMe, CH_3, $OSiMe_3$, Et_3SiO	(80)
B		(±) HC≡C, CH_3, $OSiMe_3$	O, (±), COOMe, CH_3, $OSiMe_3$, Et_3SiO	(73)

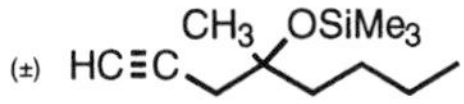

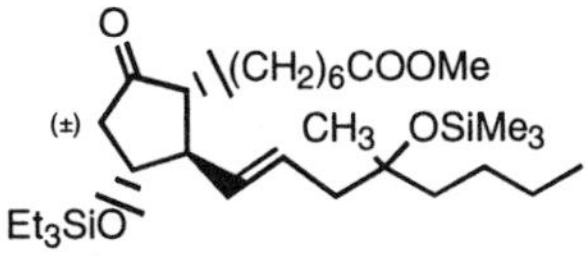

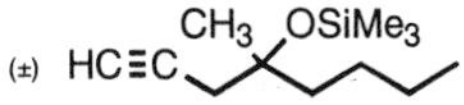

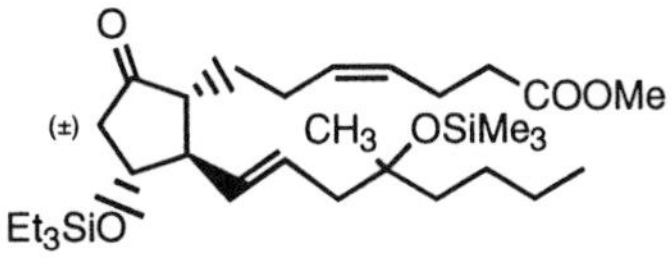

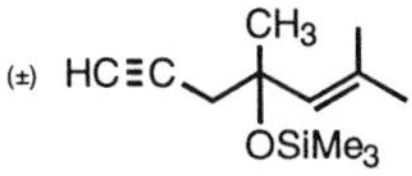

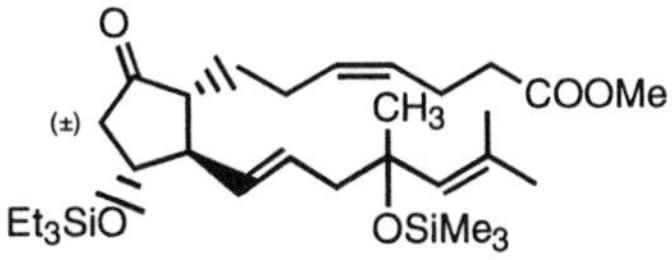

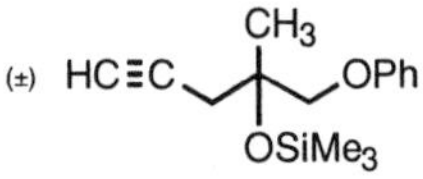

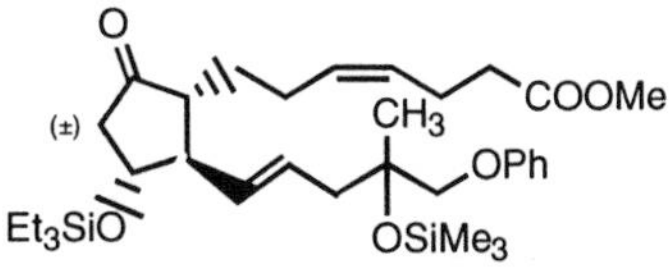

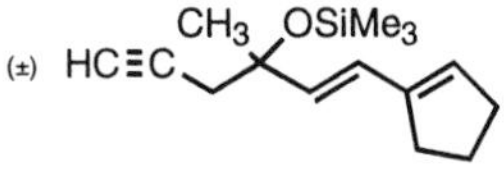

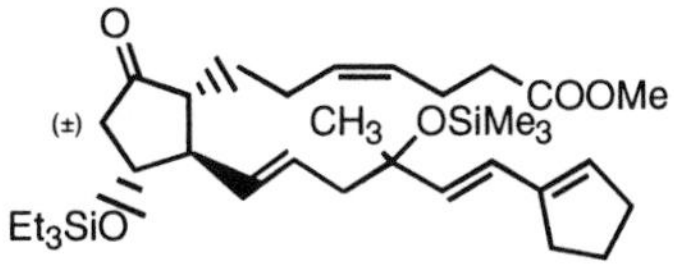

O, (±), $(CH_2)_6COOMe$, Et_3SiO **A**

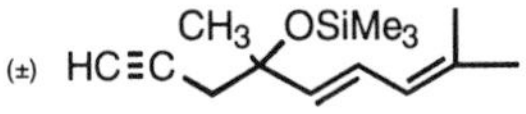

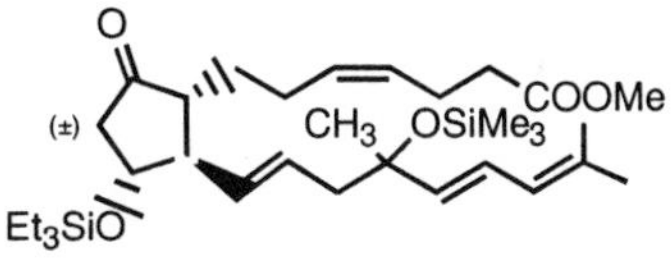

O, (±), COOMe, Et_3SiO **B**

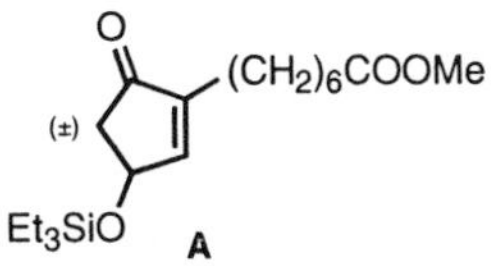

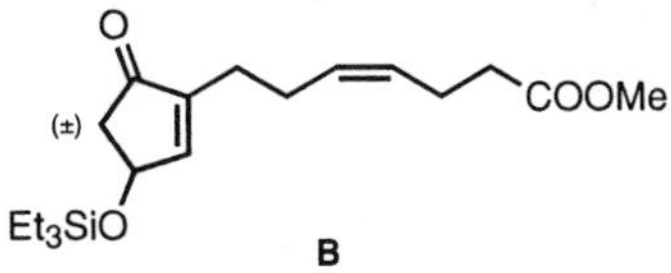

Zirconocene hydride chloride (4.5 g, 0.0178 m) was weighed into a previously flamed dried flask under an argon atmosphere. THF (20 mL, freshly distilled from sodium benzophenone ketyl) was added and the resulting slurry was treated with the terminal acetelene **15** (3.78 g, 0.0178 m) in THF (20 mL) at room temperature. The resulting slurry was stirred for 40 min at room temperature under an argon atmosphere during which time it became homogeneous and dark green in color. In the meantime, copper cyanide (1.59 g, 0.0177 m, flame dried under argon) was slurried in cold (–78 °C) THF (20 mL, freshly distilled) under argon in a separate flask. The solution containing the vinyl zirconate **35** was added to the copper cyanide slurry via a cannula and then methyllithium (0.0536, 38 mL, 1.44 m in diethyl ether) was added via syringe. The resulting dark reaction mixture was stirred at –78 °C for one hour followed by the addition of enone **6** (3.3 g, 0.0089 m) in THF (20 mL, freshly distilled). The reaction mixture was stirred cold (–78 °C) for 30 min and then quenched into a 9:1 mixture of saturated ammonium chloride: ammonium hydroxide. The resulting blue mixture was filtered through a pad of celite and extracted with two portions of ethyl acetate (200 mL). The organic extracts were combined and washed with saturated sodium chloride (200 mL), dried over sodium sulfate, filtered and concentrated to an oil. The residue was dissolved in diethyl ether (50 mL) and filtered through silica gel (10 g). The solvent was removed by evaporation to provide an oily residue which was dissolved in 25% aqueous acetone (100 mL) and treated with a catalytic amount of PPTS (300 mg). The reaction mixture was stirred for 4 h after which the acetone was removed by evaporation and the aqueous residue was extracted with ethyl acetate (200 mL). The organic layer was washed with saturated sodium chloride solution (50 mL) and evaporated to an oily residue which was purified by medium pressure chromatography on silica gel using an ethyl acetate/hexane gradient as eluant. Combination of the appropriate fractions provided the product 14 (2.75 g, 91%) as a colorless oil.

This convergent approach has become our preferred method for the introduction of the ω-chain of prostaglandins. The convenience of vinyl cuprate preparation and subsequent conjugate addition using the *in-situ* approach has provided methodology that is ultimately usable for the production scale synthesis of prostaglandins.

V. SYNTHESIS OF THE SINGLE BIOACTIVE ISOMER OF MISOPROSTOL

Misoprostol is a mixture of two racemates or four stereoisomers. Even though the conjugate addition reaction is stereospecific and controls the stereochemical outcome at C8 and C12, the use of racemic cyclopentenone and ω-chain precursors results in the production of two pairs of diastereomers. Only one of the stereoisomers of misoprostol, the (11*R*),(16*S*)-isomer, is biologically active,[7] and one of the

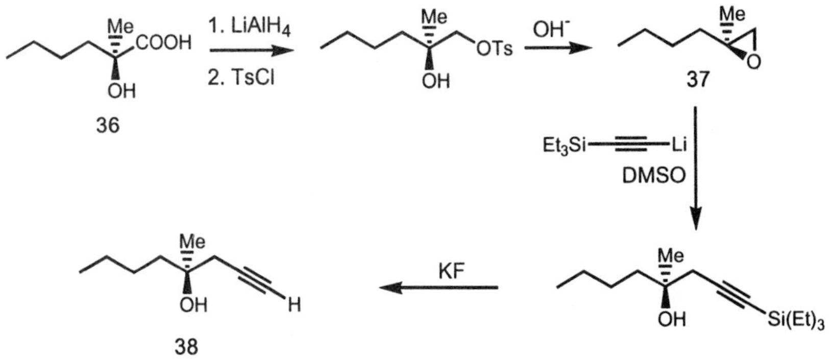

Scheme 21.

long-term goals at Searle has been to efficiently produce this isomer. The preparation of the single isomer is complicated by the fact that chromatographic separation of the misoprostol diastereomers is extremely difficult. Thus, the most effective methodology for the production of the single isomer was the preparation of the individual optically active conjugate addition components.

Our initial approach to obtain optically pure cyclopentenone[53] was to derivatize the racemic material with (*S*)- or (*R*)-2-aminoxy-4-methylvaleric acid, separate the resulting diastereomeric oximes by chromatography, and finally, liberate the optically active product with titanium trichloride. This methodology, while acceptable on a laboratory scale, suffered from a lengthy preparation of the resolving agent, a low isolated yield from the oxime cleavage reaction, and the use of chromatography rendering it unacceptable for large scale synthesis.

The first synthesis of chiral lower side chain was accomplished using a rather circuitous route (Scheme 21). The hydroxy acid **36** was resolved via its napthylethylamine salt and then converted to resolved acetylenic alcohol **38** by hydride reduction to the corresponding diol, tosylation of the primary alcohol, ring closure to the oxirane **37**, ring opening with lithium triethylsilyl acetylide, and finally, removal of the protecting group with potassium fluoride. This acetylenic alcohol was protected as a trimethylsilyl ether, converted to the corresponding pentynyl cuprate, and added to the chiral cyclopentenone in a conjugate addition reaction to provide the first sample of the bioactive isomer of misoprostol.

Given the difficulty and number of steps required to produce the single isomer prostaglandin using this "double resolution" approach, a much more efficient synthesis of the chiral precursors was needed. With this in mind, we sought to design an asymmetric approach that was applicable to large-scale synthesis of the chiral product.

A total asymmetric synthesis of cyclopentenone **39** was undertaken and was shown[54] to produce the chiral product as a 95:5 mixture of enantiomers. The synthesis, starting from cyclopentadiene, required multiple transformations and chromatographic purifications. Therefore, the synthesis was deemed unsuitable for

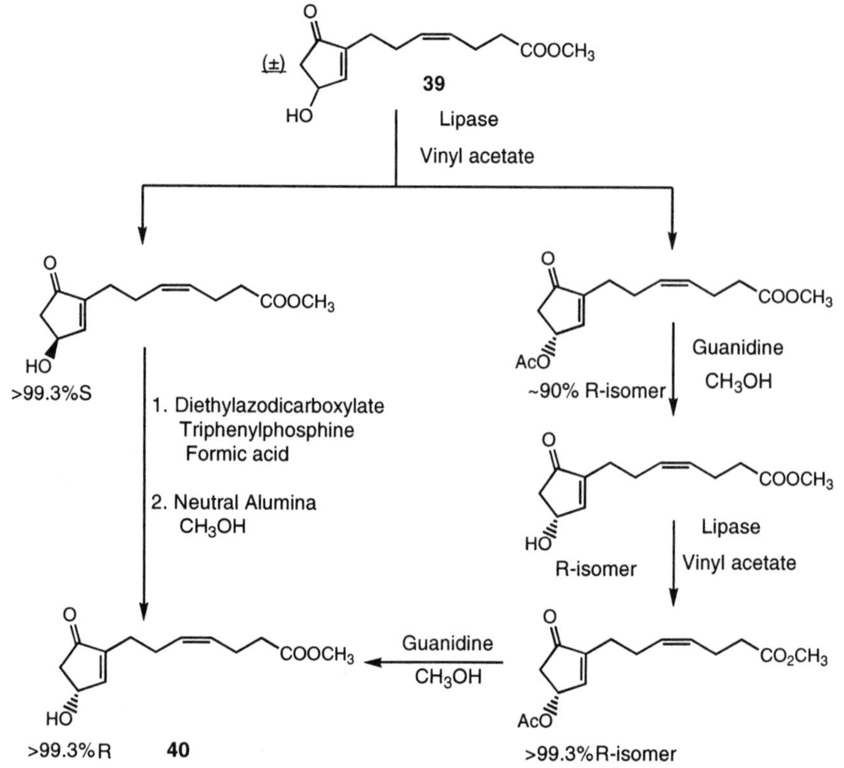

(±) 39 $COOCH_3$ HO

Lipase
Vinyl acetate

HO $COOCH_3$
>99.3%S

1. Diethylazodicarboxylate
Triphenylphosphine
Formic acid
2. Neutral Alumina
CH_3OH

AcO $COOCH_3$
~90% R-isomer

Guanidine
CH_3OH

HO $COOCH_3$
R-isomer

Lipase
Vinyl acetate

AcO CO_2CH_3
>99.3%R-isomer

Guanidine
CH_3OH

HO $COOCH_3$
>99.3%R 40

Scheme 22.

scale-up despite the fact that it could be used to provide significant quantities of the chiral precursor.

A much simpler approach capable of generating multikilogram quantities of the required chiral cyclopentenone has recently been reported.[55] This method involved the lipase catalyzed resolution of racemic cyclopentenone **39** via stereoselective transesterification with vinyl acetate (Scheme 22). The resolution is extremely efficient in that both antipodes of **39** can be converted, in very high yield, to a single enantiomer (99% ee) by employing sequential lipase catalyzed transesterification and Mitsunobu alcohol inversion technologies. This approach has been used to provide large quantities (100 g) of the optically pure prostaglandin intermediate in high yield.

The chiral lower side chain was prepared using the Sharpless asymmetric epoxidation[56] as the source of chirality (Scheme 23). Hexanal was reacted with dimethylamine hydrochloride and formaldehyde to provide 2-methylenehexanal[57] in 81% yield. After sodium borohydride reduction of the aldehyde, the 2-

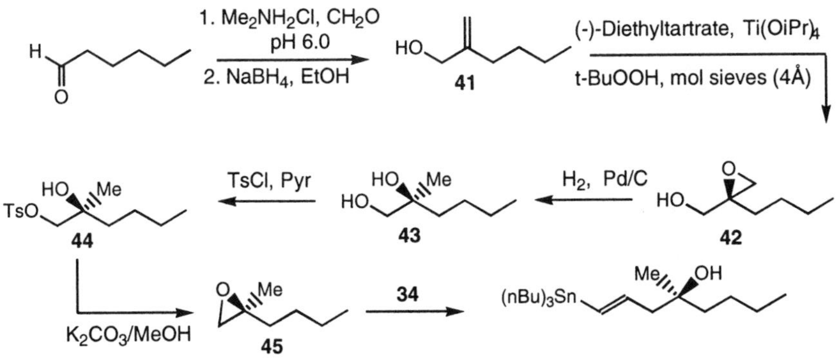

1. Me_2NH_2Cl, CH_2O pH 6.0
2. $NaBH_4$, EtOH

41

(-)-Diethyltartrate, $Ti(OiPr)_4$
t-BuOOH, mol sieves (4Å)

42

H_2, Pd/C

43

TsCl, Pyr

44

K_2CO_3/MeOH

45

34

$(nBu)_3Sn$

Scheme 23.

methylene-1-hexanol **41** was subjected to catalytic Sharpless epoxidation conditions to give chiral epoxide **42** in 63% chemical yield and good optical purity (96% as determined by NMR analysis of the Mosher esters). Reductive opening of **42** provided diol **43**. Selective tosylation of the primary alcohol, followed by treatment of tosylate **44** with tetrabutylammonium bisulfate and potassium hydroxide under phase transfer conditions generated the desired chiral epoxide **45** in a 77% yield from **43**. Epoxide **37** was treated with 2-thienyl-[2(*E*)-tributylstannylethenyl]-cyano cuprate in the *in-situ* fashion[28,37] producing the required chiral vinylstannane in an overall yield of 35%. Coupling of these two pieces provided the single bioactive isomer of misoprostol (95% ee).

VI. SUMMARY

Since the discovery of misoprostol at Searle, we have sought to develop chemistry that was amenable to the large scale general synthesis of prostaglandins. The strategies have centered around the conjugate addition approach for the introduction of a vinyllic omega side chain to an appropriately functionalized cyclopentenone. This research has resulted in extremely efficient processes useful for the manufacturing scale synthesis of misporostol as well as other structurally complex prostaglandin analogs.

ACKNOWLEDGMENTS

The authors wish to acknowledge the scientific contributions of their co-workers, most of whom are cited in the references. The pivotal contributions of Drs. R. Pappo, A. Campbell, K. Babiak, J. Dygos, Mr. A. Gasecki and Ms. K. McLaughlin to the prostaglandin program at Searle are especially appreciated.

REFERENCES AND NOTES

1. Pappo, R.; Collins, P.W. *Tetrahedron Lett.* **1972**, *26*, 2627.
2. Kluge, A.F.; Untch, K.G.; Fried, J.H. *J. Am. Chem. Soc.* **1972**, *94*, 9256.
3. Sih, C.J.; Solomon, R.G.; Price, P.; Perruzzoti, G.; Sood, R. *Chem. Commun.* **1972**, 240.
4. Floyd, M.B.; Weiss, J.J. *Prostaglandins* **1973**, *3*, 921.
5. Collins, P.W. *J. Med. Chem.* **1986**, *29*, 437.
6. Collins, P.W. *Med. Res. Rev.* **1990**, *10*, 149.
7. Pappo, R.; Collins, P.W.; Bruhn, M.S.; Gasiecki, A.F.; Jung, C.F.; Sause, H.W.; Schulz, J.A. In *Chemistry Biochemistry and Pharmocology Activity of Prostanoids*; Roberts, S.M.; Scheinmann, F. Eds.; Pergamon Press: New York, 1979, p. 17.
8. Collins, P.W.; Jung, C.J.; Pappo, R. *Israel J. Chem.* **1968**, *6*, 839.
9. Pappo, R.; Collins, P.W.; Jung, C. *Ann. N.Y. Acad Sci.* **1971**, *180*, 64.
10. Collins, P.W.; Dajani, E.Z.; Driskill, D.R.; Bruhn, M.S.; Jung, C.J.; Pappo, R. *J. Med. Chem.* **1977**, *20*, 1152.
11. Brown, H.C.; Gupta, S.K. *J. Am. Chem. Soc.* **1972**, *94*, 4370.
12. These results are consistent with Corey's findings (ref. 16) but differ from those of Sih. Sih, C.J.; Solomon, R.G.; Price, P.; Sood, R.; Peruzzotti, G. *J. Am. Chem. Soc.* **1975**, *97*, 857.
13. Kluge, A.F.; Untch, K.G,; Fried, J.H. *J. Am. Chem. Soc.* **1972**, *94*, 7827.
14. Zweifel, G.; Whitney, C.C. *J. Am. Chem. Soc.* **1967**, *89*, 2753.
15. Collins, P.W.; Dajani, E.Z.; Bruhn, M.S.; Brown, C.H.; Palmer, J.R.; Pappo, R. *Tetrahedron Lett.* **1975**, *48*, 4217.
16. Corey, E.J.; Beams, D.J. *J. Am. Chem. Soc.* **1972**, *94*, 7210.
17. Corey, E.J.; Wollenberg, R.H. *J. Org. Chem.* **1975**, *40*, 2265.
18. Collins, P.W.; Jung, C.J.; Gasiecki, A.F.; Pappo, R. *Tetrahedron Lett.* **1978**, 3187.
19. Corey, E.J.; Ulrich, P.; Fitzpatrick, J.M. *J. Am. Chem. Soc.* **1976**, *98*, 222.
20. Collins, P.W. (unpublished results).
21. Lipshutz. B.H.; Wilhelm, R.S.; Floyd, D.M. *J. Am. Chem. Soc.* **1981**, *103*, 7672.
22. Bertz, S.H. *J. Am. Chem. Soc.* **1990**, *112*, 4031.
23. Lipshutz, B.H.; Sharma, S.; Ellsworth, E.L. *J. Am. Chem. Soc.* **1990**, *112*, 4032.
24. Lipshutz, B.H.; Wilhelm, R.S.; Kozlowski, J.A. *J. Org. Chem.* **1984**, *49*, 3938.
25. Lipshutz, B.H.; Kozlowski, J.A.; Wilhelm, R.S. *J. Org. Chem.* **1984**, *49*, 3943.
26. Behling, J.R.; Babiak, K.A.; Ng, J.S.; Campbell, A.L.; Moretti, R.; Koerner, M.; Lipshutz, B.H. *J. Am. Chem. Soc.* **1988**, *110*, 2641.
27. Behling, J.R.; Campbell, A.L. U.S. Patent 4,777,275, 1988.
28. Behling, J.R.; Ng, J.S.; Babiak, K.A.; Campbell, A.L.; Elsworth, E.; Lipshutz, B.H. *Tetrahedron Lett.* **1989**, *30*, 27.
29. Mesmeyanov, A.N.; Borisov, A.E. *Dokl. Akad. Nauk. SSSR.* **1967**, *174*, 96.
30. Bataro, J.C.; Hanson, R.N.; Seitz, D.E. *J. Org. Chem.* **1967**, *46*, 5221.
31. Corey, E.J.; Chaykovsky, M.J. *J. Am. Chem. Soc.* **1962**, *84*, 867.
32. Corey, E.J.; Chaykovsky, M.J. *J. Am. Chem. Soc.* **1965**, *87*, 1353.
33. Golobobov, Y.G.; Nesmeyano, A.N.; Lysenko, V.P.; Boldeskul, I.E. *Tetrahedron*, **1987**, *43*, 2609.
34. Sax, N.I. *Dangerous Properties of Industrial Materials*, 6th ed., Van Nostrand Reinhold: New York, 1984, 433.
35. *Handbook of Reactive Chemical Hazards*, 3rd ed.; Butterworths: 1985, 295.
36. Ng, J.S. *Synthetic Communications* **1990**, *20*, 1193.
37. Campbell, A.L.; Behling, J.R.; Babiak, K.A.; Ng, J.S. U.S. Patent 5,011,958, 1991.
38. Lipshutz, B.H.; Koerner, M.; Parker, D.A. *Tetrahedron Lett.*, **1987**, *28*, 945.
39. Behling, J.R.; Medich, J.R, unpublished results (subject of U.S. patent application).
40. Hart, D.W.; Blackburn, T.F.; Schwartz, J. *J. Am. Chem. Soc.* **1975**, *97*, 679.
41. Grieco, P.A.; Ohfune, Y.; Yokoyama, Y.; Owens, W. *J. Am. Chem. Soc.* **1979**, *101*, 4749.

42. Larock, R.C.; Kondo, F.; Narayanan, K.; Sydnes, L.K.; Hsu, M.F.H. *Tetrahedron Lett.* **1989**, *30*, 5737.
43. Kalish, V.J.; Shone, R.L.; Kramer, S.W.; Collins, P.W.; Babiak, K.W.; McLaughlin, K.T.; Ng, J.S. *Synthetic Communications* **1990**, *20*, 1641.
44. Babiak, K.A.; Behling, J.R.; Dygos, J.H.; McLaughlin, K.T.; Ng, J.S.; Kalish, V.J.; Kramer, S.W.; Shone, R.L. *J. Am. Chem. Soc.* **1990**, *112*, 7441.
45. Yoshifuji, M.; Loots, M.; Schwartz, J. *Tetrahedron Lett.* **1977**, 1303.
46. Loots, M.; Schwartz, J. *J. Am. Chem. Soc.* **1977**, *99*, 8045.
47. Schwartz, J. *J. Organomet. Chem. Libr.* **1976**, *1*, 480.
48. Negishi, E.; Takahashi, T. *Aldrichimica Acta* **1985**, *18*, 36.
49. Dygos, J.H.; Adamek, J.P.; Babiak, K.A.; Behling, J.R.; Medich, J.R., Ng, J.S.; Wieczorek, J.J. *J. Org. Chem.* **1991**, *56*, 2549.
50. Lipshutz, B.H.; Ellsworth, E.E. *J. Am. Chem. Soc.* **1990**, *112*, 7440.
51. Ireland, R.E.; Wipf, P. *J. Org. Chem.* **1990**, *55*, 1425.
52. Lipshutz, B.H.; Dimrock, S. *J. Org. Chem.* (in press).
53. Pappo, R.; Collins, P.; Jung, C. *Tetrahedron Lett.* **1973**, *12*, 943.
54. Babiak, K.A.; Campbell, A.L. U.S. Patent 4,952,710, 1990.
55. Babiak, K.A.; Ng, J.S.; Dygos, J.H.; Weyker, C.L.; Wang, Y.F.; Wong, C.H. *J. Org. Chem.* **1990**, *55*, 3377.
56. Gao, Y.; Hanson, R.M.; Klunder, J.M.; Ko, S.Y.; Masamune, H.; Sharpless, K.B. *J. Am. Chem. Soc.* **1987**, *109*, 5765.
57. Moronov, G.S.; Farberov, M.I.; Korshunov, M.A. *Uch. Zap. Yaroslavsk. Tekhnol. Inst.* **1962**, 33–48; *Chem Abstr.* **1964**, *61*, 568.

TRANSITION METAL-PROMOTED HIGHER ORDER CYCLOADDITION REACTIONS

James H. Rigby

Advances in Metal-Organic Chemistry
Volume 4, pages 89–127.

ISBN: 1-55938-709-2

I. INTRODUCTION

Higher order cycloaddition reactions are characterized by the interaction of π-systems that are more extensively conjugated than those normally encountered in the traditional Diels–Alder cycloaddition process. Typical examples include 4π+4π, 6π+2π, and 6π+4π combinations (Figure 1). In general, highly functionalized polycyclic arrays displaying eight- and 10-membered ring substructural units result from these transformations.[1] As a class, these cycloadditions possess many features that are often associated with synthetic utility, including: (1) high levels of stereocontrol, (2) accommodation of extensive functionalization in the reaction partners, and (3) high convergency.

We initially became interested in the application of higher order (H.O.) cycloaddition reactions to problems in organic synthesis in connection with our long-standing involvement with the total synthesis of the structurally elaborate tumor-promoting diterpene, ingenol (**1**). The prominent bicyclo[4.4.1]undecanone moiety contained within the highly oxygenated carbon framework of this class of compounds appeared to be ideally suited to construction employing a thermally allowed [6π+4π] cycloaddition of a substituted tropone (**3**) with an appropriate diene partner as depicted in Eq. 1.

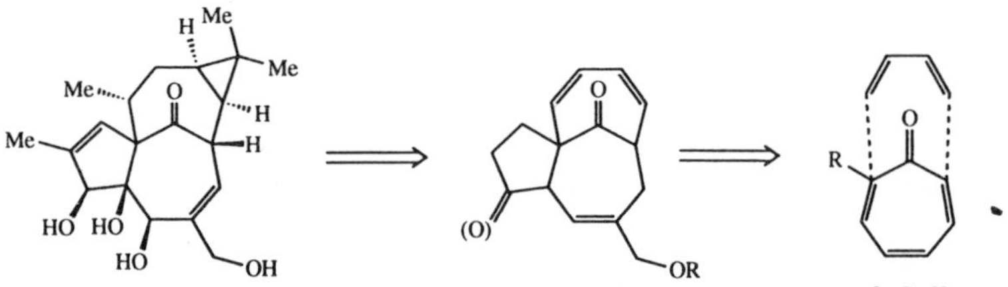

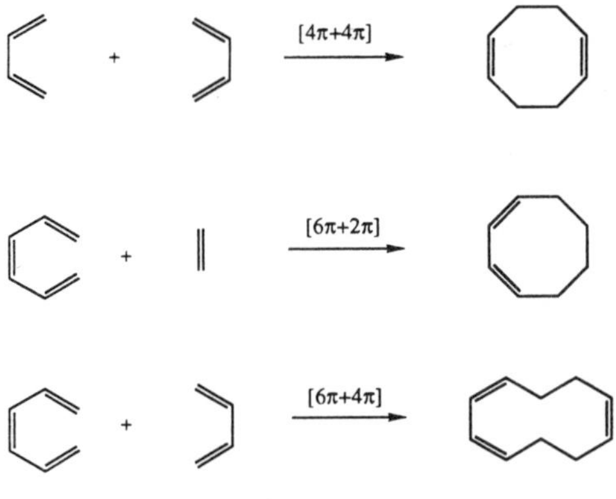

Figure 1.

The [6+4] cycloaddition chemistry of tropone and analogous 6π substrates has been the subject of sporadic investigations over the years and the basic characteristics of many of these reactions have been established.[2,3] While chemical yields tend to be modest (50–60%), the resultant cycloadducts are produced in essentially diastereomerically homogeneous form and are invariably the result of an *exo* approach of the diene to the triene π-system. The ample functionalization displayed in the resultant adducts is a particularly attractive feature of these reactions.

Our initial plan of attack was to assemble an advanced bicyclo[4.4.1]undecanone intermediate from a tropone addend substituted with the elements of the eventual A-ring assembly already installed.[4] Completion of the requisite five-membered carbocycle would follow employing a hetero-Diels–Alder sequence previously developed in our laboratory.[5] Unfortunately, in practice, heating the appropriate 2-substituted tropone **4** in the presence of 1-acetoxy-1,3-butadiene provided *none* of the expected H.O. adduct **5**. However, a small quantity of a mixture of bicyclo[3.2.2]nonane products **6** derived from a competitive [4+2] pathway was isolated (Eq. 2).

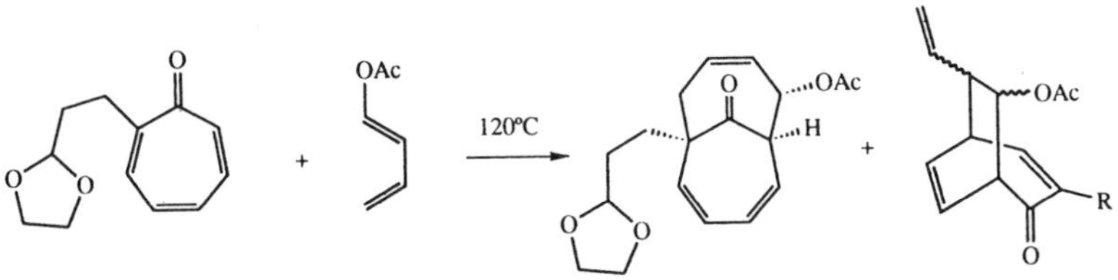

4 + OAc-butadiene —120°C→ **5** (0%) + **6** (20%) (2)

This result can be contrasted with the course of the reaction between the unsubstituted tropone with the same diene in which a 55% yield of the desired bicyclo[4.4.1]undecanone product was obtained.[4] The extreme sensitivity to addend substitution patterns as well as the general inefficiencies of these reactions are some of the root causes of the relative obscurity of H.O. cycloaddition reactions as a class, particularly in terms of synthetic applications. It is worth noting, however, that despite these observations, tropone remains one of the more effective 6π participants in thermal H.O. cycloadditions. More typical examples of this characteristically poor level of periselectivity are evident in the cycloaddition chemistry of cycloheptatriene. A result reported by Woodward and Houk in the early 1970s illustrates this point quite clearly and is depicted in Eq. 3.[6]

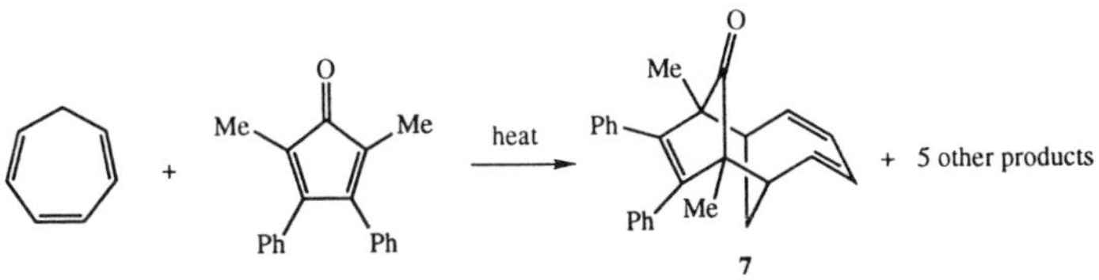

cycloheptatriene + dimethyldiphenylcyclopentadienone —heat→ **7** + 5 other products (3)

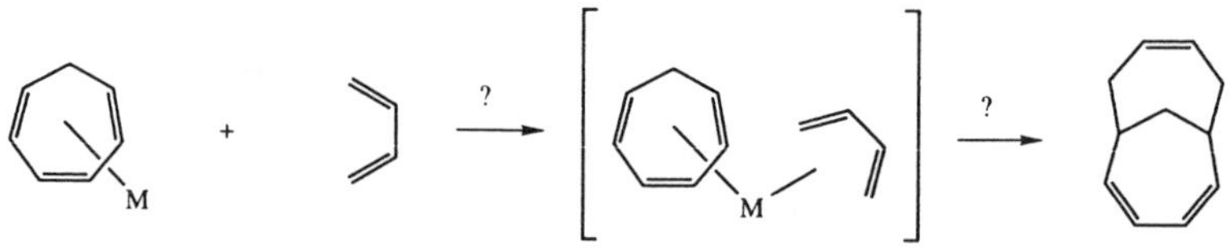

Scheme 1.

Heating cycloheptatriene with 2,5-dimethyl-3,4-diphenylcyclopentadienone provided a small amount of the *exo*-[6+4]-adduct **7** along with at least five additional products derived from a variety of competitive pericyclic pathways. The negative consequences of reacting substrates possessing extensive π-systems is clearly evident in these results. The multiple, competitive pericyclic pathways available to these reactants usually result in depressed yields of H.O. products, rendering the reactions unattractive from a synthetic perspective.[1]

Convinced as we were that H.O. cycloaddition reactions could become important additions to the arsenal of synthetically useful reactions if a means for controlling periselectivity could be identified, we sought methods for modifying the normal course of these reactions without tampering with the attractive features already in place. An intriguing strategy for achieving this objective would involve the intervention of an appropriate transition-metal template designed to preorganize reactants and render the reaction temporarily intramolecular in nature. Scheme 1 illustrates the salient features of this approach. At the outset of this investigation it was not at all obvious, however, that the desired H.O. pathway would, in fact, be the principal beneficiary of this tactic.

The concept of employing metal templates for promoting normally difficult cycloaddition reactions has been under consideration for some time. Pettit and co-workers were among the first investigators to recognize the advantages that this strategy could offer for facilitating H.O. cycloadditions.[7] Although the photoinduced cycloaddition of (η^4-1,3,5-cycloheptatriene)tricarbonyl iron(0) with dimethyl acetylenedicarboxylate gave adduct **8** in only 10% yield, the capability of effecting a [6+2] cycloaddition under these conditions was a significant advance.[8] X-ray analysis of complex **8** revealed that the acetylene partner approached the triene face that was complexed to the metal adding credence to the veracity of the template concept. Unfortunately, limited scope and low efficiency have characterized most attempted metal-mediated H.O. cycloaddition reactions.[9]

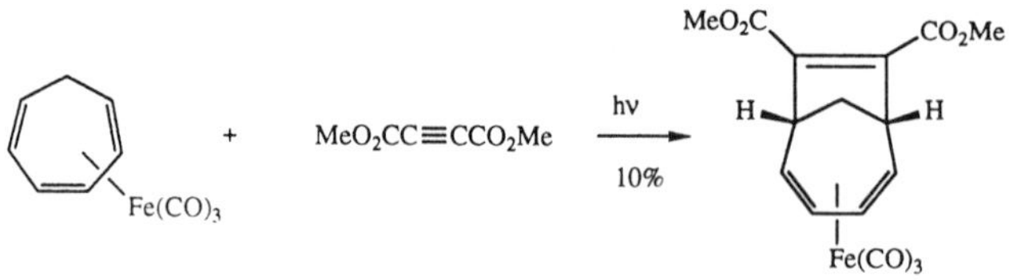

$$\text{(cycloheptatriene)Fe(CO)}_3 + MeO_2CC{\equiv}CCO_2Me \xrightarrow[10\%]{h\nu} \mathbf{8} \quad (4)$$

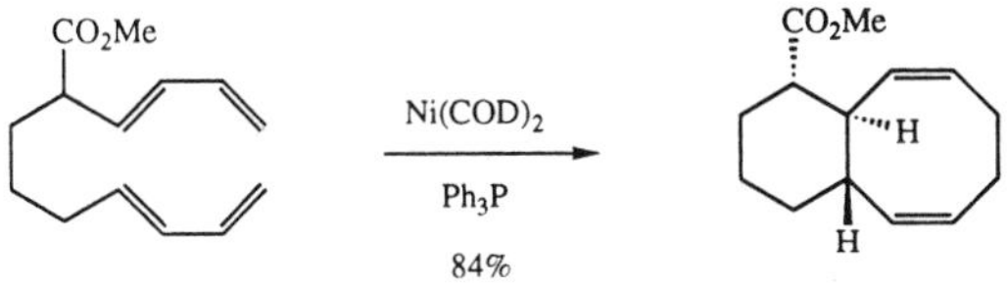

CO₂Me Ni(COD)₂ Ph₃P 84% CO₂Me H H (5)

A dramatic exception to this generalization has been the nickel(0) facilitated [4π+4π] reaction, and in particular, the intramolecular version as developed by Wender and co-workers.[10] The innovation of tethering the diene reactants with three or four atom spacers transformed the well-known nickel(0) mediated butadiene dimerization reaction into a controllable and stereorational process (Eq. 5). Applications of variations of this powerful technology to the synthesis of natural products including taxol have been reported.[10b] A modification of this protocol has also provided a concise and stereocontrolled route to the sesquiterpene lactone, (+)-asteriscanolide (**9**).[10d]

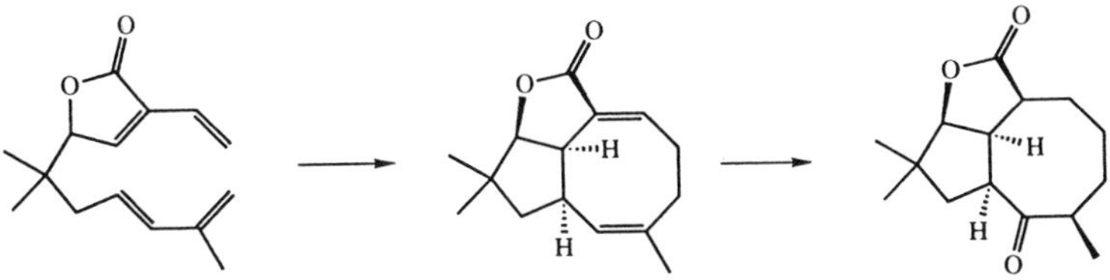

9

In a provocative series of papers, Kreiter and his co-workers at Kaiserlautern have demonstrated that a variety of cyclic (triene)tricarbonylchromium(0) complexes can undergo light-induced [6+4] cycloaddition with simple hydrocarbon diene partners in modest yields.[11] These observations were particularly pertinent to our objectives and a thorough investigation of the cycloaddition chemistry of (η-1,3,5-cycloheptatriene) tricarbonylchromium(0) (**10**) and its derivatives from a synthetic perspective was initiated in our laboratory.[12]

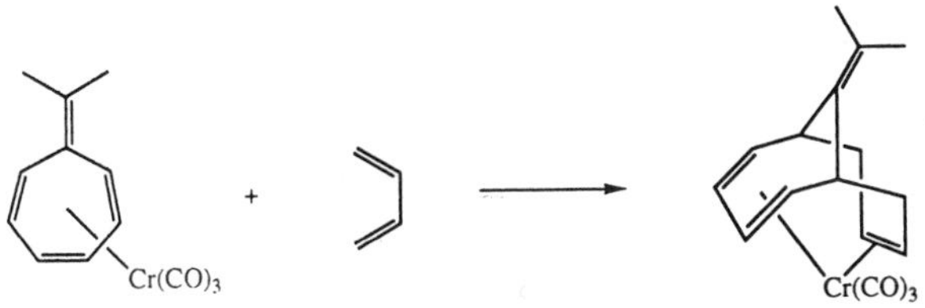

$Cr(CO)_3$ + $Cr(CO)_3$

II. CHROMIUM(0) PROMOTED [6π+4π] CYCLOADDITION REACTIONS

A. Basic Reaction Characteristics

Complex **10**, which has served as the generic triene complex throughout our investigations, is conveniently available as a stable red-orange solid by heating cycloheptatriene with either $Cr(CO)_6$ in refluxing diglyme[13a,b] or with $(MeCN)_3$

$Cr(CO)_6$ or $(MeCN)_3Cr(CO)_3$ → $Cr(CO)_3$ **10** (6)

$Cr(CO)_3$[13c] in refluxing THF. The latter reagent is a particularly useful "$Cr(CO)_3$" transfer agent by virtue of its exceptional reactivity and has emerged as our most important source of chromium.

Our initial foray into the photoinitiated cycloaddition chemistry of complex **10** provided bicyclo[4.4.1]undecane products in modest, but serviceable yields. Typically, a mixture of the complex and excess diene was irradiated (450-W Canrad–Hanovia medium pressure Hg vapor lamp, quartz filter) for a number of hours followed by treatment with excess trimethylphosphite to remove the metal. A number of intriguing features of these reactions became apparent early in our studies. Representative results are displayed in Eq. 7. In each case examined, only

10 + Y–diene–X 1) hν, (quartz) 2) $(MeO)_3P$ → (7)

11a, X=OTMS, Y=H (56%)
11b, X, Y=CO_2Me (59%)

the diastereomer derived from an *endo* transition state was produced. No evidence of the alternative *exo*-isomer was detected in any of these experiments.[14] It is particularly noteworthy that the corresponding thermal, metal-free [6+4] cycloadditions proceed exclusively via an *exo* transition state.[1] This stereocomplementarity, while greatly extending the potential synthetic versatility of these cycloadditions also provided a means for proving the *endo* nature of the metal-mediated process (Eq. 8).

+ Me-diene 140°C 63% → **12**

13 + Me-diene 1) hν (quartz) 2) $(MeO)_3P$ 3) H_3O^+ 56% → **14** (8)

Heating tropone with 1,3-pentadiene provided the known *exo*-adduct **12** and the readily available 7,7-dimethoxycycloheptatriene complex **13**[15] gave, under photochemical conditions, the bicyclic product **14** after hydrolysis. This material exhibited spectral and chromatographic characteristics which were similar, but not identical, to **12**, thus confirming the configurational difference at C7.

One of the more intriguing aspects of these cycloadditions is the virtual absence of electronic effects on reaction efficiency. The results depicted in Eq. 7 illustrate this point quite clearly. Both electron-rich and electron-deficient 4π addends engage complex **10** in cycloaddition with essentially identical outcomes. This phenomenon can be contrasted with the dominant influence of reactant electronics that characterizes the Diels–Alder reaction, wherein carefully matched reaction partners are a prerequisite for effective reaction. This unusual feature of the metal-mediated [6+4] cycloaddition affords significant flexibility when selecting potential reaction partners.

B. Mechanistic Aspects and Yield Optimization

Early in our investigations, a number of interesting observations were made that had important mechanistic implications as well as providing insight into these processes that ultimately led to dramatic improvements in the general efficacy of the reactions. Equation 9 presents one of the more interesting and pertinent events.

10 + TMSO-diene —hν→ [16: OTMS, $Cr(CO)_2$] —1) CO, 2) $(MeO)_3P$, 3) H^+, 82%→ 15 (9)

Treatment of complex **10** with 2-trimethylsilyloxy-1,3-butadiene under standard conditions did not lead to the anticipated bicyclo[4.4.1]undecane adduct **15**. Instead, a very labile and structurally ill-defined species was produced for which structure **16** was tentatively assigned based on what little spectral and chemical evidence that could be accumulated.[16] Significantly, only two carbonyl absorbances were evident in the infrared spectrum of this transient intermediate and exposure to excess carbon monoxide resulted in rapid and complete collapse to the desired cycloadduct **15** after decomplexation and silylenol ether hydrolysis. Remarkably, the yield of this multistep process was an impressive 82%, a result far superior to any previously obtained in our laboratory at that juncture of our investigation.

While these observations offered information on possible mechanistic schemes, they provided an opportunity to improve the efficiencies of the reactions as well. Indeed, the yields of many cycloadditions were improved by 10–20% by stirring the photolysis reaction mixtures under a blanket of carbon monoxide prior to

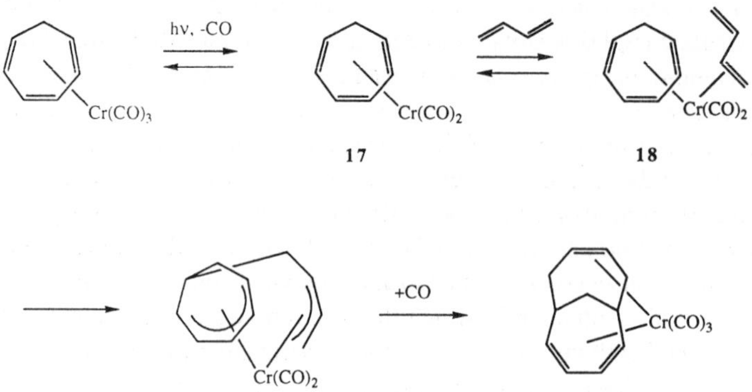

Scheme 2.

decomplexation. It is presumed that this operation intercepts intermediates such as **16** that may have been formed during photolysis. Interestingly, the reaction involving 2-trimethylsilyloxybutadiene was the only example in which the formation of this putative intermediate was the exclusive pathway. An interesting trend became apparent during this phase of our studies. Only those reactions involving *electron rich* diene partners benefited from treatment with excess CO. A corresponding yield enhancement was not observed when electron-deficient dienes were employed. Consequently, reaction mixtures were routinely treated with excess CO subsequent to photolysis whenever electron-rich diene partners were involved, however, this protocol has not been routinely used with electron-deficient diene reactions.

All of the evidence accumulated to date indicates that these cycloaddition reactions are stepwise in nature and the possible intermediacy of species such as **16** was suggestive of a mechanistic profile as shown in Scheme 2. Production of coordinatively unsaturated metal centers by CO ejection through photoactivation is well established[17] and the observations described above are fully consistent with this scheme. In this scenario, light-induced ejection of a CO ligand would produce a coordinatively unsaturated species **17**, which could then engage the diene to give complex **18**. Bond reorganization would follow to afford intermediate **19**, which must then recapture the dissociated CO and collapse to the observed cycloadduct complex **20**.

Recently, Stufkens and co-workers have demonstrated that the sequence of steps outlined in Scheme 2 is the likely reaction pathway for these cycloaddition processes in low-temperature matrices and in liquid noble gas solutions.[18] A crucial, and apparently necessary step in this scheme is the recapture of the dissociated CO by species **19** to produce the observed cycloadduct complex **20**. In support of this contention, Stufkens noted that no cycloaddition occurred in his examples when the CO gas was allowed to escape from the reaction mixture. In addition, all cycloadduct complexes isolated to date have had three CO ligands bound to the metal center, so recapture appears to be an essential element of the overall cycloaddition process.

One practical implication of the mechanism depicted in Scheme 2 is that the modest yields that characterize these transformations may be due to inefficient or incomplete recapture of the ejected CO ligand. More efficient trapping of the ligand, it was reasoned, should provide for better cycloadduct yields. To further probe this issue, experiments were performed to compare yields obtained under conventional conditions with a situation in which the reactions were intentionally deprived of the dissociated carbonyl ligand by vigorously flushing the solution with an inert gas during photolysis. The results are shown in Eq. 10. Surprisingly, the yield of

10 + Me–⁄⁄–CO_2Me → 1) hv; 2) $(MeO)_3P$; 49% → 21 (Me, CO_2Me, H, H)

10 + " " " " → 1) hv, "Ar purge"; 2) $(MeO)_3P$; 74% → " " " "

(10)

adduct **21** increased rather substantially when argon gas was bubbled through the reaction mixture during photolysis. A dramatic decrease in yield would have been anticipated if free CO was being produced during these reactions. While no effort has been made to detect CO evolution in these reactions, the purging conditions employed were sufficient to remove at least a substantial portion of this modestly soluble gas. These observations, while in no way conclusive, are consistent with the notion that initial CO dissociation may not be a necessary step for successful cycloaddition in certain cases. The increased yield may be due to removal of adventitious molecular oxygen, since the presence of this gas is known to labilize metal triene complexes in solution.[19]

In light of the findings described above, an alternative mechanism, the basic features of which were first suggested by Kreiter,[11d] may also merit consideration in certain circumstances (Scheme 3). The principal features of this scheme are: (1) all three CO ligands remain bound to the metal center throughout the reaction, and (2) the requisite coordinatively unsaturated species **22** is produced by a light-induced "hapticity slippage" from η^6 to η^4 (**10** → **22**).[20] The observations in Eq. 9 can also be accommodated in this mechanism by invoking intermediate **19** in those

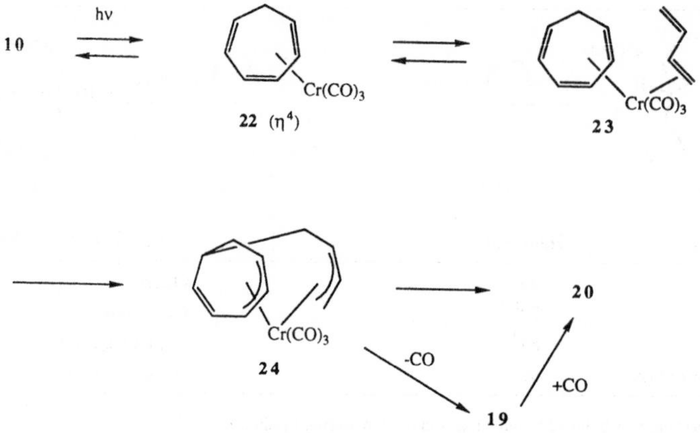

Scheme 3.

cases involving electron-rich dienes. This intermediate can then be shuttled to the product complex by exposure to excess CO.

Regardless of the actual mechanistic steps that may be involved in these cycloadditions, consideration of these issues has been instrumental in developing reaction conditions that routinely deliver cycloadducts in yields dramatically improved over those initially obtained in our early investigations. The evolution of conditions for optimizing the yields of cycloaddition products in reactions involving both electron-rich and electron-deficient dienes is displayed in Table 1. The apparent influence of the wavelength of light used in these cycloadditions is noteworthy.[21] It is not clear at what stage of the stepwise process this influence is exerted; however, control experiments reveal that cycloadduct metal complexes are stable for extended irradiation periods (>3 h) with uranium glass and pyrex filtered light, while relatively rapid decomposition was evident when quartz filtered light was employed.

Equation 11 depicts some typical yields for metal-promoted cycloaddition reactions employing optimized conditions. Yields in the vicinity of 90% are not

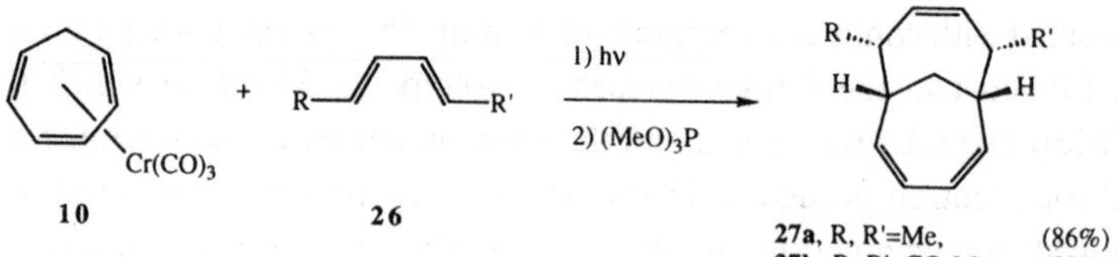

(11)

27a, R, R'=Me, (86%)
27b, R, R'=CO_2Me, (89%)
27c, R=CO_2Me, R'=H, (83%)

Table 1. Optimization of Cycloaddition Reaction Conditions

Conditions	*Yield (%)*	*Conditions*	*Yield (%)*
Quartz	49	Quartz	46
Ar, quartz	74[a,b]	CO, quartz	56[c]
Ar, pyrex	84[a]	N_2, CO, quartz	67[a,c]
Ar, uranium glass	96[a]	N_2, CO, pyrex	86[a,c,d]

Notes: [a]Inert gas was bubbled through reaction mixture during irradiation.
[b]No influence was noted in reactions involving electron deficient dienes when treated with excess CO.
[c]Stirring under a blanket of CO after termination of irradiation.
[d]No additional yield enhancement was noted using a uranium glass filter.

uncommon when the cycloadditions are conducted using uranium glass or pyrex filtered light accompanied by argon or nitrogen purging throughout the photolysis.

7α-Methoxycarbonyl-10α-methyl-(1Hβ,6Hβ)-bicyclo[4.4.1]undeca-2,4,8-triene (**21**). A solution of (η^6-1,3,5-cycloheptatriene)tricarbonylchromium(0) (**10**) (570 mg, 2.5 mmol) and methyl sorbate (284 mg, 2.25 mmol) in hexanes (320 mL) was placed in a dry photochemical reaction vessel equipped with a pyrex immersion well fitted with a uranium glass sleeve. Vigorous bubbling of argon gas through the orange-red solution was initiated. Irradiation (Canrad–Hanovia 450W medium pressure Hg vapor lamp) of the solution was conducted with continuous purging for 30 min. The resultant orange solution was filtered and the filtrate reduced in volume to 50 mL at which time $P(OMe)_3$ (10 mL) was added and the solution stirred at room temperature until the solution became colorless (10 h). The remaining volatiles were removed *in vacuo* and the product was purified by flash chromatography (silica gel, hexanes/ether, 19:1) to afford 469 mg (96%) of a colorless oil.

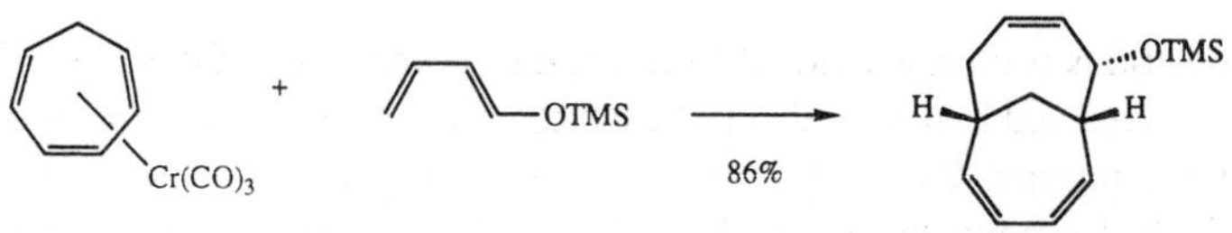

7α-Trimethylsilyloxy-(1Hβ, 6Hβ)-bicyclo[4.4.1]undeca-2,4,8-triene (**25**). A solution of complex **10** (310 mg, 1.36 mmol) and 1-trimethylsilyloxy-1,3-butadiene (1.21 g, 8.5 mmol) in hexanes (350 mL) was placed in a dry photochemical reaction vessel equipped with a pyrex immersion well. Vigorous bubbling of nitrogen gas was begun and the orange-red solution was irradiated (Canrad–Hanovia 450W medium pressure Hg vapor lamp) with continuous purging for 30 min. The reaction mixture was then saturated with CO and stirred under a blanket of this gas at room temperature for 15 h at which time the mixture was concentrated, treated with $(MeO)_3P$ and purified as described above to provide 273 mg (86%) of a colorless oil.

We have also had occasion to examine the effectiveness of other group 6 metals for promoting H.O. cycloadditions. The corresponding CHT-$Mo(CO)_3$[19] and CHT-$W(CO)_3$[22] were prepared in excellent yields by heating the hydrocarbon with either $M(CO)_6$ or $(MeCN)_3$ $M(CO)_3$. The complexes derived from the second and third row metals were then reacted with a series of dienes employing the conditions found to be optimal for the chromium complexes. The superiority of Cr(0) as a mediator

Table 2. Comparison of the Photoinduced [6+4] Cycloadditions of a Series of Group 6 Metal Tricarbonyl Complexes

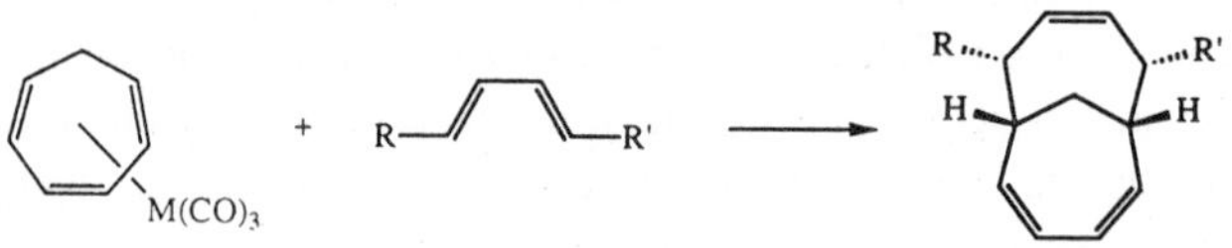

Entry	*M*	*Diene*	*Yield (%)*
1	Cr	R = OTMS, R′ = H	86
2	Mo	R = OTMS, R′ = H	51
3	W	R = OTMS, R′ = H	0
4	Cr	R = Me, R′ = H	70
5	Mo	R = Me, R′ = H	32
6	W	R = Me, R′ = H	0
7	Cr	R = Me, R′ = CO_2Me	96
8	Mo	R = Me, R′ = CO_2Me	27
9	W	R = Me, R′ = CO_2Me	0

of these reactions is clearly evident from the data compiled in Table 2. Each metal complex was reacted in turn with a typical electron-rich, electron-poor, and hydrocarbon diene partner. Cycloaddition was observed in each example employing Cr(0) and Mo(0); however, no evidence for cycloaddition was apparent with the corresponding tungsten complex. Within each set of reactions, the Cr(0) mediated entry was superior. The total absence of cycloaddition in the tungsten case may be reflective of the ligand–metal bond strengths in this complex, which are known to be the strongest in the group 6 series.[23] Molybdenum complexes are known to be the most labile toward ligand exchange and, as a consequence, may be too prone to decomplexation to be effective participants in these cycloaddition reactions.[24]

An important and intriguing implication of the mechanisms described above is the possibility of effecting these cycloadditions employing thermal activation conditions. Considerable precedent for thermally induced exchange of carbonyl ligands in the chromium (0) series is available[25] and thermal hapticity slippages formally related to those suggested in Scheme 3 are also well established for the group 6 metals.[20,24,26] These considerations prompted an examination of thermally induced [6+4] cycloadditions of the chromium cycloheptatriene complex **10**.

10 + R–(diene)–R' $\xrightarrow[\text{reflux}]{n\text{-Bu}_2\text{O}}$ (12)

R=Me, R'=CO_2Me, (60%)
R=H, R'=OAc, (59%)
R, R'=Me, (70%)

Heating complex **10** with a series of diene partners in refluxing n-Bu_2O provided the corresponding cycloadducts in unoptimized yields in the 55–70% range. The products of the thermal variant were identical in every respect to those obtained in the corresponding photochemical reactions and, significantly, the adducts from the thermal reactions were isolated metal-free in each case. In an effort to maximize reaction efficiency, a series of solvents were examined to evaluate the role of nucleophilicity and temperature on these reactions. Dibutyl ether emerged as the solvent of choice even though it is generally regarded as a weakly nucleophilic solvent in other ligand exchange processes.[25] In some instances *tert*-butyl methyl ether was an effective solvent as well. Further studies on thermally induced H.O. cycloadditions will be detailed in the section on [6+2] reactions.

Preparation of 7α-Methoxycarbonyl-10α-methyl-(1Hβ,6Hβ)-bicyclo[4.4.1]-undeca-2,4,8-triene via thermal activation. To a flame-dried flask equipped with a condenser and inert gas inlet was added complex **10** (270 mg, 1.1 mmol) and methyl sorbate (190 mg, 1.5 mmol) in freshly distilled (from $LiAlH_4$) n-butyl ether (12 mL). The solution was heated at reflux for 44 h at which time the reaction mixture was cooled, the solvent removed *in vacuo*, and the residue purified by flash chromatography (silica gel, hexanes/ether, 9:1). This afforded 156 mg (60%) of a colorless oil identical in all respects with the corresponding compound produced photochemically.

C. Cycloaddition of Substituted Cycloheptatriene Complexes (Regiochemical Issues)

The capacity for constructing functionally elaborate bicyclic structures is potentially one of the most appealing and important features of the metal-mediated cycloaddition technology. A wide range of substituted cycloheptatriene complexes are readily available using procedures originally developed by Pauson and co-workers,[27] and a systematic examination of the cycloaddition chemistry of a number of these substituted complexes has been conducted. It appears that the cycloaddition process can accommodate extensive functionalization in both addends, which is a distinct advantage when compared with the rather limited scope of the corresponding thermal, metal-free process.[3a]

The reactions of complex **28**[27a] (Eq. 13) demonstrate that extensive functionalization in the 6π partner does not result in suppressed adduct yields. The somewhat lower yield for **30b** is due to the relative lability of diene **29b**[28] and does not appear to reflect any intrinsic limitations of the reaction itself.

$(CO)_3Cr$ OMe **28** + R' R' R R **29** 1) hv, pyrex 2) CO 3) $(MeO)_3P$ → R' R' R R H H OMe (13)

30a, R=Me, R'=H, (90%)
30b, R=OTBDMS, R'=H, (70%)
30c, R=H, R'=Me, (93%)

The 2-methoxycycloheptatriene complex **31** also undergoes smooth cycloaddition with both electron-rich and electron-deficient diene partners. However, when unsymmetrically substituted dienes are employed there is no detectable discrimination between the formation of the two possible regioisomeric products in each case. This trend is repeated with all other 2- and 3-substituted cycloheptatriene complexes studied to date. Additional regiochemical results that further define the trends in this area are compiled in Table 3.

$(CO)_3Cr$ OMe **31** + Y X → (X)Y X(Y) H H OMe (14)

32a, X, Y=CO_2Me, Me, (89%)
32b, X, Y=H, OTMS, (89%)

Table 3. Cycloadditions of 2- and 3-Substituted Cycloheptatriene Complexes with Selected Dienes

Entry	Complex	Diene	Isomer Ratio	Yield (%)
1	R′ = CO_2Me, R = H	X = OTMS, Y = H	1:1	75
2	R = CO_2Me, R′ = H	X = CO_2Me, Y = Me	1:1	90
3	R = OMe, R′ = H	X = OTMS, Y = H	1:1	79
4	R′ = CO_2Me, R = H	X = Y = Me	1:0	74
5	R = CO_2Me, R′ = H	X = Y = Me	1:0	90

The results presented in Table 3 reinforce the observations made with the unsubstituted complex **10**. Once again, the electronic nature of the reactants appears to have little or no impact on reaction efficiency. For example, entries 2 and 3 involve the reaction of dienes and complexes possessing substituents of similar electronic character and yet the yields of the cycloadducts remain high. Also consistent with the insensitivity of these transformations to electronic effects is the total absence of regioselectivity when unsymmetrically substituted reactants engage in cycloaddition. Entries 4 and 5 involve a symmetrically substituted diene but further demonstrate that virtually all types of dienes are able to undergo efficient cycloaddition.

As discussed previously, the presence of substitution at one or more of the triene bond-forming centers in the thermal [6+4] cycloaddition of tropone has been shown to completely suppress the H.O. pathway (see Eq. 2). We were delighted to find that the metal-promoted reaction can overcome this obstacle and can be employed with reasonable effectiveness to produce a range of bicyclo[4.4.1]undecane systems displaying bridgehead substituents. Equation 15 illustrates the key features of this important development. Readily available complex **33**[27b] undergoes photoinduced cycloaddition with 2-*tert*-butyldimethylsilyloxy-1,3-butadiene to provide adduct **34**. While the chemical yield for this example is somewhat lower than with the

OMe, Cr(CO)$_3$, **33** + TBDMSO-diene —57%→ TBDMSO, H, OMe, **34** (15)

unsubstituted complex, only a single regioisomer is produced. In light of the absence of electronic effects in most of the corresponding cycloadditions, this result is somewhat surprising and a rationale for this selectivity remains obscure at this time. It is noteworthy, however, that dienes that bear an oxygen substituent at the 2 position have been shown to behave somewhat differently than other 4π partners throughout these investigations (see Eq. 9), and unique stabilization of a crucial intermediate in the reaction process by this substitution pattern may be responsible for this regioselectivity. Interestingly, the reaction between complex **33** and isoprene displays a similar, although, less pronounced regioselectivity profile (Eq. 16).

33 + Me — 47% → H, OMe (3) + H, Me, OMe (1) (16)

As would be expected, 1-substituted dienes add to complex **33** in such a fashion so as to generate the less hindered cycloadduct. Once again, only the endo diastereomer is produced (Eq. 17).

33 + X → X, H, OMe (17)

X= Me, (45%)
X= OTMS, (38%)

Cycloheptatriene complexes substituted at the 7-position on the organic ligand also participate in these cycloadditions. These reactions are particularly powerful because adducts possessing as many as five contiguous stereogenic centers can be conveniently accessed. Furthermore, a number of interesting C7-substituted complexes can be easily prepared using methodology originally described by Pauson allowing for the ready preparation of bicyclo[4.4.1]undecane systems rich in stereochemical information.[27]

A typical example of the capabilities of these transformations is the formation of cycloadduct **36**, which can be constructed in essentially quantitative yield by photoinduced cycloaddition of substituted complex **35** with methyl sorbate. We have demonstrated unambiguously, via single crystal X-ray analysis of a related

Me, $Cr(CO)_3$ (**35**) + Me—CO_2Me — 97% → Me, CO_2Me, H, H, Me (**36**) (18)

adduct, that the relative stereochemistry at C7 in the complex is retained during the cycloaddition. This observation supports the notion that the diene engages the complex exclusively from the triene face that is complexed to the metal center.

D. Asymmetric Induction Studies

Control of the absolute stereogenicity of newly formed carbon–carbon bonds is a significant contemporary issue in organic synthesis. Cycloadditions of all types have been shown to be amenable to chiral induction,[29] and a number of strategies can be envisioned for inducing asymmetry during the [6+4] cycloaddition process. An attractive and experimentally convenient approach would employ a chiral diene component. We have examined a number of candidate chiral auxiliaries during this study including chiral terpene alcohols and chiral oxazolidinones,[30] but the most effective agent has been the so-called Oppolzer sultam.[31] The corresponding *N*-sorbyl sultam derived from *D*-(–)-2,10-camphorsultam reacts smoothly with complex **10** to produce an easily separable mixture of diastereomeric cycloadducts **37** in a 74% chemical yield with an 84% de (Eq. 19). The absolute configuration

10 + [N-sorbyl camphorsultam: Me, Me, H, O, N, SO_2, Me] → (74%, 84%de) [Me, COX_c, H, H] **37** (19)

of the major diastereomer produced during this reaction has been determined to be that depicted in structure **37** by single crystal X-ray analysis of a related adduct in the enantiomeric series.

An alternative strategy for inducing chirality in H.O. cycloadducts would be to incorporate chiral auxiliaries in the coordination sphere of the metal center. This approach is particularly appealing because, in principle, it does not place structural limitations on either organic reaction partner. The commercial availability of many types of chiral phosphines recommends this class of compounds as a good starting point for this investigation. In view of the dramatically different electronic natures of phosphine and carbonyl ligands, it was regarded as prudent to initially establish in simple systems that exchange of carbonyl ligands for phosphine ligands would not adversely effect the course of the H.O. cycloaddition process.

To test this premise, complex **38** was prepared using the method of Anderson[32] and subjected to standard cycloaddition conditions in the presence of both electron-rich and electron-deficient dienes. To our delight, the cycloadditions proceeded

[cycloheptatriene $Cr(CO)_3$] + PPh_3 →(hν) [cycloheptatriene $Cr(CO)_2PPh_3$] **38** (20)

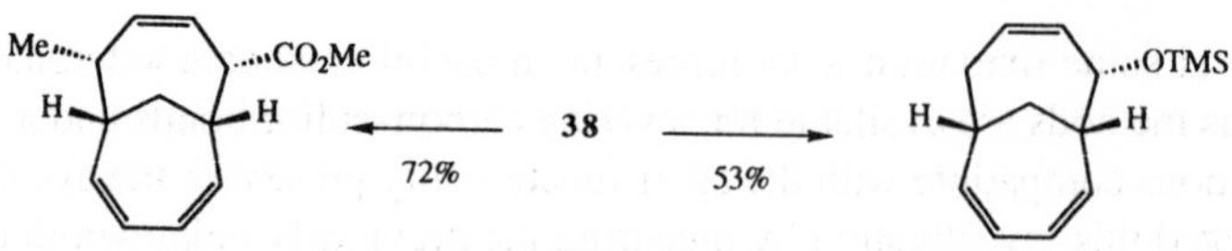

(21)

without incident in both cases, although the yields (not optimized) were somewhat lower than in the corresponding reactions with the tricarbonyl complex **10**.

With these encouraging results in hand, we then turned to a preliminary examination of a simple chiral induction using the commercially available monodentate ligand (+)-neomenthyldiphenylphosphine. Although the preparation of the phosphine complex was executed without difficulty, no induction could be detected in the resultant cycloadduct using methyl sorbate as the 4π partner. More extensive study of this very promising strategy for chiral induction is currently underway in our laboratories.

E. Cycloadditions of Heterocyclic Triene Complexes

One important feature of the [6+4] cycloaddition process as executed in these studies is that the resultant products contain a potentially substituent-rich 10-membered carbocyclic substructure embedded within the larger bicyclo[4.4.1]undecane framework. Cycloaddition reactions in general are among the most effective means for assembling ring structures of a variety of sizes; however, 10-membered and larger rings are particularly difficult to access via this methodology. In light of the growing number of natural products displaying ten-membered ring systems, a rapid, efficient and stereoselective cycloadditive strategy into these species would be a welcome addition to the synthetic armory. The [6+4] cycloaddition could provide an attractive technology for addressing this issue if a suitable protocol could be identified for excising the one-atom bridge.

A tactic, as depicted in Scheme 4, in which a heteroatom X replaces the usual carbon at position 7 in the triene partner represents an attractive potential solution to this problem. In view of the high degree of stereocontrol afforded by the metal-promoted [6+4] methodology as well as the facial bias intrinsic to the resultant bicyclo[4.4.1]undecane system,[4] stereochemically elaborate 10-membered carbocycles could become readily available with the successful implementation of this strategy.

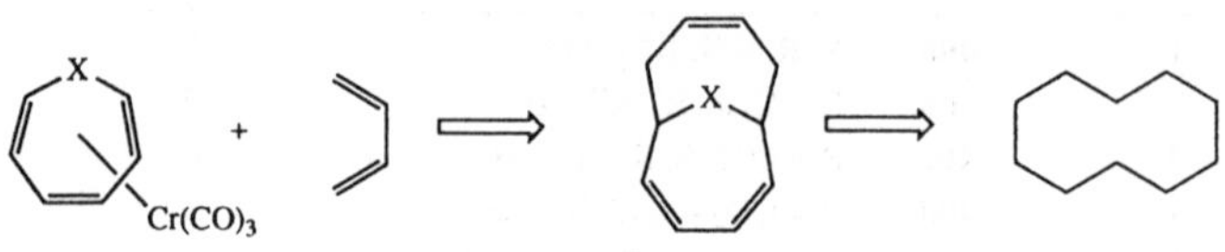

Scheme 4.

Sulfur at some oxidation level represents a useful candidate heteroatom since numerous methods are available for severing carbon–sulfur bonds under a variety of conditions compatible with the other functionality present in the cycloadducts. We initiated this investigation by preparing the previously unknown thiepin-1,1-dioxide complex **39**[34] by treatment of the readily available heterocycle[35] with $(MeCN)_3Cr(CO)_3$ at room temperature.

$(MeCN)_3Cr(CO)_3$ 86% $Cr(CO)_3$ **39**

(η^6-Thiepin-1,1-dioxide)tricarbonylchromium(0) (**39**). Trisacetonitrile tricarbonylchromium(0)[13c] [from ($Cr(CO)_6$ (2.05 g, 9.28 mmol)] and thiepin-1,1-dioxide[35] (0.67 g, 4.64 mmol) were stirred in THF (75 mL) at room temperature for 12 h. The solvent was removed *in vacuo* and chromatography (silica gel, hexanes/EtOAc, 1:1) afforded 1.11 g (86%) of a red solid: mp: 173–4 °C dec.

The cycloaddition chemistry of complex **39** has been examined in some detail. The reactions, in general, were well behaved and proceeded, for the most part, in a fashion similar to the all-carbon series. A single, *endo*-diastereomer was produced in each case, but unlike previous [6+4] cycloadditions, the electronic character of the diene partners appeared to have a modest influence on reaction yields and rates. Product recoveries were uniformly lower for reactions involving electron-deficient dienes when compared with those employing electron-rich 4π partners. Some typical examples are compiled in Table 4.

Table 4. [6+4] Cycloadditions of Complex **39**

39 + R, R', R" diene; 1) hν (uranium glass); 2) O_2/Et_2O → **40**

Entry	*Product*	*Substituents*	*Yield (%)*
1	**40a**	R, R′ = H, R″ = Me	77
2	**40b**	R = OAc, R′, R″ = H	78
3	**40c**	R = OTMS, R′, R″ = H	65
4	**40d**	R = CO_2Me, R′, R″ = H	38
5	**40e**	R = CO_2Me, R′ = Me, R″ = H	21
6	**40f**	R, R′ = CO_2Me, R′ = H	0

Entries 4–6 demonstrate that electron-deficient dienes are inferior participants in these reactions. Indeed, no cycloadduct could be detected when the very electron-poor diene, dimethyl muconate, was employed as the 4π addend. This result can be contrasted with the virtually quantitative yield obtained from the reaction of (cycloheptatriene)tricarbonylchromium(0) with methyl sorbate. Several other points are worth noting, including the use of molecular oxygen as the decomplexing agent in these reactions. Oxygen is known to labilize certain metal-π-complexes[19] and achieves this end quite admirable in the heterocyclic series. Interestingly, similar efficiencies have not been realized in the all-carbon analogs.

> *7α-Acetoxy-(1Hβ,6Hβ)-11-thiabicyclo[4.4.1]undeca-2,4,8-triene-11,11-dioxide* (**40b**). A solution of complex **39** (150 mg, 0.54 mmol) and 1-acetoxy-1,3-butadiene (E/Z isomer mixture, 0.64 mL, 5.4 mmol) in hexanes/CH_2Cl_2 (1:1) (350 mL) was irradiated (Canrad–Hanovia 450 W Hg vapor lamp, uranium glass sleeve) with constant purging with Ar gas for 30 min. The volatiles were removed *in vacuo* and the adduct complex was suspended in dry diethyl ether (25 mL) and the mixture purged with oxygen. The resultant mixture was stirred under a blanket of oxygen for 36 h at which time the mixture was filtered, the solvent removed *in vacuo* and the residue purified by flash column chromatography (silica gel, hexanes/EtOAc, 2:1) affording 180 mg (78%) of a white solid: mp: 151–2 °C.

In a fashion paralleling the all-carbon series, the complex **39** can also be induced to undergo [6+4] cycloaddition by thermal activation. Unfortunately, the yields tend to be a bit lower than in previous cases apparently due to loss of sulfur dioxide from the starting material complex during thermolysis. Equation 22 displays a typical thermal cycloaddition result.

39 + OAc —(n-Bu_2O, reflux, 47%)→ **40b** (H, O_2S, H, OAc) (22)

Among the more attractive methods available for excising the elements of sulfur dioxide from these cycloadducts is thermally or photochemically initiated cheletropic extrusion.[36] Successful implementation of these transformations would provide 10-membered rings exhibiting multiple unsaturations with complete control of resultant geometry. We have briefly examined this methodology in the context of revealing the 10-membered carbocycle residing within the adduct's bicyclic array with very encouraging results. In the event, cycloadducts **40b** and **40e** were irradiated through a quartz filter to effect SO_2 extrusion resulting in the production of the carbocycles **41a** and **41b**, respectively.

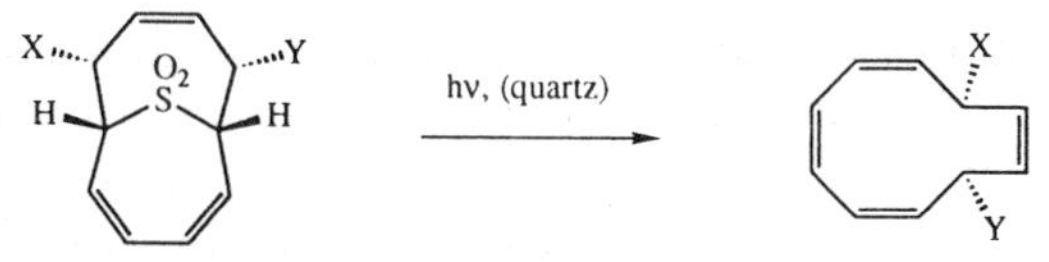

40b, X=H, Y=OAc
40e, X=Me, Y=CO_2Me

hv, (quartz)

41a, (54%)
41b, (41%)

(23)

7-Acetoxy-1,3,5,8-decatetraene (**41a**). A solution of **40b** (75 mg, 0.295 mmol) in a mixture of dichloromethane (150 mL) and hexanes (200 mL) was placed in a dry photochemical reaction vessel equipped with a quartz immersion well. Irradiation over a 13 min period was carried out by means of a Canrad–Hanovia 450W medium pressure Hg lamp. The solution was concentrated in vacuo and the residue purified by flash chromatography (silica gel, CH_2Cl_2/hexanes, 1:1) to afford 30 mg (54%) of a colorless oil.

That the anticipated all (*Z*)-cyclodecatetraenes were produced in these reactions was easily discernable by ^{1}H NMR decoupling experiments. The corresponding thermal extrusion process, which could provide the complementary (*E*,*Z*,*Z*)-isomer, was briefly examined without a successful conclusion. Molecular modelling clearly revealed that the requisite bond rotations for thermal extrusion were precluded by the presence of the two sp^2 carbons making up the isolated double bond across the 10-membered ring (Eq. 24). Efforts to selectively reduce this unsaturation are currently underway.

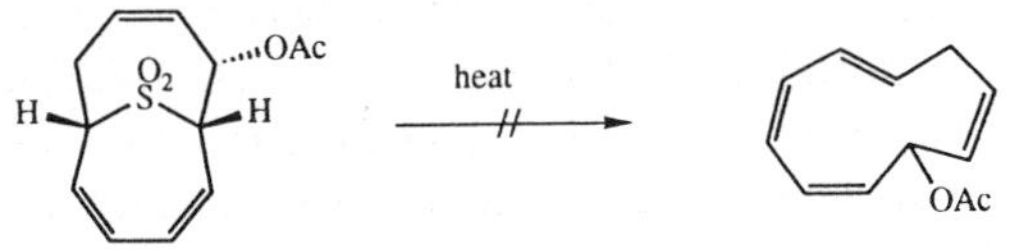

heat

(24)

The all (*Z*)-tetraenes that are available via this protocol are ideally suited as templates from which to assemble a variety of (*Z*,*Z*)-sesquiterpene lactones such as 15-desoxy-(*Z*,*Z*)-artemisiifolin (**42**).[37]

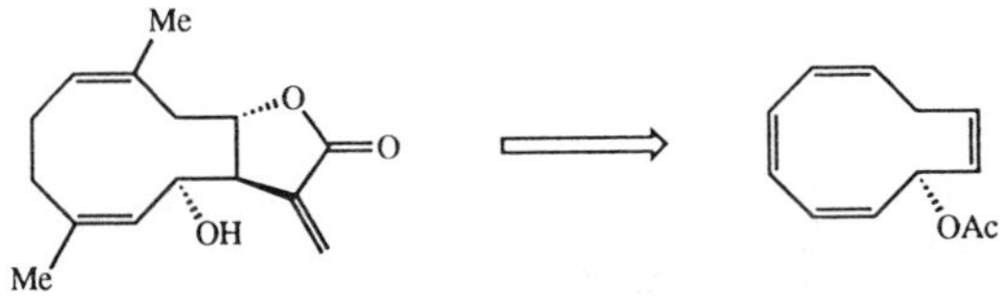

42

(25)

Members of a second set of heterocyclic trienes that have also proven to be quite useful participants in metal-promoted cycloaddition reactions are the *N*-alkoxycarbonylazepines. The parent (*N*-methoxycarbonylazepine)tricarbonylchromium(0) (**43**) is easily prepared employing the method of Kreiter and Özkar.[38] Representative photochemical [6+4] cycloaddition reactions of azepine complex **43** are collected in Table 5.

Table 5. Typical [6+4] Cycloaddition Reactions of Complex **43**

43 + diene → 1) hν (pyrex); 2) O_2/Et_2O → **44**

Entry	*Product*	*Substituents*	*Yield (%)*
1	**44a**	R, R′ = H, R″ = OTMS	87[a]
2	**44b**	R = OAc, R′, R″ = H	75
3	**44c**	R = OTMS, R′, R″ = H	79
4	**44d**	R = Me, R′ = CO_2Me, R″ = H	83

Note: [a]Product isolation was accompanied by some of the corresponding ketone.

The cycloaddition chemistry of azepine complex **43** parallels quite closely that for the all-carbon series. For example, both electron-rich and electron-deficient dienes engage in cycloaddition with essentially equal effectiveness (Entries 3 and 4). Once again the use of molecular oxygen to decouple the metal from the photoadducts proved superior to $(MeO)_3P$ for that purpose. The *endo* nature of the adducts in this series was further supported by means of a single crystal X-ray

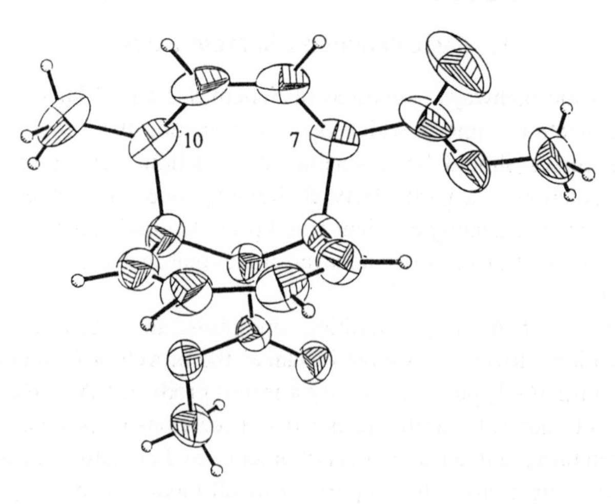

Figure 2. X-ray crystal structure of cycloadduct **44d**.

CO_2Et N 45 + MeO_2C O CO_2Me Ph Ph → Ph E O Ph E N CO_2Et 46 (18%) + EN Ph E Ph E O (82%)

structure of **44d**. The ORTEP diagram for this determination is presented in Figure 2 and the anticipated pseudoequatorial disposition of the C7 and C10 substituents is clearly evident from this solid-state structure.

The clean H.O. cycloaddition pathway followed by complex **43** can be contrasted with the normal behavior of metal-free azepines in the presence of diene reaction partners. In most instances those combinations tend to display predominantly [4π + 2π] cycloaddition chemistry wherein the azepine can participate as either the 4π or 2π partner.[39] There have, however, been sporadic reports of metal-free azepine derivatives that provide small quantities of [6+4] adducts with reactive dienes under thermal activation.[40] For example, *N*-ethoxycarbonylazepine (**45**) yielded a minor amount of a [6+4] adduct when heated in the presence of 2,5-dimethoxycarbonyl-3,4-diphenylcyclopentadienone.[40b,c] It is noteworthy that the resultant [6+4] adduct **46** exhibited exclusive exo stereochemistry, consistent with other metal-free, thermal [6+4] cycloaddition reactions.

We are currently in the process of extending the metal-promoted [6+4] cycloaddition chemistry of heterocyclic triene complexes to applications in natural product synthesis, and additional chemistry displayed by these intriguing complexes will be described in subsequent sections.

III. CHROMIUM(0) PROMOTED [6π+2π] CYCLOADDITION REACTIONS

A. Basic Reaction Characteristics

The mechanistic pathways suggested in Schemes 2 and 3 have a number of intriguing implications, among which is the expectation that other higher order pericyclic processes should be amenable to transition metal promotion. The [6π+2π] combination is a particularly challenging one for synthetic chemists, which, if rendered sufficiently efficient could provide rapid access to eight-membered carbocycles with all of the attendant attributes that characterize H.O. cycloaddition reactions.

A few examples of [6+2] cycloadditions are known to occur in the absence of metal intervention. However, in most instances the bicyclo[4.2.1]nonane adduct resulting from the [6+2] pathway is only a minor product.[41] An exception is the fascinating photoinduced, intramolecular [6+2] reactions of protonated tropone, which has been brought to a useful level of practice by Feldman and co-workers.[42] Prior to our investigations, a few reports of metal-mediated [6+2] cycloaddition reactions have appeared. However, these have generally been of somewhat limited

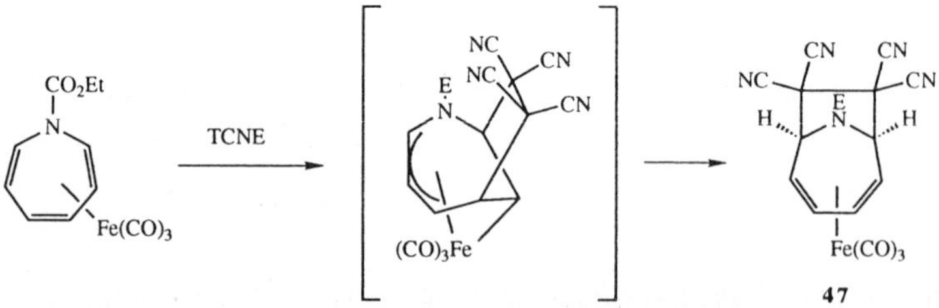

(26)

47

scope.[43] A noteworthy example of these earlier investigations is the cycloaddition of a (η^4-azepine)tricarbonyl iron(0) complex with TCNE to give adduct **47** (Eq. 26).[43c] The reaction was shown to proceed by an initial 1,3 pathway which then rearranged to the net [6+2] adduct. In this particular transformation the 2π addend approached the triene from the face distal to the metal center which is in direct contrast to the situation in the related reactions reported by Pettit[7] (Eq. 4).

In an effort to develop a more general and versatile cycloadditive entry into the important bicyclo[4.2.1]nonane system, we reasoned that the all carbon 4π partner in either of the mechanisms described previously could be replaced with an α,β-unsaturated carbonyl component which might be expected to react as a 2π partner and deliver the desired adducts. To examine the feasibility of this scheme, the parent chromium(0) complex **10** was irradiated (15 min, pyrex filter) in the presence of a slight excess of ethyl acrylate. To our delight this process provided adduct **48** as a single *endo*-diastereomer in 92% isolated yield.[44,45]

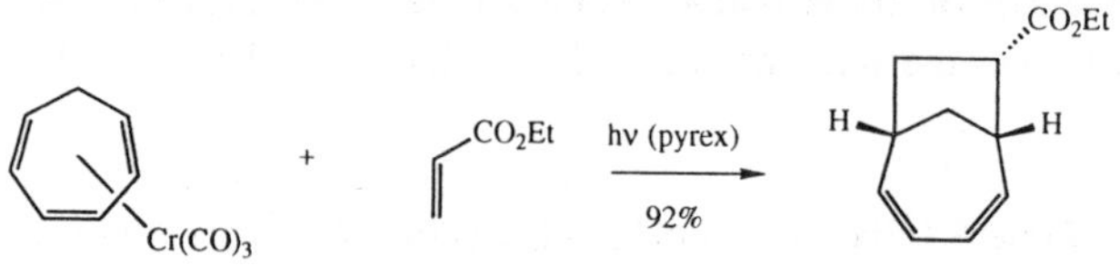

(27)

10 48

7α-Ethoxycarbonyl-(1Hβ,6Hβ)-bicyclo[4.2.1]-nona-2,4-diene (**48**). A solution of complex **10** (228 mg, 1.0 mmol) and ethyl acrylate (150 mg, 1.5 mmol) in hexanes (320 mL) was irradiated [Canrad–Hanovia medium pressure Hg vapor lamp, uranium glass sleeve (15 min)]. The resultant yellow solution was stirred in an open vessel exposed to air for 10 min and then filtered. The colorless solution was concentrated and the residue purified by flash chromatography (silica gel, hexanes/diethyl ether, 1:1) to give 177 mg (92%) of a colorless oil.

In contrast to the metal-mediated [6+4] cases in which a second decomplexation step was required to isolate the organic moiety, the product in Eq. 27 was isolated metal-free directly. The *endo* nature of the adduct was established by equilibration studies and comparison with known bicyclo[4.2.1]nonane systems.[41a,e] Further support for the stereochemical assignments came from NOE experiments on

OMe, $Cr(CO)_3$, **50** + CO_2Et —hν, 68%→ MeO, 9, H, 7, CO_2Et, **49** (28)

substituted adduct **49**. Irradiation of the 9-methoxy signal resulted in a 7.6% enhancement of the signal for the *exo*-proton at C7. The results of this experiment also suggests, at least circumstantially, that the acrylate approaches the complex **50** from the face proximate to the metal center since the stereochemical integrity of the 7-*endo*-methoxy complex starting material used in this transformation has been assured.[46] Several representative examples of the photochemical [6+2] cycloaddition are collected in Table 6.

The generally high level of reaction efficiency is one of the most notable features of the metal-promoted [6+2] cycloaddition process. In contrast, however, to the [6+4] cycloaddition, the electronic nature of the 2π partner plays a crucial role in the outcome of the [6+2] reaction as evidenced by the total absence of any product when butyl vinyl ether was reacted with complex **10** (Entry 6, Table 6). Once again, quartz filtered light results in lower product yields (Entry 4).

A number of substituted cycloheptatriene complexes have also been examined in this reaction and regioselectivity results vary with the placement and nature of the substituent on the triene partner. For example, a single regioisomer was obtained in the reaction of the 1-methoxycycloheptatriene complex **33** with ethyl acrylate.

Table 6. Photochemical [6+2] Cycloaddition of Complex **10**

$Cr(CO)_3$ + R′ R —hν→ R′, R, H, H

Entry	*Substituents (R,R′)*	*Yield (%)*[a]
1	R = COMe, R′ = H	97
2	R, R′ = CO_2Et	80
3	R = SOPh, R′ = H	39
4	R = CO_2Et, R′ = H	58[b]
5	R = CO_2Et, R′ = H	92
6	R = OBu, R′ = H	0

Notes: [a]All cycloadditions were preformed by irradiation using a pyrex or uranium glass filter unless otherwise noted.

[b]Irradiation through a quartz filter.

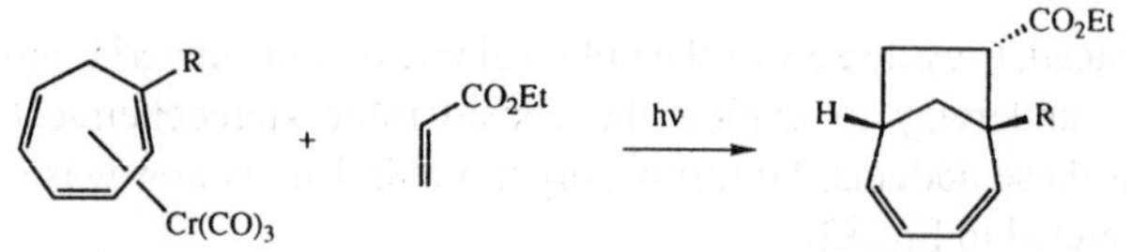

(29)

33, R=OMe
52, R=Me

51, R=OMe, (81%)
53, R=Me, (87%)

Interestingly, the cycloadduct isomer displaying vicinal substitution (presumably the more hindered option) prevails. The remarkably high reaction efficiency of triene complexes with substituents at the bond-forming centers is also noteworthy *vis-a-vis* the corresponding [6+4] examples. Similar levels of regioselectivity and reaction efficiency were observed in the photochemical reaction of the 1-methyl substituted complex **52**.

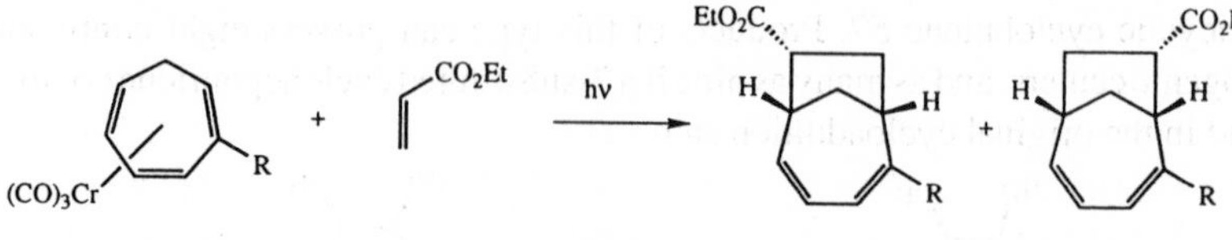

(30)

31, R=OMe	64%	1	1
54, R=CO_2Et	67%	1	1

No detectable regioselectivity was exhibited in reactions involving complexes having substituents at the 2-position of the triene ligand (Eq. 30); however, good selectivity was observed with the corresponding 3-substituted species (Eq. 31). A fully satisfactory rationale for these contrasting observations is not obvious at this point.

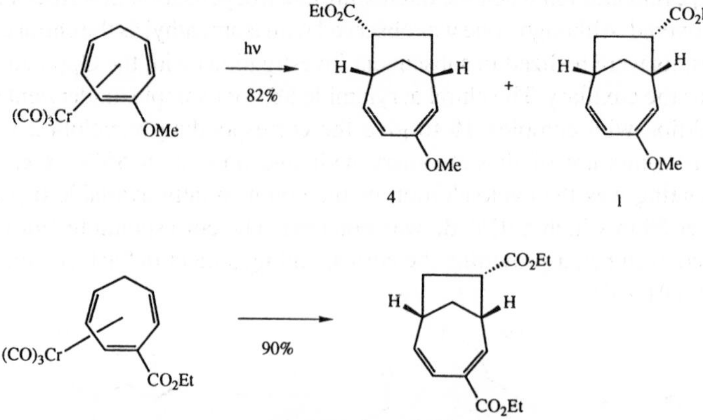

(31)

A consequence of the facile decomplexation noted for the photochemical [6+2] cycloadditions is that further photoinduced chemistry can occur with the metal-free cycloadducts present in the reaction mixtures. In an interesting case in point, extended irradiation (quartz) provided good yields of cyclobutene products in many cases. For example, irradiation of complex **10** in the presence of (+)-isomenthyl acrylate for a 2 hour duration provided none of the expected bicyclo[4.2.1]nonane product. Instead, tricycle **55** was the only isolable organic material produced. Work is currently underway to exploit the considerable stereochemical information available in these adducts. Unfortunately, no chiral induction was evident in the reaction depicted in Eq. 32.

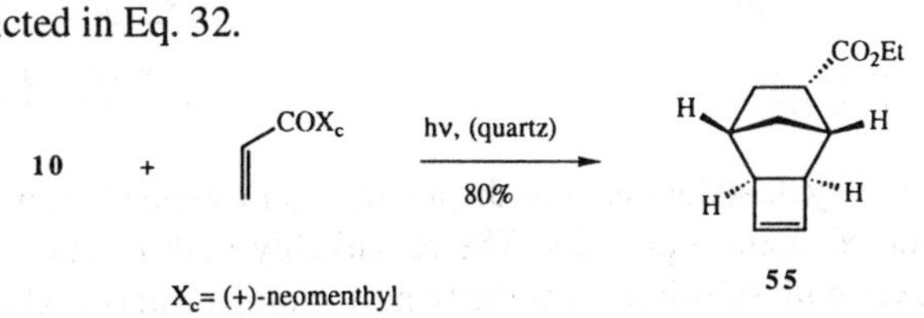

X_c= (+)-neomenthyl

10 + COX$_c$ → hν, (quartz), 80% → **55** (32)

A similar transformation can also be effected in the metal-free [6+4] cycloadducts, which can be conveniently irradiated after metal decomplexation. For example, photolysis of bicyclo[4.4.1]undecane derivative **56** provided the interesting tricyclic cyclobutene **57**. Products of this type can possess eight contiguous stereogenic centers, and as many as nine if a 7-substituted cycloheptatriene complex is used in the original cycloaddition step.

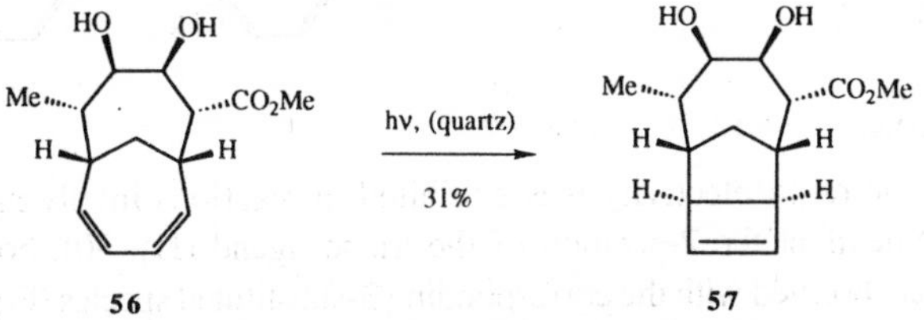

56 → hν, (quartz), 31% → **57** (33)

We were actually attempting to explore the possibilities for chiral induction in the [6+2] cycloaddition when the interesting electrocyclization described in Eq. 32 was discovered. Although none was observed with isomenthyl as the chiral element, better results were realized in subsequent investigations with the Oppolzer sultam serving as the auxiliary. The chiral acrylamide **58**, for example, underwent smooth cycloaddition with complex **10** to give the corresponding bicyclo[4.2.1]nonane adduct as a mixture of diastereomers with one present in 55% excess. More discriminating was the cycloaddition of the commercially available α,β-unsaturated ester **59** in which a 91% de was obtained. The corresponding *trans*-isomer performed even better providing the corresponding adduct in 94% chemical yield and with a 91% de.

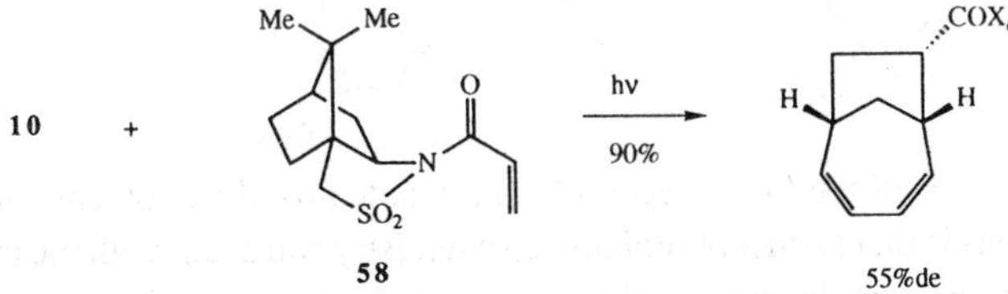

10 + **58** → hν, 90% → 55%de (34)

To date, in addition to the use of chiral auxiliaries attached to the 2π and 4π components, we have also briefly explored the effects of chiral ligands in the coordination sphere of the metal as described previously. A third approach to

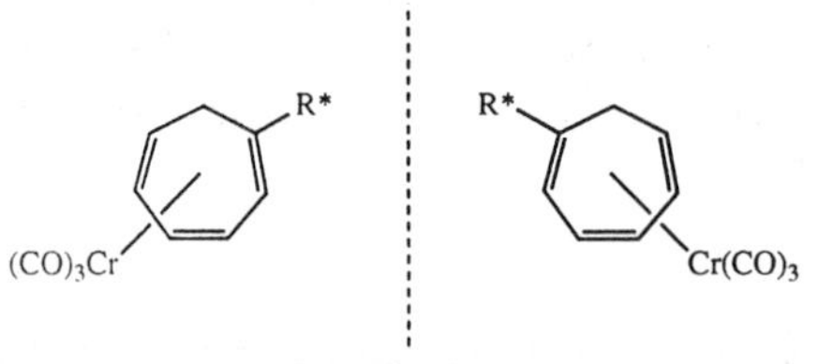

(CO)₃Cr R* | R* Cr(CO)₃

Figure 3.

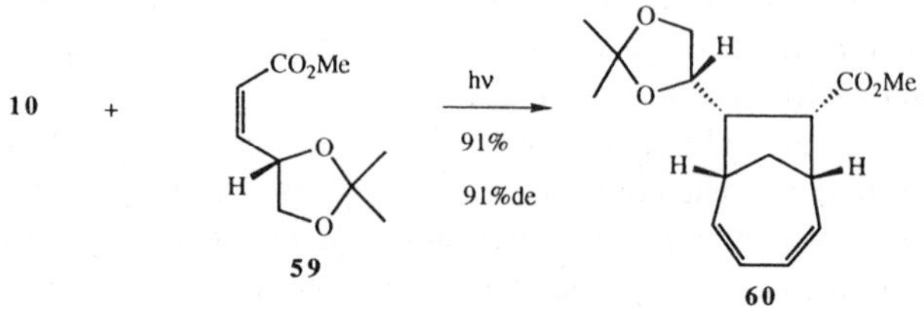

10 + 59 —hν, 91%, 91%de→ 60 (35)

solving the problem of inducing chirality during these cycloadditions is to append an auxiliary to the triene moiety. It was reasoned that the maximum induction would be observed if the center of asymmetry was as close to the bond-forming centers as possible; consequently 1-substituted cycloheptatrienes were deemed to be particularly promising candidates. However, this induction strategy would trigger additional complications, since a chiral appendage at this position would render the two triene faces diastereotopic. This potential predicament would, none-the-less, afford us the opportunity to explore diastereofacially selective complexation, an objective not easily accomplished in systems with little rigidly enforced facial bias. Figure 3 depicts some of the salient issues involved with planar chirality in the chromium(0)-cycloheptatriene series. Most important, if R* is chiral, the enantiomers shown become diastereomers. Unfortunately, precedent for diastereoselective complexation in systems related to ours is quite slim.[48] For example, Potter and McCague reported only a 7% de in the complexation shown in Eq. 36.[48a]

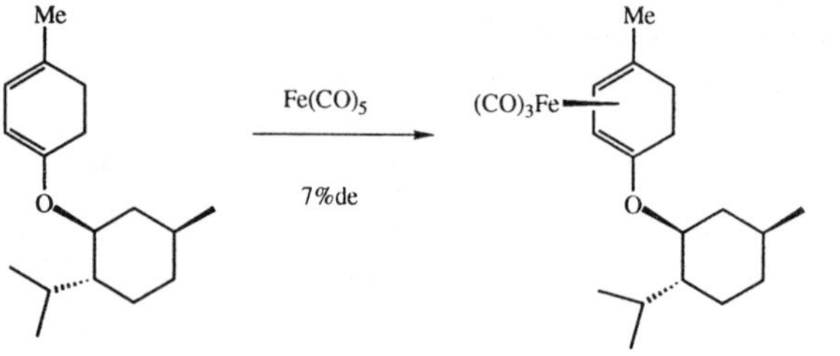

$Fe(CO)_5$, 7%de (36)

Nevertheless, we have briefly examined the viability of this projected chemistry in the chromium series with some simple examples. Treatment of the chiral cycloheptatriene **61**, prepared from 1-*endo*-(+)-fenchol and tropylium fluoroborate[27b] with $(MeCN)_3Cr(CO)_3$ at room temperature, provided complex **62** as a mixture of two easily separated diastereomers with one produced in an excess of

FenchylO — $(MeCN)_3Cr(CO)_3$, 89%, 40%de → FenchylO, $Cr(CO)_3$

61 → 62 (37)

40%. The absolute configuration of the major complex has not been established with any rigor; consequently, the depiction is drawn arbitrarily. We were particularly delighted with this rather unexpected outcome and are currently exploring other auxiliaries that may provide superior facial discrimination as well as affording better opportunities for eventual removal.

In an effort to glean additional information about the behavior of these chiral complexes, the major diastereomer of **62** was exposed to ethyl acrylate under standard photochemical conditions to give adduct **63** in optically pure form. None of the product arising from the alternative diastereomeric complex was detected in this reaction. This result strengthens our belief that the metal center remains on the same face of the triene throughout the stepwise photocycloaddition process.

62 + CO_2Et — hν, 86%, 100%de → CO_2Et, H, OFenchyl

63 (38)

The [6+2] cycloaddition protocol is also amenable to heteroatom substitution in the triene ligand. Treatment of complex **43** with a slight excess of ethyl acrylate under standard photochemical conditions leads to the corresponding azabicyclo[4.2.1]nonane **64** in excellent yield.[49] The relationship of compounds such as **64**, which are readily available using this protocol, and the potent nerve depolarizing agent, anatoxin-a is noteworthy.[50]

CO_2Me, N, $Cr(CO)_3$ + CO_2Et — hν, 88% → CO_2Et, H, E, N, H ⟹ H, H, N, H, O, Me

43 — 64 — anatoxin-a

7α-Ethoxycarbonyl-9-methoxycarbonyl-(1Hβ,6Hβ)-9-azabicyclo[4.2.1]-nona-2,4-diene (**64**). To a solution of complex **43** (266 mg, 0.927 mmol) in Et_2O (5 mL) and hexanes (345 mL) was added ethyl acrylate (138 mg, 1.38 mmol). The resultant mixture was irradiated (Canrad–Hanovia 450W medium pressure Hg vapor lamp, pyrex immersion well) for 35 min. The reaction mixture was then concentrated *in vacuo* and purified by flash column chromatography (hexanes/ethyl acetate, 9:1) to give 206 mg (88%) of compound **64**.

Substituted cyclooctenes can be easily accessed from the azepine adducts obtained in these reactions. An illustration of this capability has been achieved with adduct **64**. Saturation of the diene system in **64** followed by reduction of the carbamate to the corresponding *N*-methyl derivative **65** sets the stage for exhaustive methylation and subsequent Hofmann elimination. Although eliminations of this type proceed smoothly at moderate temperatures in the closely related tropane alkaloid series, temperatures in excess of 325 °C were required to effect a similar fragmentation in the homotropane **65**. In this instance the reaction provided an equimolar mixture of isomers **66** and **67**.

64 — 1) H_2-Pd/C; 2) $LiAlH_4$; 92% → 65 (H, N, Me, H, OH) → 66 (Me_2N, HO) + 67 (HO, Me_2N) (39)

With these results in hand, one of the original goals of this research program—to assemble structurally elaborate 8- and 10-membered carbocycles via a cycloaddition process—has been successfully brought to practice.

B. Thermal and Catalytic Cycloadditions

The success of the thermal version of the [6+4] cycloaddition as described previously prompted a similar investigation in the [6π+2π] cycloaddition series.[51] Heating complex **10** in the presence of excess ethyl acrylate in refluxing n-Bu_2O provided the anticipated bicyclic adduct in 80% isolated yield (Eq. 40). Once again the adduct from the thermal cycloaddition was isolated metal-free and proved to be

10 + (CO_2Et) — n-Bu_2O, reflux, 80% → (H, H, CO_2Et) (40)

identical in every respect with the corresponding product available from the photochemically initiated reaction. Interestingly, heating complex **10** with a chiral 2π partner as in Eq. 41 provided the resultant adduct with virtually the same level and sense of stereoinduction as obtained from the corresponding photochemical transformation.

The efficiency of the thermal process in the [6+2] series and the metal-free nature of the adducts suggested the possibility of effecting this transformation employing only a small quantity of a "$Cr(CO)_3$" source. Previous observations concerning the lability of the cycloadduct–metal complexes in this series were also suggestive of the viability of a practical, catalytic version of these reactions. Initial investigations

10 + CO_2Me … H … n-Bu_2O 88% 90%de … CO_2Me H H (41)

employed complex **10** as the catalyst since it is readily available, reasonably stable, and easy to manipulate. The first generation of catalytic reactions were encouraging but not particularly efficient. After one or two turnovers the chromium(0) catalyst was deactivated by oxidation [presumably to a Cr(III) species] as evidenced by the development of a green solid in the reaction mixture. Performing the cycloaddition in a thoroughly degassed solution in a sealed tube allowed for efficient conversion to product. Equation 42 depicts a typical result. Heating cycloheptatriene and excess ethyl acrylate in the presence of complex **10** (15 mol %) provided a 99% isolated yield of a mixture consisting of the expected bicyclo[4.2.1]nonane **48** and adduct **68** in a 10:1 ratio, respectively. The influence of the catalytic metal in this transformation is remarkable since heating the same two reactants together under identical conditions *in the absence of the metal* gave a 65% yield of a product mix in which **68** was virtually the exclusive component (Eq. 43).[52] *The chromium(0) catalyst completely altered the course of the cycloaddition.* Interestingly, employing only 2 mol % of catalyst resulted in even greater discrimination between the two competing pericyclic pathways, but with somewhat lower yields.

+ CO_2Et — **10**, (15 mol%), n-Bu_2O, 160°C, sealed tube, 99% → **48**, (90%) + **68**, (9%) (42)

7α-Ethoxycarbonyl-(1Hβ,6Hβ)-bicyclo[4.2.1]nona-2,4-diene (**48**) *via catalytic cycloaddition.* To an oven-dried Carius tube was added complex **10** (160 mg, 0.7 mmol), ethyl acrylate (1 mL, 9 mmol), cycloheptatriene (0.55 mL, 5 mmol) and freshly distilled *n*-Bu_2O (10 mL). The contents of the tube were degassed several times by the freeze-pump-thaw technique, sealed under vacuum and then heated at 160 °C for 15 h. After cooling to room temperature, the green suspension was filtered through celite and the filtrate concentrated *in vacuo*. The residue was purified by flash column chromatography (silica gel, chloroform/hexanes, 10:1) to give 948 mg (90%) of compound **48** and 95 mg (9%) of compound **68**.

The *N*-alkoxycarbonylazepines also undergo efficient catalytic [6+2] cycloaddition. In these cases (η^6-naphthalene)tricarbonylchromium(0) proved to be a superior catalyst for effecting the reactions (Eq. 44).

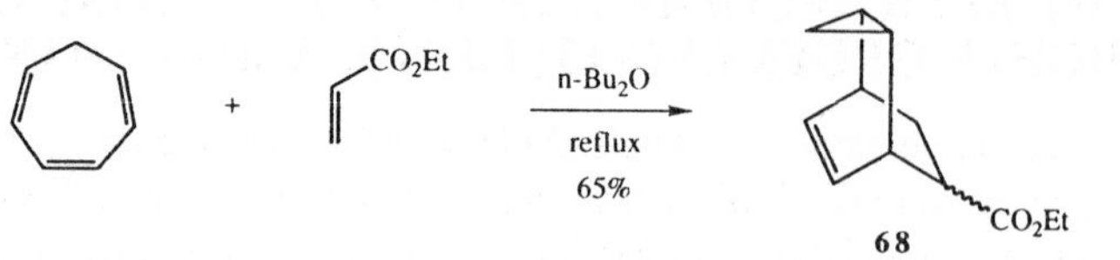

(43)

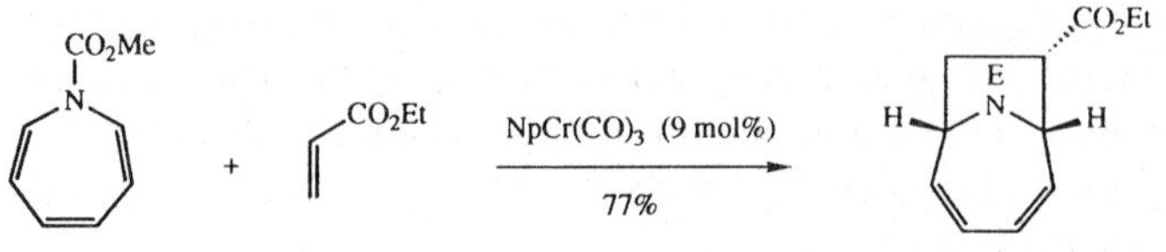

n-Bu_2O, reflux, 65%, CO_2Et, **68**

(44)

CO_2Me, CO_2Et, $NpCr(CO)_3$ (9 mol%), 77%, E, N, H, H

Np=naphthalene

The cycle depicted in Figure 4 can satisfactorily explain many of the observations made to date in these catalytic reactions. A coordinating solvent such as n-Bu_2O can serve both to initiate ligand exchange at the beginning of the process and to decomplex the resultant cycloadduct metal tricarbonyl complex and recycle the metal in the form of a very reactive "$Cr(CO)_3$" transfer agent. It is worth noting that the putative "$Cr(CO)_3$" carrier **69** has also been invoked as the actual catalytic agent in the facile chromium(0) mediated 1,4-reduction of dienes in coordinating solvents.[53]

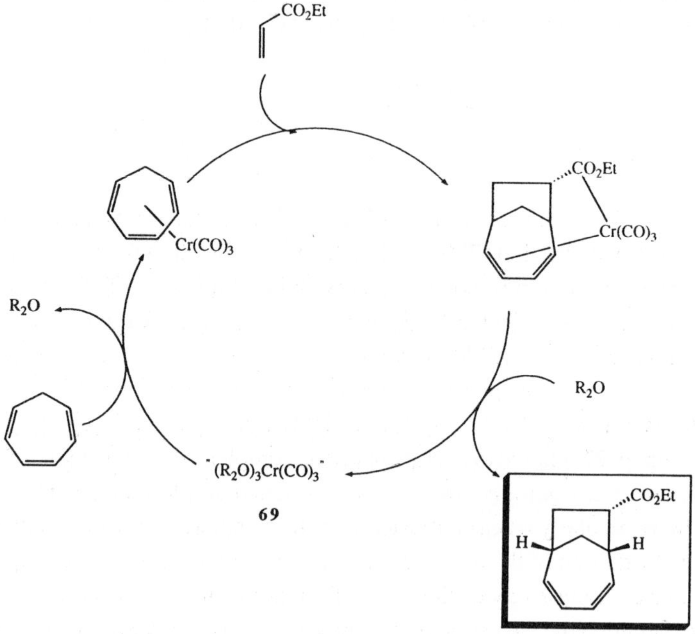

CO_2Et, $Cr(CO)_3$, R_2O, "$(R_2O)_3Cr(CO)_3$" **69**, CO_2Et, $Cr(CO)_3$, R_2O, CO_2Et, H, H

Figure 4.

IV. INTRAMOLECULAR CHROMIUM(0) PROMOTED HIGHER ORDER CYCLOADDITION REACTIONS

The intramolecular version of the Diels–Alder reaction has emerged as a powerful method in organic synthesis.[54] In contrast, relatively few examples of H.O. cycloadditions are known in which the interacting π-systems are connected by a tether.[55] The capability of effecting both [6+4] and [6+2] cycloadditions in an intramolecular fashion would be an important extension of the metal-promoted H.O. cycloaddition technology that would afford additional opportunities for application to natural product synthesis. Equation 45 depicts the salient features of the intramolecular version of the [6+4] reaction.

X
$Cr(CO)_3$
1) hv (pyrex)
2) $(MeO)_3P$
X
H
H

70a, X=O
70b, $X=CH_2$

71a, (52%)
71b, (85%)

(45)

The reaction of complex **70a** is interesting because it provides rapid access to a tricyclic intermediate suitable for attacking, from a unique perspective, the ingenane problem alluded to earlier in this chapter. The essential *endo* nature of adduct **71a** was confirmed by comparison with the known *exo*-adduct derived from the metal-free thermal intramolecular tropone cycloaddition (Eq. 46).[4]

O
heat
O
H
H

(46)

The lower yield for the conversion of **70a** into **71a** relative to the corresponding hydrocarbon transformation (**70b** $\rightarrow$ **71b**) can be attributed to the relative lability of the (tropone)chromium(0) complex, as established during preliminary investigations in our laboratory on the parent complex. Prior to embarking on the study of the intramolecular cycloaddition, we briefly examined the general behavior of the known (η-2,4,6-cycloheptatrien-1-one)tricarbonylchromium(0) (**72**)[27b] in H.O. cycloaddition reactions. Uniformly lower yields were obtained in the cycloadditions of complex **72** relative to cycloheptatriene complex **10**. However, the reasonably efficient cycloaddition observed for 2-methyl-substituted complex **73** is noteworthy, particularly when contrasted with the complete failure of the attempted thermal cycloaddition of the related *metal-free*, substituted tropone (see Eq. 2).

At the outset of our study of the intramolecular version of the metal-promoted [6+4] cycloaddition there was some concern about the ability to selectively prepare the η^6 complexes **70a,b** in the presence of the potentially competitive diene ligand.

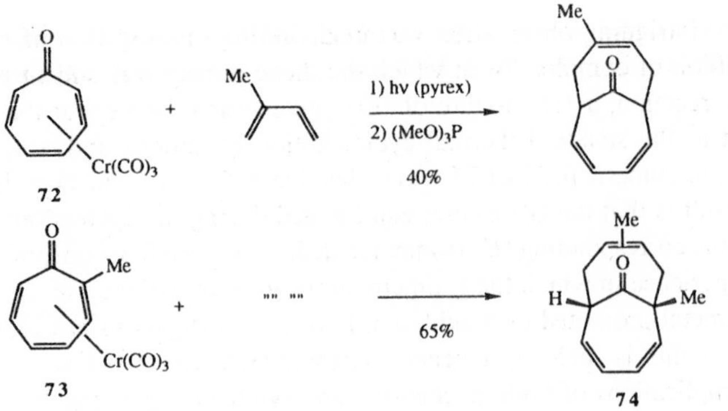

(47)

Precedent suggested,[56] however, that, in general, η^4-chromium(0) complexes of conjugated dienes are relatively unstable and, in the event, we were able to prepare the required complexes with complete haptoselectivity.

Subsequent observations served to further dispel the notion that "haptoselective" complexation in these systems would be an obstacle to the development of the intramolecular [6+4] cycloaddition reaction. Heating readily available 7-*exo*-substituted complex **75** in a sealed tube led, presumably via the intermediacy of **70b**, to the bicyclo[4.4.1]undecane **71b** in 60% yield. The ability to utilize 7-substituted

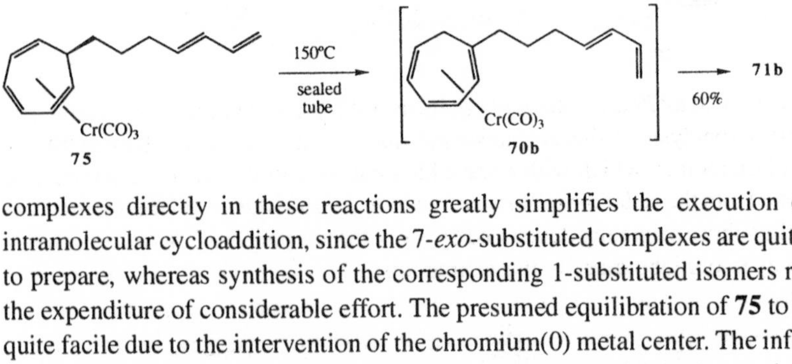

(48)

complexes directly in these reactions greatly simplifies the execution of the intramolecular cycloaddition, since the 7-*exo*-substituted complexes are quite easy to prepare, whereas synthesis of the corresponding 1-substituted isomers require the expenditure of considerable effort. The presumed equilibration of **75** to **70b** is quite facile due to the intervention of the chromium(0) metal center. The influence of the metal center on the 1,5-migrations of the *endo*-C7 hydrogen in these complexes is well known.[27a,d] Cycloaddition occurs, in these cases, exclusively via isomer **70b** because all other species that may be present in solution are geometrically precluded from engaging in cyclization. As a consequence, although a number of triene regioisomers are probably involved at one point or another during these reactions, only one of them can actually undergo cycloaddition.

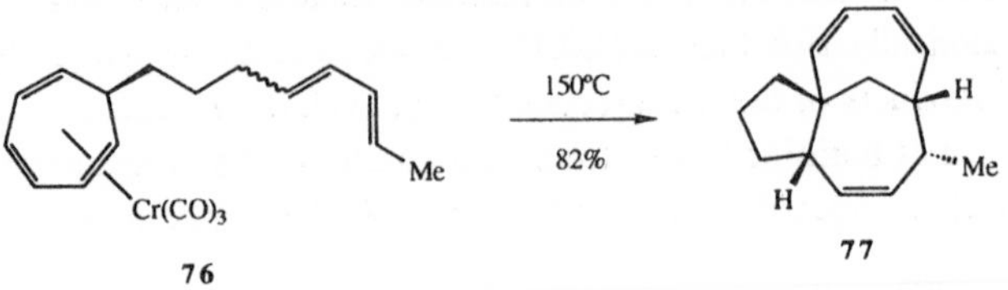

(49)

Another intriguing observation was made during investigation of the thermal cycloaddition of complex **76** in which the diene moiety was substituted. In this particular reaction, a 1:1 mixture of *(E/Z)*-geometrical isomers at the diene was subjected to the standard thermal cycloaddition conditions and, surprisingly, a single diastereomeric product **77** was isolated in 82% yield. The clear implication of this result is that the (*Z*)-isomer equilibrated during the cycloaddition process and only the corresponding (*E*)-isomer reacted. This is the first confirmed example in our experience in which the geometrical integrity of a diene was compromised during a metal-promoted cycloaddition. Further investigations are in progress to establish if this is, indeed, a general phenomenon associated with (*Z*)-dienes. Several implications of both mechanistic and synthetic significance would result from a positive verdict on this question.

In a related development, we have established that an intramolecular cycloaddition can be effected in the [6+2] series as well. Surprisingly, however, unactivated double bonds undergo the reaction cleanly (Eq. 50), a result which is quite unlike the intermolecular version. Normally, only electron-deficient 2π species are acceptable reaction partners in the latter case.

$Cr(CO)_3$ **78** —hν (pyrex) (63%) or 140°C dioxane (65%)→ **79** (50)

Adducts such as **79** are potentially quite useful for natural product synthesis. For example, ozonolysis of this substance provided the diquinane dialdehyde **80** as a single diastereomer, which with some additional structural modifications could be elaborated into the cedar oil component, cedrene, via a sequence of transformations paralleling those originally developed by Stork for this purpose.[57] Work is currently underway in our laboratory in these directions.

79 —O_3, 40%→ **80** ⟹ cedrene (51)

V. CONCLUSIONS

Many of the initial objectives of our investigation of the metal-promoted higher order cycloaddition reaction have been realized. Access to structurally elaborate and stereochemically rich bicyclo[4.4.1]undecane species is now quite routine. Many of the products of the metal-promoted cycloaddition reactions are particularly well suited as building blocks for the construction of biologically significant diterpenes, such as ingenol, as well as a number of other candidates.

Furthermore, the synthesis of extensively substituted 8- and 10-membered carbocycles via these transformations is easily achieved in a stereochemically rational fashion. The resultant products offer potential entries into a wide range of important natural products. The characteristics of the metal-promoted cycloaddition reactions suggest numerous applications to synthesis which we are now vigorously pursuing in our laboratory.

ACKNOWLEDGMENTS

I wish to express my sincerest gratitude to all of the co-workers who have contributed so much to the development of the chemistry discussed in this review. These very capable investigators include: Humy Ateeq, James Henshilwood, Nelly Charles, Cyprian Ogbu, Kevin Short, Mark Ferguson, Chris Krueger, Stephane Cuisiat, Vincent Sandanayaka, and Priyantha Sugathapala. I also wish to acknowledge the National Institutes of Health for their generous support of this research program.

REFERENCES AND NOTES

1. For a general overview of higher-order cycloaddition chemistry, see: Rigby, J.H. In *Comprehensive Organic Synthesis*; Trost, B.M.; Fleming, I.; Eds., Pergamon Press: Oxford, 1991, Vol. 5, pp. 617–643.
2. (a) Fujise, Y.; Saito, H.; Ito, S. *Tetrahedron Lett.* **1976**, 1117. (b) Ito, S.; Ohtani, H.; Narita, S.; Honma, H. *Ibid.* **1972**, 2223; (c) Cookson, R.C.; Drake, B.V.; Hudec, J.; Morrison, A. *J. Chem. Soc., Chem. Commun.* **1966**, 15; (d) Ito, S.; Fujise, Y.; Okuda, T.; Inoue, Y. *Bull. Chem. Soc. Jpn.* **1966**, *39*, 1351.
3. (a) Garst, M.E.; Roberts, V.A.; Houk, K.N.; Rondan, N.G. *J. Am. Chem. Soc.* **1984**, *106*, 3882; (b) Garst, M.E.; Roberts, V.A.; Prussin, C. *Tetrahedron* **1983**, *39*, 581; (c) Garst, M.E.; Roberts, V.A.; Prussin, C. *J. Org. Chem.* **1982**, *47*, 3969.
4. Rigby, J.H.; Moore, T.L.; Rege, S. *J. Org. Chem.* **1986**, *51*, 2398.
5. (a) Rigby, J.H. *Tetrahedron Lett.* **1982**, *23*, 1863; (b) Rigby, J.H.; Wilson, J.Z. *J. Am. Chem. Soc.* **1984**, *106*, 8217.
6. Houk, K.N.; Woodward, R.B. *J. Am. Chem. Soc.* **1970**, *92*, 4143.
7. (a) Davis, R.E.; Dodds, T.A.; Hseu, T.-H.; Wagnon, J.C.; Devon, T.; Tancrede, J.; McKennis, J.S.; Pettit, R. *J. Am. Chem. Soc.* **1974**, *96*, 7562; (b) Ward, J.S.; Pettit, R. *Ibid.* **1971**, *93*, 262.
8. In the absence of a metal center virtually no bicyclo[4.2.1]nonane products can be isolated from related reactions, see: (a) Jenner, G.; Papadopoulos, M. *J. Org. Chem.* **1986**, *51*, 585; (b) Rigby, J.H.; Denis, J.-P. *Synth. Commun.* **1986**, *16*, 1789; (c) Bellus, D.; Helfrich, G.; Weis, C.D. *Helv. Chem. Acta* **1971**, *54*, 463.
9. For representative examples of metal mediated cycloaddition reactions that work well in a variety of situations, see: (a) Wilson, A.M.; Waldman, T.E.; Rheingold, A.L.; Ernst, R.D. *J. Am. Chem. Soc.* **1992**, *114*, 6252; (b) Lautens, M.; Tam, W.; Edwards, L.G. *J. Org. Chem.* **1992**, *57*, 8; (c) Kreiter, C.G.; Lehr, K.; Leyendecker, M.; Sheldrick, W.S.; Exner, R. *Chem. Ber.* **1991**, *124*, 3; (d) Jolly, R.S.; Luedtke, G.; Sheehan, D.; Livinghouse, T. *J. Am. Chem. Soc.* **1990**, *112*, 4965; (e) Slegeir, W.; Case, R.; McKennis, J.S.; Pettit, R. *Ibid.* **1974**, *96*, 287; (f) Onoue, H.; Moritani, J.; Murahashi, S.-I. *Tetrahedron Lett.* **1973**, 121; (g) Wristers, J.; Brener, L.; Pettit, R. *J. Am. Chem. Soc.* **1970**, *92*, 7499.
10. (a) Wender, P.A.; Ihle, N.C. *J. Am. Chem. Soc.* **1986**, *108*, 4678; (b) Wender, P.A.; Snapper, M.L. *Tetrahedron Lett.* **1987**, *28*, 2221; (c) Wender, P.A.; Ihle, N.C. *Ibid.* **1987**, *28*, 2451; (d) Wender,

P.A.; Ihle, N.C.; Correia, C.R.D. *J. Am. Chem. Soc.* **1988**, *110*, 5904; (e) Wender, P.A.; Tebbe, M.J. *Synthesis* **1991**, 1089.

11. (a) Özkar, S.; Kurz, H.; Neugebauer, D.; Kreiter, C.G. *J. Organomet. Chem.* **1978**, *160*, 115; (b) Kreiter, C.G.; Kurz, H. *Chem. Ber.* **1983**, *116*, 1494; (c) Michels, E.; Kreiter, C.G. *J. Organomet. Chem.* **1983**, *252*, C1; (d) Michels, E.; Sheldrick. W.S.; Kreiter, C.G. *Chem. Ber.* **1985**, *118*, 964; (e) Kreiter, C.G.; Michels, E. *J. Organomet. Chem.* **1986**, *312*, 59; (f) Kreiter, C.G. *Adv. Organomet. Chem.*, **1986**, *26*, 297.
12. For a preliminary account of our work in this area, see: Rigby, J.H.; Ateeq, H.S. *J. Am. Chem. Soc.* **1990**, *112*, 6442.
13. (a) Abel, E.W.; Bennett, M.A.; Burton, R.; Wilkinson, G. *J. Chem. Soc.* **1958**, 4559; (b) Munro, J.D.; Pauson, P.L. *Ibid.* **1961**, 3475; (c) Tate, D.P.; Knipple, W.R.; Augl, J.M. *Inorg. Chem.* **1962**, *1*, 433.
14. The stereochemistry of these adducts was intitally assigned based on the small coupling constants (1-2 Hz) observed for the protons on C7 and/or C10 (compounds **11a, b**). This was consistent with a pseudoequitorial disposition for the non-hydrogen substituents at C7 and C10. A subsequent X-ray structure of a related adduct confirmed the veracity of this analysis.
15. Pauson, P.L.; Todd, K.H. *J. Chem. Soc.* (C) **1970**, 2638.
16. Kreiter and co-workers have noted what may be an analogous intermediate during their investigations on the reaction of (heptafulvene)tricarbonylchromium with certain hydrocarbon dienes, ref. 11d.
17. (a) Wrighton, M. *Chem. Rev.* **1974**, *74,* 401. For some recent studies on the photochemistry of polyene-M$(CO)_3$ complexes, see: (b) Astley, S.T.; Churton, M.P.V.; Hitam, R.B.; Rest, A.J. *J. Chem. Soc., Dalton Trans.* **1990**, 3243; (c) Bloyce, P.E.; Hooker, R.H.; Rest, A.J.; Bitterwolf, T.E.; Fitzpatrick, N.J.; Shade, J.E. *Ibid.* **1990**, 833; (d) Hooker, R.H.; Rest, A.J. *Ibid.* **1982**, 2029; (e) Hitam, R.B.; Mahmoud, K.A.; Rest, A.J. *Coord. Chem. Rev.* **1984**, *55*, 1; (f) Perutz, R.N.; Turner, J.J. *J. Am. Chem. Soc.* **1975**, *97*, 4800.
18. Van Houwelingen, T.; Stufkens, D.J.; Oskam, A. *Organometallics* **1992**, *11,* 1146.
19. Cotton, F.A.; McCleverty, J.A.; White, J.E. *Inorg. Synth.* **1967**, *9*, 121.
20. For discussion of other "hapticity slippages," see: (a) Schuster-Woldan, H.G.; Basolo, F. *J. Am. Chem. Soc.* **1966**, *88,* 1657; (b) Dixon, D.T.; Burkinshaw, P.M.; Howell, J.A.S. *J. Chem. Soc., Dalton Trans.* **1980**, 2237; (c) O'Connor, J.M.; Casey, C.P. *Chem. Rev.* **1987**, *87*, 307; (d) Cheong, M.; Basolo, F. *Organometallics* **1988**, *7*, 2041; (e) Arthurs, M.; Piper, C.; Morton-Blake, D.A.; Drew, M.G.B *J. Organomet. Chem.* **1992**, *429*, 257.
21. For recent examples of wavelength effects on CO extrusion in metal complexes, see: (a) Merlic, C.A.; Xu, D. *J. Am. Chem. Soc.* **1991**, *113*, 7418; (b) Song, J.-S.; Bullock, R.M.; Creutz, C. *Ibid.* **1991**, *113*, 9862.
22. King, R.B.; Fronzaglia, A. *Inorg. Chem.* **1966**, *5*, 1837.
23. Brown, D.L.S.; Connor, J.A.; Demain, C.P.; Leung, M.L.; Martinho-Simoes, J.A.; Skinner, H.A.; Zafarani Moatlar, M.T. *J. Organomet. Chem.* **1977**, *142*, 321.
24. Al-Kathumi, K.M.; Kane-Maguire, L.A.P. *J. Chem. Soc., Dalton Trans.* **1974**, 428.
25. (a) Mahaffy, C.A.L.; Pauson, P.L. *Inorg. Synth.* **1979**, *19*, 154; (b) Mahaffy, C.A.L.; Pauson, P.L. *Ibid.* **1990**, *28*, 36.
26. Cais, M.; Fraenkel, D.; Weidenbaum, K. *Coord. Chem. Reviews* **1975**, *16*, 27.
27. (a) Pauson, P.L.; Smith, G.H.; Valentine, J.H. *J. Chem. Soc. (C)* **1967**, *1057*, 1061; (b) Pauson, P.L.; Todd, K.H. *Ibid.* **1970**, 2315; (c) Pauson, P.L.; Todd, K.H. *Ibid.* **1970**, *2636*, 2638; (d) Munro, J.D.; Pauson, P.L. *J. Chem. Soc.* **1961**, 3475.
28. Duke, R.K.; Richard, R.W. *J. Org. Chem.* **1984**, *49*, 1898.
29. Oppolzer, W. *Angew Chem. Intl. Ed., Engl.* **1984**, *23*, 876.
30. (a) Evans, D.A.; Chapman, K.T.; Bisaha, J. *J. Am. Chem. Soc.* **1988**, *110*, 1238; (b) Evans, D.A.; Chapman, K.T.; Bisaha, J. *Ibid.* **1984**, *106* , 4261.

31. (a) Oppolzer, W.; Chapuis, C.; Bernardinelli, G. *Helv. Chem. Acta.* **1984**, *67*, 1397; (b) Oppolzer, W.; Poli, G.; Kingma, A.J.; Starkemann, C.; Bernardinelli, G. *Ibid.* **1987**, *70*, 2201.
32. Anderson, W.P.; Blenderman, W.G.; Drews, K.A. *J. Organomet. Chem.* **1972**, *42*, 139.
33. Rigby, J.H.; Ateeq, H.S.; Krueger, A.C. *Tetrahedron Lett.* **1992**, *33*, 5873.
34. To the best of our knowledge this is the first report of a chromium complex of thiepin-1,1-dioxide, however, the corresponding iron tricarbonyl complex was recently described: Nishino, K.; Takagi, M.; Kawata, T.; Murata, I.; Inanga, J.; Nakasuji, K. *J. Am. Chem. Soc.* **1991**, *113*, 5059.
35. (a) Mock, W.L. *J. Am. Chem. Soc.* **1967**, *89*, 1281; (b) Mock, W.L.; McCausland, J.H. *J. Org. Chem.* **1976**, *41*, 242.
36. For an informative review of cheletopic extrusions, see: Mock, W.L. In *Pericyclic Reactions*; Marchand, A.P.; Lehr, R.E., Eds.; Academic Press: New York, 1977, Vol. II, pp. 141–179.
37. Bohlmann, F.; Zdero, C. *Phytochemistry* **1977**, *16*, 1243.
38. Kreiter, C.G.; Özkar, S. *Z. Naturforsch.* **1977**, *32b*, 408.
39. (a) Wiseman, J.R.; Chong, B.P. *Tetrahedron Lett.* **1969**, 1619; (b) Paquette, L.A.; Kuhla, D.E.; Barrett, J.H.; Leichter, L.M. *J. Org. Chem.* **1969**, *34*, 2888.
40. (a) Saito, K.; Iida, S.; Mukai, T. *Bull. Chem. Soc., Jpn.* **1984**, *57*, 3783; (b) Harano, K.; Yasuda, M.; Ban, T.; Kanematsu, K. *J. Org. Chem.* **1980**, *45*, 4455; (c) Harano, K.; Ban, T.; Yasuda, M.; Kanematsu, K. *Tetrahedron Lett.* **1979**, 1599.
41. For some representative examples, see ref. 8 and: (a) Wu, T.-C.; Houk, K.N. *J. Am. Chem. Soc.* **1985**, *107*, 5308; (b) Yang, N.-C.; Chiang, W.L. *Ibid.* **1977**, *99*, 3163; (c) Tabushi, I.; Yamada, H.; Yoshita, Z.; Kuroda, H. *Tetrahedron Lett.* **1971**, 1093; (d) Mori, A.; Takeshita, H. *Chem. Lett.* **1975**, 599.
42. (a) Feldman, K.S.; Come, J.H.; Freyer, A.J.; Kosminder, B.J.; Smith, C.M. *J. Am. Chem. Soc.* **1986**, *108*, 1327; (b) Feldman, K.S.; Come, J.H.; Fegley, G.J.; Smith, B.D.; Parvez, M. *Tetrahedron Lett.* **1987**, *28*, 607; (c) Feldman, K.S.; Come, J.H.; Kosmider, B.J.; Smith, P.M.; Rotella, D.P.; Wu, M.-J. *J. Org. Chem.* **1989**, *54*, 592; (d) Feldman, K.S.; Wu, M.-J.; Rotella, D.P. *J. Am. Chem. Soc.* **1990**, *112*, 8490.
43. (a) Green, M.; Heathcock, S.; Wood, D.C. *J. Chem. Soc., Dalton Trans.* **1973**, 1564; (b) Green, M.; Heathcock, S.M.; Turney, T.W.; Mingos, D.M.P. *Ibid.* **1977**, 204; (c) Chopra, S.K.; Moran, G.; McArdle, P. *J. Organomet. Chem.* **1981**, *214*, C36; (d) Kreiter, C.G.; Michels, E.; Kurz, H. *Ibid.* **1982**, *232*, 249; (e) Mack, K.; Antropiusova, H.; Sedmera, P.; Hanus, V.; Turecek, F. *J. Chem. Soc., Chem. Commun.* **1983**, 805; (f) Mack, K.; Antropiusova, H.; Petrusova, L.; Hanus, V.; Turecek, F. *Tetrahedron* **1984**, *40*, 3295; (g) Bourner, D.G.; Brammer, L.; Green, M.; Moran, G.; Orpen, A.G.; Reeve, C.; Schaverien, C.J. *J. Chem. Soc. Chem. Commun.* **1985**, 1409.
44. For a preliminary account of this work, see: Rigby, J.H.; Henshilwood, J.A. *J. Am. Chem. Soc.* **1991**, *113*, 5122.
45. Subsequent to the publication of our [6+2] cycloaddition (ref. 44), two reports on the photochemical [6+2] cycloaddition of substituted acetylenes to (η^6-cycloheptatriene)tricarbonylchromium(0) appeared: (a) Fischler, I.; Grevels, F.-W.; Leitich, J.; Özkar, S. *Chem. Ber.* **1991**, *124*, 2857; (b) Chaffee, K.; Sheridan, J.B.; Aistars, A. *Organometallics* **1992**, *11*, 18.
46. Reynolds, S.D.; Albright, T.A. *Organometallics* **1985**, *4*, 980 and references cited therein.
47. For a review of planar chirality in transition metal complexes, see: Schlögl, K. In *Topics in Stereochemistry*; Allinger, N.L.; Eliel, E.L., Eds.; John Wiley & Sons: New York, 1967, Vol. 1, pp. 39–92.
48. (a) Potter, G.A.; McCague, R. *J. Chem. Soc., Chem. Commun.* **1990**, 1172; (b) Uno, M.; Ando, K.; Komatsuzaki, N.; Takahashi, S. *Ibid.* **1992**, 964.
49. To date we have been unable to effect a [6+2] cycloaddition with (η^6-thiepin-1,1-dioxide)tricarbonylchromium(0) (**39**).
50. (a) Malpass, J.R.; Smith, C. *Tetrahedron Lett.* **1992**, *33*, 273; (b) For a synthesis of the homotropane anatoxin, see: Javier Sardina, F.; Howard, M.H.; Morningstar, M.; Rapoport, H. *J. Org. Chem.* **1990**, *55*, 5025.

51. For a preliminary account of this work, see: Rigby, J.H.; Short, K.M.; Ateeq, H.S.; Henshilwood, J.A. *J. Org. Chem.* **1992**, *57*, 5290.
52. It is well known that cycloheptatriene cycloadds almost exclusively via the norcaradiene tautomer under thermal conditions. For examples, see ref. 8.
53. Yagupsky, G.; Cais, M. *Inorganica Chim. Acta* **1975**, *12*, L27.
54. Roush, W.R. In *Comprehensive Organic Synthesis*; Trost, B.M.; Fleming, I., Eds.; Pergamon Press: Oxford, 1991, Vol. 5, pp. 513–550.
55. For recent examples, see ref. 4 and: (a) Funk, R.L.; Bolton, G.L. *J. Am. Chem. Soc.* **1986**, *108*, 4655; (b) Wu, T.-C.; Mareda, J.; Gupta, Y.N.; Houk, K.N. *Ibid.* **1983**, *105*, 6996; (c) Gupta, Y.N.; Doa, M.J.; Houk, K.N. *Ibid.* **1982**, *104*, 7336.
56. See, for example: Davis, R.; Kane-Maguire, L.A.P. In *Comprehensive Organometallic Chemistry*; Wilkinson, G.; Stone, F.G.; Abel, W.E., Eds.; Pergamon Press, Oxford, Vol. 3, 1982, p. 958.
57. Stork, G.; Clarke, Jr.; F.H. *J. Am. Chem. Soc.* **1961**, *83*, 3114.

ACYCLIC DIENE TRICARBONYLIRON COMPLEXES IN ORGANIC SYNTHESIS

René Grée and Jean Paul Lellouche

Advances in Metal-Organic Chemistry
Volume 4, pages 129–273.

ISBN: 1-55938-709-2

I. GENERAL DATA ON THE SYNTHESIS, RESOLUTION, AND REACTIVITY OF ACYCLIC DIENE TRICARBONYLIRON COMPLEXES

The first [η^4-diene tricarbonyliron] complex has been obtained by heating butadiene with $Fe(CO)_5$ as early as 1930[1a] and the first complex of a cyclic diene was synthesized by Pauson in 1958.[1b] The strong development in organometallic chemistry lead to the preparation of a very large number of such complexes, together with basic studies dealing with their structure and reactivity.[2] However, it was only during the last 15 years that [η^4-diene tricarbonyliron] complexes have been considered as potentially useful intermediates in organic chemistry and asymmetric synthesis.[3,4] The main reasons explaining the growing interest in such derivatives can be summarized in the following way:

1. Their preparation is easy: in most cases, they are obtained by reaction of (1,3)-dienes with various (carbonyliron) derivatives which transfer the $Fe(CO)_3$ unit. These complexes are then accessible on a relatively large scale (10–100 g), even if caution must be taken in using the toxic starting material $Fe(CO)_5$. Furthermore, these organometallic derivatives can usually be handled in the laboratory

a : $Fe(CO)_5$ or $Fe_2(CO)_9$ or $Fe_3(CO)_{12}$ or Ph-(enone)$Fe(CO)_3$ or (COD) $Fe(CO)_3$

b : $Ce(NH_4)_2(NO_3)_6$ or $FeCl_3$ or Me_3NO or $H_2O_2/OH^{\ominus}$ or CrO_3, Py...

under standard conditions.

2. These complexes are stable towards a large variety of chemical reagents, as described later. Thus the $Fe(CO)_3$ unit can be used as a very efficient, temporary protecting group for a conjugated dienyl ligand.

3. The decomplexation process involves oxidations which are in almost every case chemioselective; it is then possible to recover chiral and polyfunctionalized dienes or polyenes which are of much interest in further synthesis.

4. They stabilize, as their tricarbonyliron complexes, very labile molecules. This property opens new routes to organic synthesis as will be described later, for instance in the case of pentadienyl cations.

5. Most importantly these complexes, when they are unsymmetrically substituted, are chiral. They can be obtained in optically active form using various resolution methods; furthermore, the thermal racemization barrier (around 120 kJ mol^{-1})[5] is high enough to allow reactions to be performed over a wide range of temperature. Thus, it is not surprising that these complexes have attracted recently a great interest in stoichiometric asymmetric synthesis.

This review, especially devoted to the *acyclic* [η^4-dienyl tricarbonyliron] complexes,[6] will emphasize recent progress in the chemistry of these organometallic derivatives and their use in organic synthesis.

In this section, we will first consider the preparation of the "parent" optically active complexes of general type **A**, as well as methods used to establish their absolute configuration.

Next, we will briefly examine the scope and limitations of the main synthetic methods used to manipulate these parent compounds in order to prepare new related chiral complexes of "second generation". Finally, we will consider some novel complexes of cumulenic type which have been developed very recently.

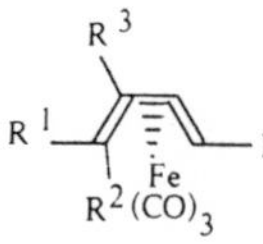

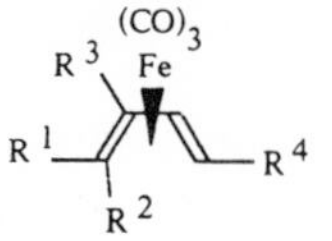

A. "Parent" Chiral Complexes

The complexes **A** are chiral, except if $R^1 = R^4$ and $R^2 = R^3 = –H$. They can be resolved by classical methods including fractional crystallization or chromatography of diastereoisomeric mixtures. Kinetic resolutions by chemical or enzymatic ways have also been used. Table 1 indicates the most significant examples of these resolution processes with the corresponding references.

In most cases, satisfactory to excellent isolated yields have been obtained. Furthermore, the optical purity of these complexes is usually very high (routinely >> 95%) and is easily established, for instance by NMR[17] or chiral phase HPLC analysis.[18]

It is interesting to point out that optically pure complex **1** has been used recently for the resolution of chiral secondary alcohols related to fatty acids metabolites.[7b]

The preceding complexes have an (*E*,*E*)-stereochemistry at the dienyl system. However, it is interesting to point out that (*E*,*Z*)-complexes can also be prepared, even in optically pure form. This has been done by Friedel–Crafts reaction in the case of **17**[4a,19] or through the complexed pentadienyl cation chemistry for **18**, **19**,[20] **20**, and **21** for instance.[21]

Unexpectedly, these (*E*,*Z*) complexes proved to be stable enough to be used in multistep syntheses without isomerization of the dienyl system; several examples will be described later.

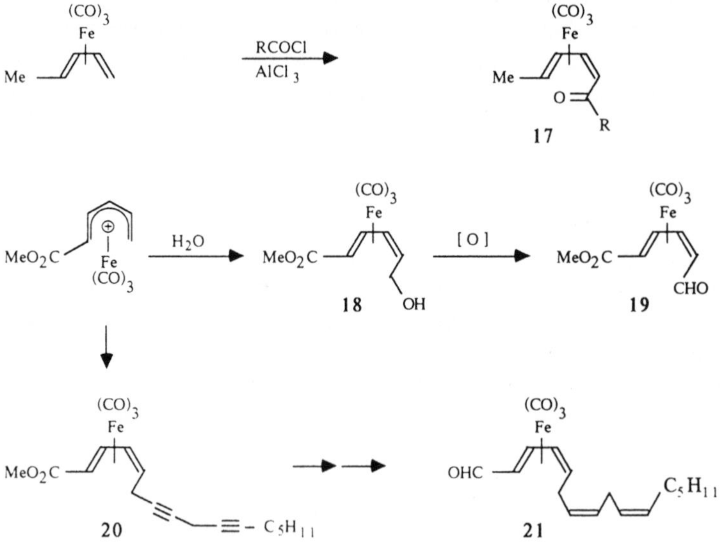

Table 1

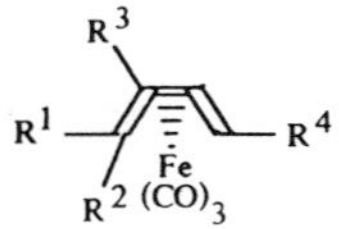

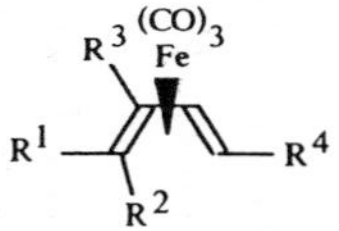

Compound	R^1	R^2	R^3	R^4	Method of Resolution of the Complex Itself or a Related Complex	Structure of the Diastereoisomeric Adducts[α]	Reference
1	$-CH_3$	–H	–H	$-CO_2H$	Fractional crystallization of their α-methyl benzyl amine salts	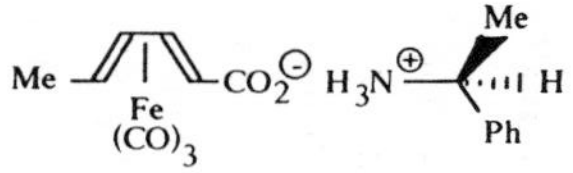 Me, Fe, $(CO)_3$, $CO_2^{\ominus}$ $H_3N^{\oplus}$, Me, H, Ph	7
2	–H	–H	–H	$-CO_2H$	Fractional crystallization of their α-methyl benzyl amine salts	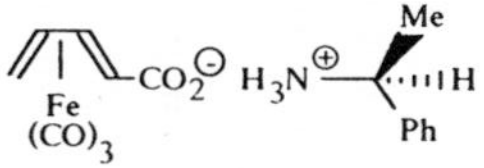 Fe, $(CO)_3$, $CO_2^{\ominus}$ $H_3N^{\oplus}$, Me, H, Ph	7a
3	$-CH_3$	–H	–H	–CHO	Fractional crystallization of the corresponding pyrrolidinium salts having *d*-camphor sulfonate as the optically active anion	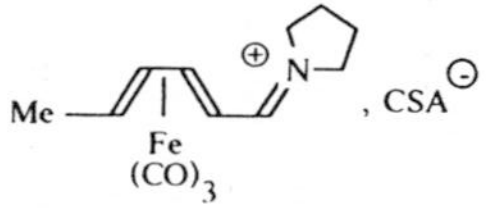 Me, Fe, $(CO)_3$, $N^{\oplus}$, $CSA^{\ominus}$	7a
3	$-CH_3$	–H	–H	–CHO	Chromatographic separation of diasteroisomeric semi-oxamazones	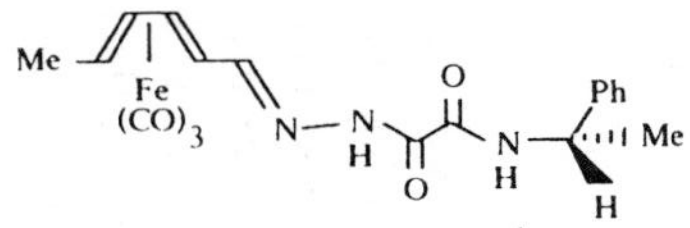 Me, Fe, $(CO)_3$, N–NH, O, O, NH, Ph, Me, H	8
3	$-CH_3$	–H	–H	–CHO	Kinetic resolution using C2-symmetric chiral allylboronates β or baker yeast reduction		9

Table 1. (Continued)

Compound	R^1	R^2	R^3	R^4	Method of Resolution of the Complex Itself or a Related Complex	Structure of the Diastereoisomeric Adducts[α]	Reference
4	$-CO_2CH_3$	–H	–H	–CHO	Fractional crystallization of diasteroisomeric oxazolidines[γ,δ].		10
4	$-CO_2CH_3$	–H	–H	–CHO	Fractional crystallization of related acidic complex diasteroisomeric-brucine salts		5
4	$-CO_2CH_3$	–H	–H	–CHO	Kinetic resolution using C2-symmetric chiral allylboronates[β,γ].		9
5	–CHO	–H	–H	–CHO	Kinetic resolution using C2-symmetric chiral allylboronates[β,γ].		9
6	–H	–H	–CHO	–H	Chromatographic separation of diasteroisomeric semi-oxamazones		11
7	–H	–H	–H	–CHO	Chromatographic separation of diasteroisomeric semi-oxamazones		4a
8	$-CH_2OSiPh_2tBu$	–H	–H	–CHO	Chromatographic separation of diasteroisomeric imidazolidines		12

9	$-nC_4H_9$	$-H$	$-H$	$-CHO$	Chromatographic separation of diasteroisomeric imidazolidines	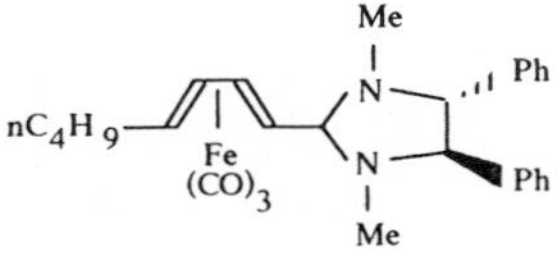 nC_4H_9; Fe $(CO)_3$; N–Me; N–Me; Ph; Ph	13
10	$-H$	$-H$	$-CO_2Et$	$-H$	Enzymatic resolution using Pig Liver esterase		14
11	$-H$	$-H$	$-CO_2H$	$-H$	Enzymatic resolution using Pig Liver esterase		14
12	$-H$	$-H$	$-H$	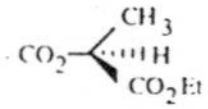 $-CO_2-C(CH_3)(H)(CO_2Et)$	Chromatographic separation of diastereoisomeric esters.		15
13	$-CH_3$	$-H$	$-H$	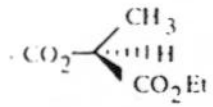 $-CO_2-C(CH_3)(H)(CO_2Et)$	Chromatographic separation of diastereoisomeric esters.		15
14	$-CH_3$	$-CH_3$	$-H$	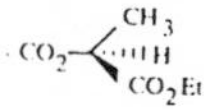 $-CO_2-C(CH_3)(H)(CO_2Et)$	Chromatographic separation of diastereoisomeric esters.		15
15	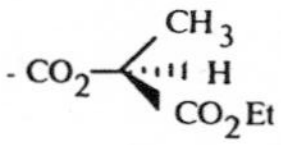 $-CO_2-C(CH_3)(H)(CO_2Et)$	$-H$	$-Et$	$-H$	Chromatographic separation of diastereoisomeric esters.		15
16	$-H$	$-H$	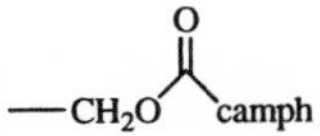 $-CH_2O-C(=O)-camph$	$-H$	Fractionalization recrystallization of diasteroisomeric camphanoate esters.		16

Notes: [α]The structures of the diastereosiomeric adducts are indicated whenever usual chemical manipulations regenerate the described chiral functionalized complexes. It emphasizes also on the resolution methods which furnish the *two* optically active complexed enantiomers in *one* reactional sequence.

[β]This highly efficient kinetic resolution using the tartrate modified (*R,R*)-allylboronate (allyl–B dioxaborolane with CO_2iPr, H, CO_2iPr, H) affords only *one* chiral complex of (2*S*)-absolute chirality.

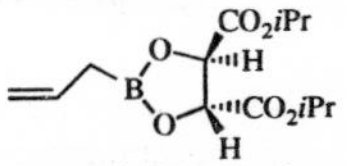

Of course, formally the use of the (*S,S*)-allyl boronate reagent would afford the chiral complex of (2*R*)-absolute chirality.

[γ]This complex is near base-line resolved by high performance liquid chromatography on a β-chiral cyclodextrin column. Nevertheless, this resolution method is interesting only on an analytical point of view.[18]

[δ]This complex has been used recently for the synthesis of new chiral liquid crystals (Ziminski L.; Malthête J. *J. Chem. Soc., Chem. Commun.* **1990**, 1495).

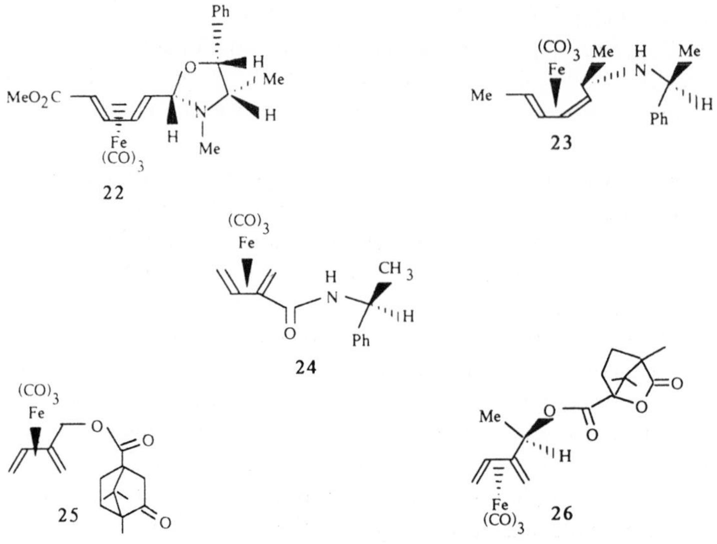

Figure 1.

The rational design of chiral, nonracemic target molecules starting from complexes of this general type involves an unambiguous knowledge of their absolute configuration. For that purpose the most reliable method is X-ray analysis, but this has been done only in five cases (Figure 1):

1. The oxazolidine **22** is one of the diastereoisomers obtained by condensation of (±)-**4** with (–)-ephedrine. The X-ray analysis established the (2*S*,5*R*)-absolute configuration of this complex as well as the corresponding aldehyde (+)-**4** and all the derivatives prepared from it since no racemization was detected during the different reactions.[10,22] It is the only reported example of an (1*E*, 4*E*)-disubstituted complex.
2. The amine **23** is one of the two diastereoisomers obtained by reaction of the (*S*)-(–)-α-methyl benzylamine with the achiral (1,5) dimethylpentadienyl tricarbonyliron cation.[23] The X-ray analysis indicates a (3*R*,6*R*)-absolute configuration for the complex and this is the only example of an (*E*,*Z*)-dienyl tricarbonyliron complex.
3. The acid corresponding to amide **24** has been obtained, in high optical purity, by an enantioselective hydrolysis catalyzed by pig liver esterase. Since the (*S*)-(–)-α-methyl benzylamine has been used to prepare amide **24**, this establishes its (*S*)-absolute configuration on C2.[14]
4. The ester **25** was isolated by classical resolution using (–)-(1*S*,4*R*)-camphanoyl chloride as the auxiliary agent. The X-ray analysis establishes the (2*S*)-absolute configuration.[16]

5. During studies dealing with the stereoselectivity of ionic hydrogenation processes, Gerlach et al. prepared the alcohol corresponding to **26**.[24] The X-ray data of this ester, obtained again by reaction with (1*R*,4*S*)-camphanoyl chloride, established unambiguously the (2*R*)-absolute configuration of this complex.

Musco et al. suggested very early, by analogy with other transition metal derivatives, that the sign of the circular dichroism (CD) maximum at long wavelengths attributable to d–d transitions should be related to the absolute configuration of these complexes.[7] Several recent studies, including also cyclohexadienyl derivatives,[25] have established that such correlations indeed can be useful in the prediction of the absolute configurations.[22] However, great caution must be taken in using these empirical correlations since the *number*, the *nature*, and the *position* of the substituents on the complex clearly have a strong influence on the position and the intensity of the CD bands between 250 and 400 nm. The number of examples is presently too small to give *general* rules. Thus, it is of much importance to correlate the data of the complex under study with those of closely related derivatives having well established structures such as the complexes of Figure 1 and derivatives thereof.

B. "Second Generation" Chiral Complexes

The number of the "first generation" functionalized and chiral complexes which are readily available by direct resolution (see Table 1) is limited. Thus, in order to develop the use of complexes of this type in organic synthesis, it was of much importance to prepare, starting from this first series, new related chiral complexes of "second generation"; furthermore, the new carbon–carbon bonds formed during these reactions should be created without loss of the initial chirality. Scheme 1 indicates the main routes which have been used presently; they involve either electrophilic or nucleophilic additions on various type of complexes. It is interesting to comment briefly on the present scope and limitations of each of these routes.

Route 1

R^1 R^2 CHO Fe $(CO)_3$ **27** → R^1 R^2 OH R^3 Fe $(CO)_3$ **28** ← R^1 R^2 O R^3 Fe $(CO)_3$ **29**

↓

R^1 R^2 R^3 Fe $(CO)_3$ **30**

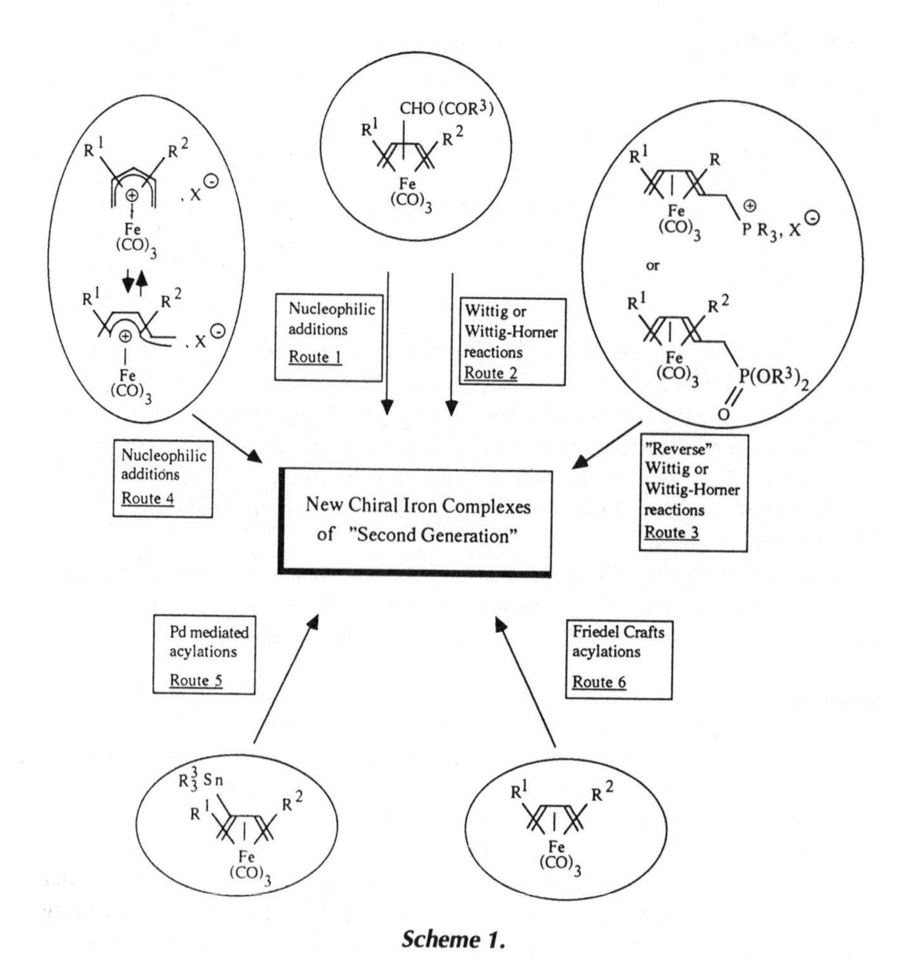

Scheme 1.

Nucleophilic addition to a carbonyl group vicinal to the complex is both very general and very efficient. In the case of the aldehyde **27**, it leads usually to two diastereoisomers (called Ψ-*exo* and Ψ-*endo*)[26] very easily separated by chromatography, while nucleophilic addition on the ketones **29** are usually highly stereoselective giving the Ψ-*endo*-diastereoisomer. No racemization has been observed during these reactions. It is also interesting to point out that the ionic deoxygenation (Et_3SiH, H^+) reaction of type **28** alcohols is highly efficient; this is a good route to prepare type **30** complexes with a CH_2–R^3 group.[16,27] Alternatively, the alane reduction ($LiAlH_4/AlCl_3$) of carbonyl derivatives has also been successfully performed.[15,19,28–30]

Routes 2 and 3

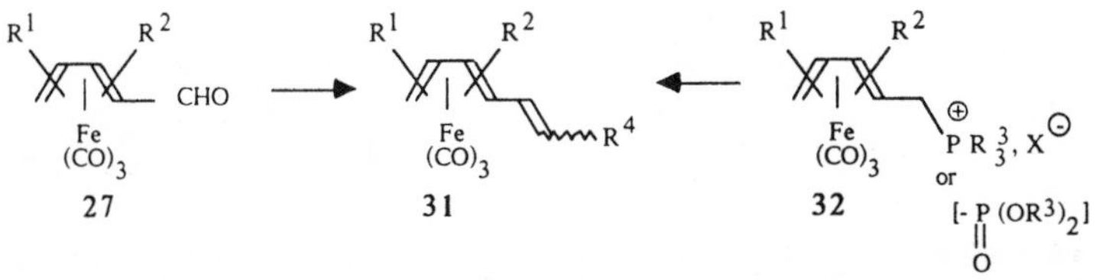

These are also very general and efficient routes to obtain trienes or polyenes selectively complexed by the $Fe(CO)_3$ group. These reactions follow the general trends concerning the stereoselectivities in Wittig or Wittig–Horner type reactions. They have been extended, without racemization, to optically active complexes. However, in a very few cases, bond-shift isomerizations have been noticed during reactions performed in a basic medium,[31,32] under Lewis acid conditions[33,34] or at elevated temperatures.[5]

Other olefination procedures, such as the Peterson's reaction[35] or the Cope–Knoevenagel condensation,[10,36,37] can be used successfully starting from type **27** aldehyde.

Route 4

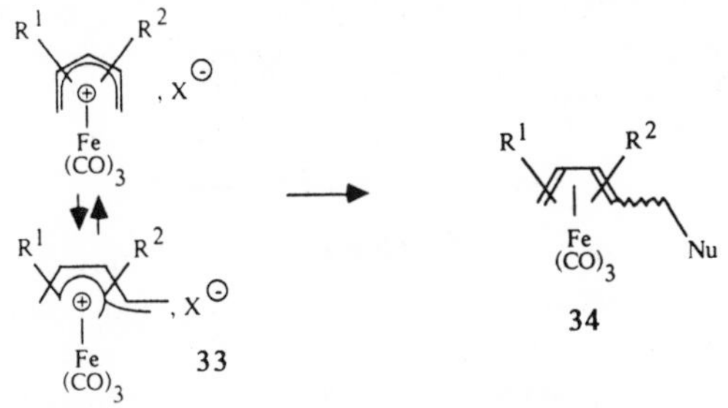

The nucleophilic addition on pentadienyl cations complexed to $Fe(CO)_3$ is also a very general route to new complexed dienes. It has been used for the synthesis of optically active derivatives but it is a very complicated reaction: it leads not only to η^4-diene complexes but also to some σ,η^3-π-allyl derivatives, as will be discussed in more detail in the next section. The nature of the substituents on the cation, the nature of the nucleophile, and the reaction conditions play a great role on the regio- and stereoselectivities of these reactions. In spite of the numerous studies already done in this field, the delicate balance between all these factors make this approach of predictive value only in homogeneous series of cations.

Route 5

R^3_3Sn ... Fe $(CO)_3$ **35** → Fe $(CO)_3$ **36**

This appears as a new and interesting route, even if the number of examples is presently small[38] and questions remain on the availability of the starting dienes of this type. However, it can be expected that other approaches of this type, involving Pd coupling reactions, will develop in the near future.

Route 6

Fe $(CO)_3$ **37** → Fe $(CO)_3$ **38**

This approach has been studied in great detail and developed in Franck-Neuman's group.[4a] It is an extremely powerful method since it is quite general, it has usually very good regioselectivities and it occurs without racemization. It is even possible, in certain cases, to prepare bisacylated derivatives.[39,40,41] It is worthy of note that among the six different routes, it is the only one which can be performed on *unfunctionalized* complexes.

C. Cumulenic-Type Complexes

Finally, the iron complexes of cumulenes or heterocumulenes should also be mentioned. New and efficient routes to these novel type of complexes have been devised recently, although their scope and limitations remain to be established; these new derivatives include the tetraenes **39**[42] and **40,**[43] the vinylallenes **43**, and the heterocumulenes **41** and **42.**[44]

These complexes will certainly open in the near future new possibilities not only with regard to structural and mechanistic studies, but also in terms of synthetic developments.

D. Conclusion

In conclusion, a wide range of "basic" functionalized [η^4-diene tricarbonyliron] complexes are now easily accessible, also in optically active form. Many types of reactions can be successfully performed with these complexes in order to create new C–C bonds and these reactions occur usually without loss of the chirality. In most cases, the absolute configuration of these complexes can be established either by chemical correlations or by spectroscopical data.

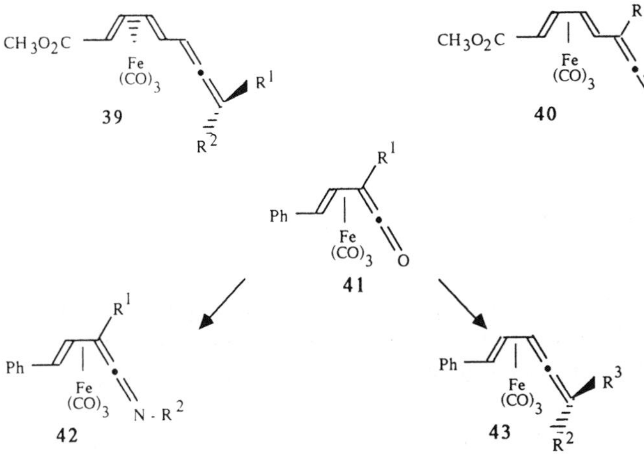

Thus, even if important improvements are to be expected in the future, these basic studies were and are still essential to the development of the use of such complexes towards the total synthesis of elaborated natural products and their structural analogs, as described later.

II. η^5-PENTADIENYL TRICARBONYLIRON COMPLEXED CATIONS

Besides the aspects of protection and/or stereodirection attached to the organometallic $Fe(CO)_3$ unit when ligated to various conjugated (1,3)-dienes to be described later, another fascinating area of chemical manipulations of these iron-complexed entities emerged when it was discovered that the dienyl iron-complexed moiety greatly stabilizes a positive charge located at the α position of the coordination site.[45]

In agreement with this statement, it is particularly relevant to note that the *isolable* cyclohexadienyl iron complexed cation **44** can be recrystallized even in water.[45a] Thus, a great deal of effort has been devoted to the reactivity of these cyclic cations and their uses in planned synthetic schemes directed towards the natural products synthesis.[2,3,4b]

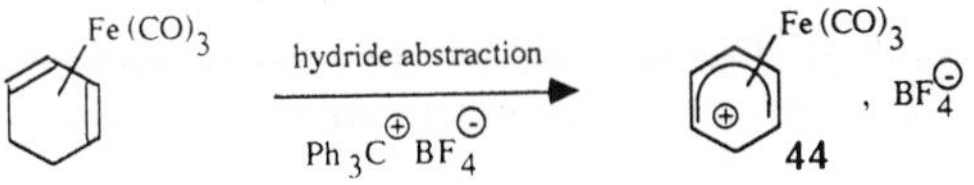

A. Preparation of η^5-Pentadienyl Tricarbonyliron Complexed Cations

The usual mode of preparation of these iron-complexed cations depends on the *acyclic* or *cyclic* nature of the dienyl ligands. Cyclic η^4-dienyl iron complexes when reacted with powerful hydride abstraction reagents like $Ph_3C\ BF_4$ or $Ph_3C\ PF_6$ lead to the corresponding crystalline iron-complexed cations:[45–47]

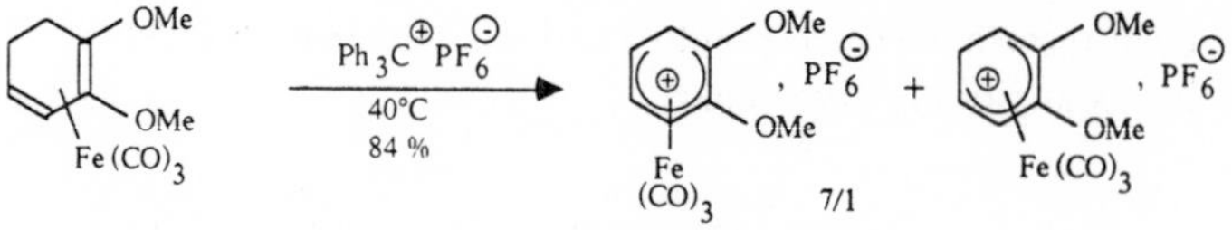

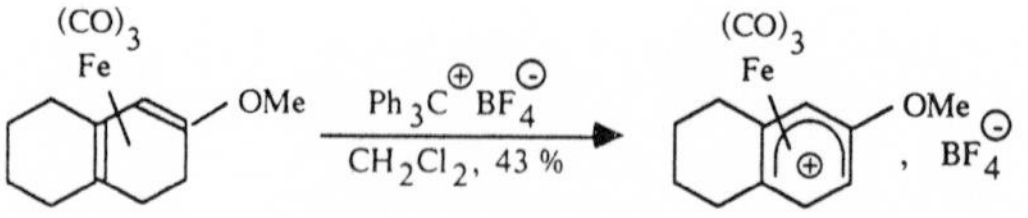

Of course, this synthetic mode of preparation is restricted to iron-complexed cations that are more stable than the triphenylmethyl cation. It is interesting to note that the cyclic nature of the ligand imposes a cisoid (U shape) nature for these cations. They are generally air- and light-stable. The hydride abstraction by the bulky cation $Ph_3C^{\oplus}$ from the *exo* face can be very chemoselective depending upon the substitution pattern of the dienyl ligand:[48,49]

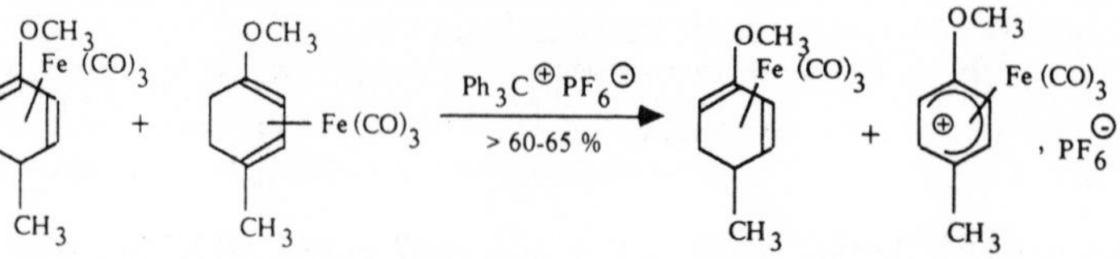

An interesting acid-catalyzed demethoxylation route[50] has been also used with the same success provided that the appropriate cyclic precursors are available;[51,52] for instance:

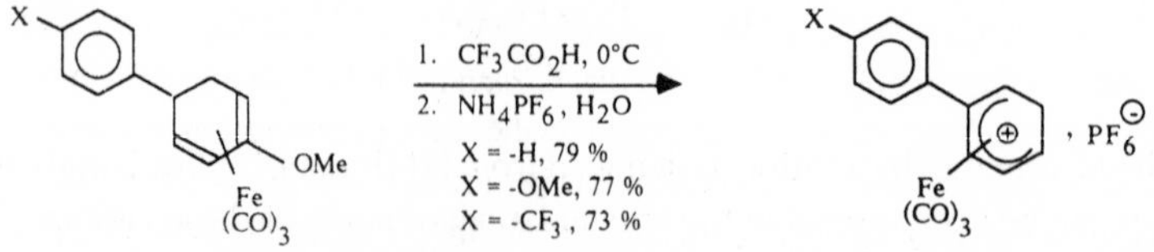

In the case of acyclic η^4-dienyl iron complexes, the *isolable* crystalline cations are best prepared via the acid-catalyzed dehydration of the corresponding alcohol precursors (Scheme 2). The complexed alcohols **45** when treated with a strong acid (HBF_4, $HClO_4$, or HPF_6) in acetic acid anhydride followed by the addition of

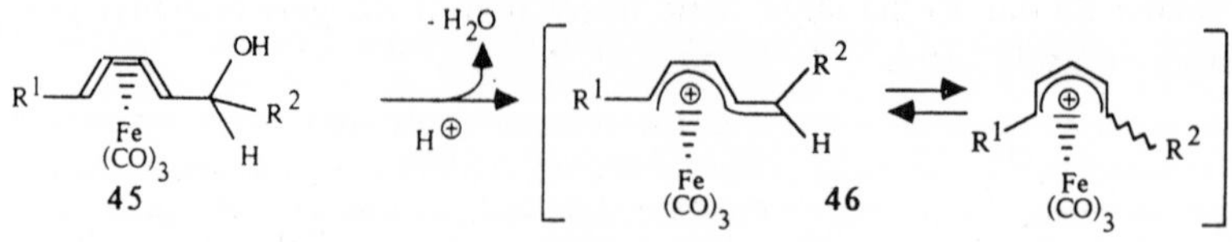

Scheme 2.

anhydrous ether furnish the corresponding cationic salts in high yield.[46,53,54] A few representative examples are given below.

1. $HClO_4/Ac_2O$
2. anhydrous ether
78 %

$Fe(CO)_3$ OH → ClO_4^- **47**

Ψ-exo **48** or Ψ-endo **49**

1. HBF_4/Ac_2O
2. anhydrous ether
94 %

→ CH_3 CH_3, BF_4^- **50**

C_nH_{2n+1}, n = 2 - 7

1. HPF_6 anhydrous ether, 0°C
2. Ether reprecipitation from a nitromethane solution
30-50 %

→ C_nH_{2n+1}, PF_6^- **51**

For the symmetric derivatives ($R^1 = R^2$), such as **47** and **50**, ^1H NMR experiments demonstrated unambiguously the cisoid (U shape) nature of these isolable cations due to the symmetry of their spectra. The cations **51** have the corresponding

51 (R, H, H, H, $Fe(CO)_3$)

predominant structure in solution as deduced from NMR data.[53] Surprisingly water molecules can be also trapped in the crystal lattice of some of these cations.[53] It is worthwhile to note that, in any case, no transoid (S shape) cation of the general formula **46** has ever been *isolated* using these syntheses. However, the presence of these transoid complexed cations must be considered, as it has been established experimentally:

1. The transoid complexed cation **52** has been characterized by ^1H NMR experiments at low temperature in an hyperacidic medium; it equilibrates with cisoid isomer **53** due to the large steric interaction of the *gem*-dimethyl group disfavoring its cisoid shape.[55]

FSO_3H, - 78°C → [**52** ⇄ **53**]

2. In contrary to Mahler and Pettit,[46] who isolated only *one* cisoid cation **50** from *each* diastereoisomeric alcohol **48** or **49**, the *two* transoid complexed cations **54** and **55** have been proposed as intermediates during the formation of the *two different* cisoid cations **50** and **56**,[55]

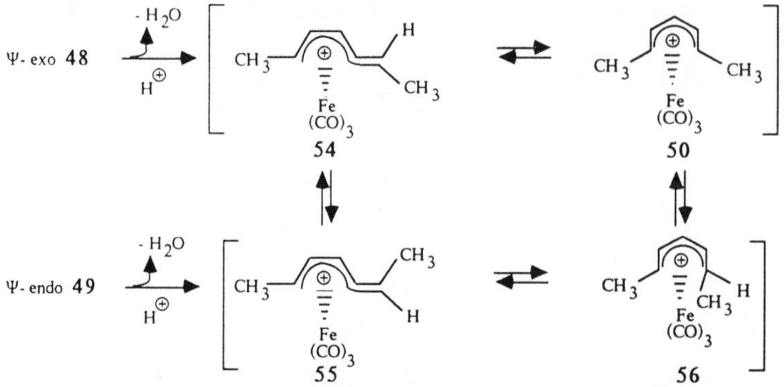

The discrepancy between these two results may be due to the temperature of the dehydration step since the conversion of **56** to **50** is easy at temperatures *above 0 °C* as shown by ^{1}H NMR. Analogous results were obtained during the dehydration of the two iron-complexed diastereoisomeric alcohols **57** and **58**:[56]

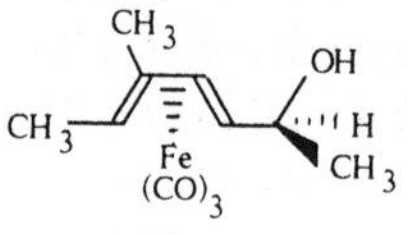

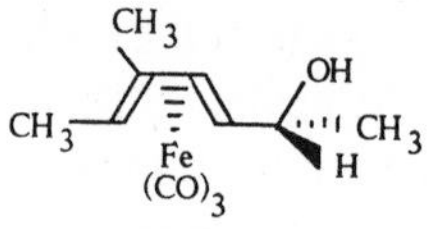

3. The solvolysis of several iron-complexed dinitrobenzoates **59** offers another direct evidence for elusive transoid cations since a complete retention of configuration is observed in all cases but one (SN_i mechanism):[57]

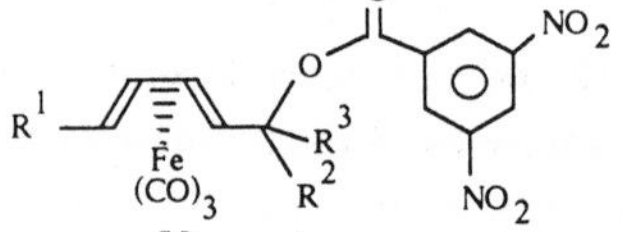

B. Bimolecular Reactions of Isolated η^5-Pentadienyl Tricarbonyliron Complexed Cations

Although less stable (and more electrophilic) than the cyclic iron-complexed cations, the acyclic ones react readily with various nucleophiles. In principle, each electronically deficient carbon of such a cation can be attacked by a nucleophile and, experimentally, this statement is confirmed for each carbon of the dienyl ligand *except one* (C3), giving rise to new η^4-dienyl (*E*,*E*)- or (*E*,*Z*)-iron complexes and σ, η^3-π-allyl derivatives as illustrated in Scheme 3:

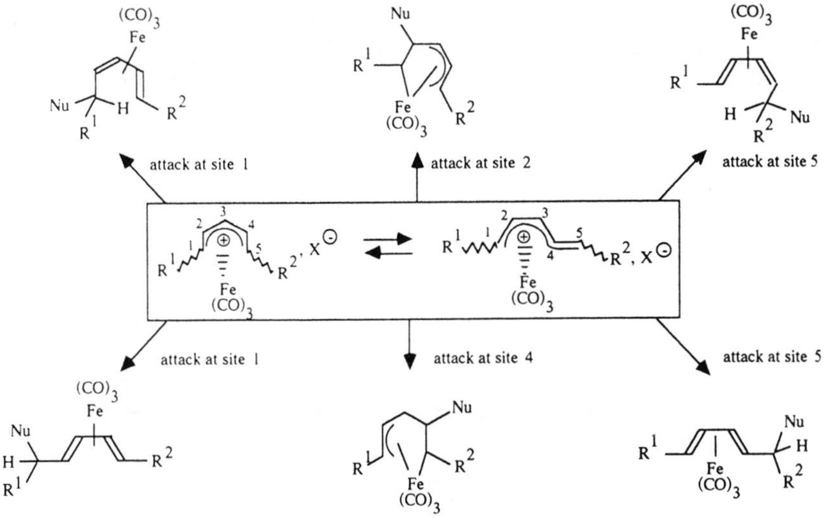

Scheme 3.

Taking into account the equilibrium between the cisoid and transoid shapes of the cations, the site of the nucleophilic attack depends not only on the nucleophile but also on the nature of the substituents for the dienyl ligand. Thus the observed regioselectivity is the result of a subtle combination of electronic and steric effects; furthermore, kinetic versus thermodynamic controls have to be considered in some cases. In summary, the numerous data available from the literature can be overviewed in the following manner.

Isolated Cations where R^1 and/or R^2 are Alkyl Groups and/or Hydrogens

In that case, the nucleophile attacks exclusively at the termini 1 and/or 5 of the cationic salt furnishing one or mixtures of the η^4-dienyl complexes. The nature of the nucleophile is the key component in the observed regioselectivity: water,[46,57b,58–60] alcohols,[61] electronically enriched aromatics,[62] and allylsilanes[60] give the (*E*,*E*)-complexes. This geometry of the complexed dienyl ligands is illustrative not only of the higher reactivity of the transoid cations but also of the hardness of the nucleophiles. Some illustrative examples are reported below:

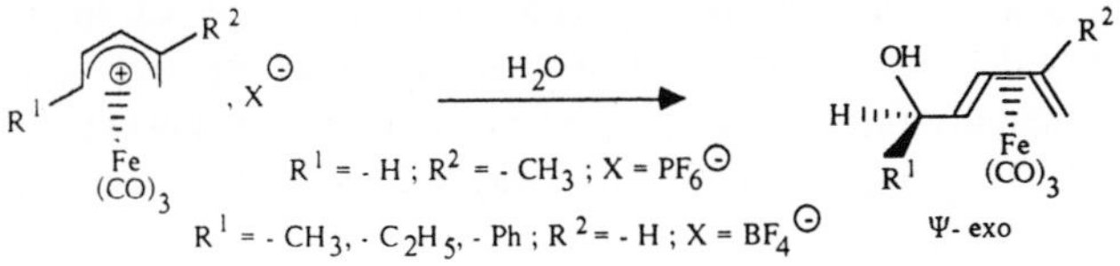

H_2O

R^1 = - H ; R^2 = - CH_3 ; X = $PF_6^{\ominus}$

R^1 = - CH_3, - C_2H_5, - Ph ; R^2 = - H ; X = $BF_4^{\ominus}$

Ψ- exo

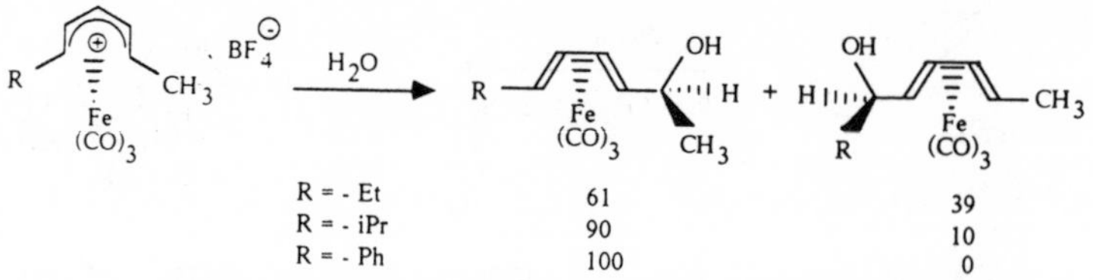

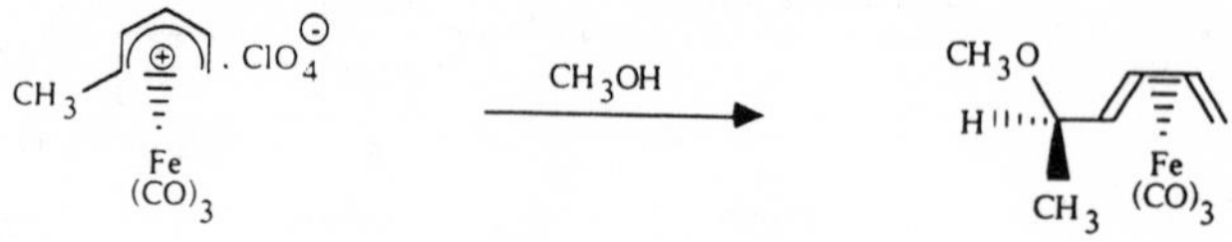

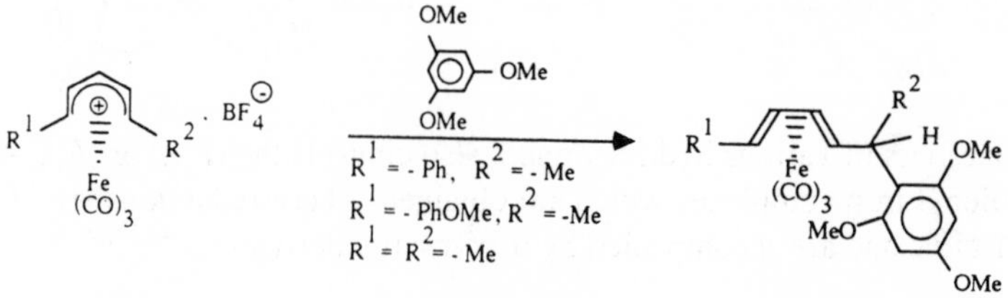

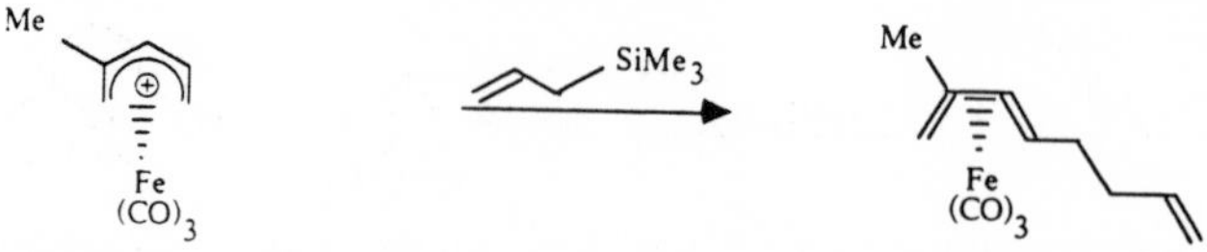

Some organometallic anions also provide the same type of (*E*,*E*)-adducts.[63]

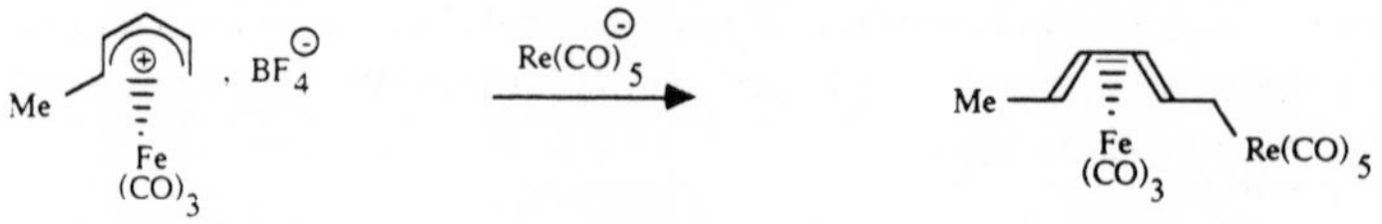

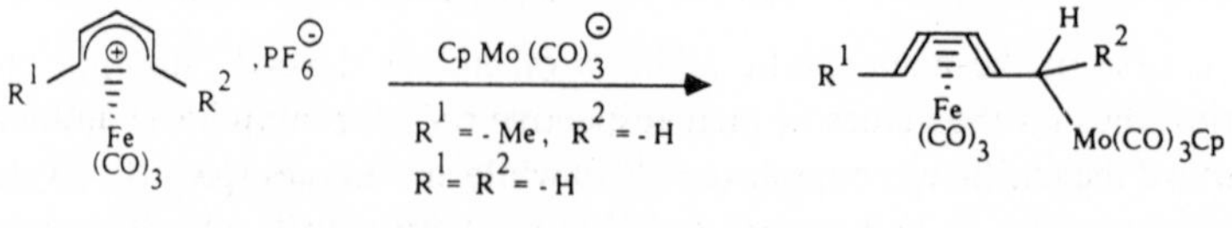

On the other hand, soft nucleophiles like phosphorous derivatives[32,60,64–66] or cadmium organometallics[67] afford the (*E*,*Z*)-dienyl iron complexes. Lack of regioselectivity during the nucleophilic addition is frequently encountered.

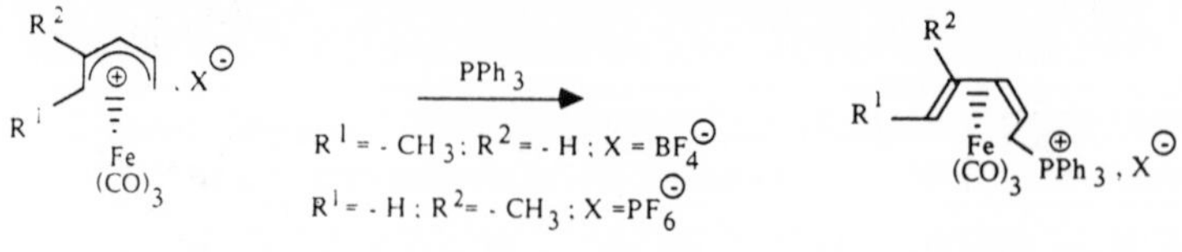

R^1 Fe $(CO)_3$. $BF_4^{\ominus}$ —$P(OCH_3)_3$; R^1 = - CH_3 ; - nC_4H_9→ R^1 Fe $(CO)_3$ P(O) OCH_3 OCH_3

CH_3 Fe $(CO)_3$. $BF_4^{\ominus}$ —$P(OC_2H_5)_3$→ CH_3 Fe $(CO)_3$ P(O) OC_2H_5 OC_2H_5 + Fe $(CO)_3$ CH_3 P(O) OC_2H_5 OC_2H_5

5 / 4

CH_3 Fe $(CO)_3$. $BF_4^{\ominus}$ —$(iPr)_2Cd$→ CH_3 Fe $(CO)_3$ + Fe $(CO)_3$ CH_3

1 / 1

The hardness of various hydride donors also controls the (*E*,*Z*) or (*E*,*E*)-type of the η^4-dienyl iron complexes which are obtained. These reductions are often not regioselective and are accompanied by σ,η^3-π-allyl derivatives.[46,54,68]

Ph Fe $(CO)_3$ Ph . $BF_4^{\ominus}$ —$NaBH_3CN$→ Ph Fe $(CO)_3$ Ph + Ph Fe $(CO)_3$ Ph

Fe $(CO)_3$ CH_3 . $BF_4^{\ominus}$ → CH_3 Fe $(CO)_3$ CH_3 + Fe $(CO)_3$ CH_3 + CH_3 Fe $(CO)_3$ CH_3 + Fe $(CO)_3$ CH_3

$NaBH_4$:	57	13	29	1
$NaBH_3CN$:	36	56	1	3
$LiBEt_3H$:	13		64	23

The course of the nucleophilic addition of amines depends strongly on their basicities, thus on the values of their respective pK_b. Strongly basic amines (pK_b 3–6) afford the (*E*,*Z*)-type complexes[23,69,70] while arylamines (pK_b 10–13) furnish the (*E*,*E*)-type ones.[70] Aniline (pK_b 9.4) and *p*-toluidine (pK_b 9.8) of intermediate pK_b lead to mixtures of the two geometrical isomeric complexes.[70]

CH_3 Fe $(CO)_3$. $BF_4^{\ominus}$ —NEt_3→ CH_3 Fe $(CO)_3$ $\overset{\oplus}{N}Et_3$ R^1

R^1 Fe $(CO)_3$ R^1 . $BF_4^{\ominus}$ —R^2NH_2→ R^1 Fe $(CO)_3$ NHR^2 R^1

R^1 = - CH_3, - H

R^2 = - CH(Me)Ph, - CH_2Ph, - iPr, - Et

$R^1 = -Br, R^2 = -H$
$R^1 = -H, R^2 = -NO_2$
$R^1 = -NO_2, R^2 = -H$

Furyl cuprate,[60] highly functionalized cupro–zinc organometallics **60**,[71a–d] alkynylcuprate **61**,[71b–d,72] and sodiomalonates **62**[60,71d–g,73] afford the retained (*E*,*Z*)-stereochemistry of the starting iron-complexed cationic salts. Like organolithiums compounds,[60,71d,e,f,74] they sometimes accompanied by-products arising via attack at C-2/C-4 giving σ,η^3-π-allyl derivatives (characterized in some cases as cyclohexenones after CO insertion[74]). The results observed with these nucleophiles deserve some comments:

- when reacting with *isolated* iron-complexed cations, cupro–zinc organometallics and alkynylcuprates are the *only* organometallics which showed a complete control of regioselectivity, and
- the obtention of σ,η^3-π-allyl derivatives is observed as a function of both the hardness of the nucleophile, and the nature of the substituents of the dienyl ligand. This point will be developed later.

R^2- Cu (CN) ZnI
60
$R^1 = -H, R^2 = -(CH_2)_3CN$
$R^1 = -H, R^2 = -(CH_2)_3CO_2Et$
$R^1 = -CH_3, R^2 = -CH_2Ph$
$R^1 = -CH_3, R^2 = -(CH_2)_3CN$

$(PhC{\equiv}C)_3CuLi_2$
61
$R^1 = -CH_3 ; -Ph$

$NaCH(CO_2R^2)_2$
62

$R = -CH_3 ; R^1 = -H ; R^2 = -Et$
$R = -Et ; R^1 = -H ; R^2 = -Et$
$R = -nPr ; R^1 = -H ; R^2 = -Et$
$R = -Ph ; R^1 = -H ; R^2 = -Et$
$R = -H ; R^1 = -CH_3 ; R^2 = -CH_3$

$R = R^1 = -H ; R^2 = -CH_3$	1	1.2
$R = R^1 = -H ; R^2 = -nBu$	1.4	1
$R = R^1 = -Ph ; R^2 = -Ph$	1	1.5

Isolated Cations Where R^1 or R^2 are Electron-Withdrawing Groups

When R^1 or R^2 are electron-withdrawing substituents like an alkoxycarbonyl group, the reactivity of η^5-pentadienyl iron complexed cations is greatly affected, most likely because of the modified electronic deficiency on the dienyl ligand. Therefore, one can expect that the course of nucleophilic additions would be diverted from the usual way whatever the nature of the control (orbital or charge control). In this case, the nucleophiles attack not only at the two termini 1 and/or 5 but also at the internal C2 and/or C4 leading to η^4-dienyl iron complexes and σ,η^3-π-allyl derivatives (see Scheme 3). Water[75,76] and alcohols[75] give the (*E,Z*)-η^4-dienyl iron complexes and not the (*E,E*)-isomers as before, while trimethylphosphite,[13a,32] lithio methylmalonate,[72] and alcynylcerium organometallics[77] afford variable mixtures of η^4-dienyl-iron complexes and σ,η^3-π-allyl derivatives.

$R^1 = -H ; R^2 = -CH_3 ; R^3 = -H$	95	5
$R^1 = -CH_3 ; R^2 = -H ; R^3 = -H$	95	5
$R^1 = -H ; R^2 = -CH_3 ; R^3 = -CH_3$	50	50
$R^1 = -CH_3 ; R^2 = -H ; R^3 = -CH_3$	63	37

Even a vinyloguous electron-withdrawing ethoxycarbonyl group has been found to promote an internal nucleophilic addition not only at the carbon near it but also distal from it.[78]

5 / 1 / 4

MeO_2C (Fe $(CO)_3$ cation), $PF_6^{\ominus}$ —$LiCH(CO_2Me)_2$→ MeO_2C, MeO_2C, CO_2Me, Fe $(CO)_3$ + MeO_2C, CO_2Me, CO_2Me, Fe $(CO)_3$

> 20 / 1

MeO_2C (Fe $(CO)_3$ cation), R, $BF_4^{\ominus}$

$Me_3Si - \equiv - CeCl_2$

MeO_2C, $SiMe_3$, Fe $(CO)_3$, R + $SiMe_3$, MeO_2C, Fe $(CO)_3$, R

R		
R = - H	60	40
R = - CH_3	40	60
R = - Ph	0	100

Fe $(CO)_3$ cation, CO_2Et, $BF_4^{\ominus}$

$NaCH(CO_2Me)_2$

CO_2Me, CO_2Me, Fe $(CO)_3$, CO_2Et + CO_2Me, MeO_2C, Fe $(CO)_3$, CO_2Et + MeO_2C, $(CO)_3$ Fe, CO_2Me, CO_2Et

39 30 / 31

In a general way, σ,η[3]-π-allyl metal complexes have been obtained previously during the C2/C4 nucleophilic attacks of relatively soft nucleophiles with open odd complexed π systems possessing an electronically enriched metal center.[79–81] It is interesting to note that the course of the nucleophilic additions of alcynylcuprates $(RC \equiv C)_3CuLi_2$ is not modified at all by the presence of a methoxycarbonyl group on the starting iron-complexed cation.[21,82]

MeO_2C (Fe $(CO)_3$ cation), $PF_6^{\ominus}$ —$(RC \equiv C)_3CuLi_2$→ MeO_2C, Fe $(CO)_3$, R

R = -Ph, - nC_5H_{11}, - $CH_2 - \equiv - nC_5H_{11}$

(*E-E*)-η^4-dienyl iron complexes can also be synthesized if the nucleophilic alkynyl residue is prepared in a different manner.[82]

MeO_2C ... $PF_6^{\ominus}$, $Fe(CO)_3$ → ($RC\equiv C\,SiMe_3$; KI, KF, CH_2Cl_2, Δ) → MeO_2C ... $Fe(CO)_3$ ... R

R = - Ph, - nC_5H_{11}

It is quite relevant to note that, in all cases, the nucleophiles attack the η^5-pentadienyl iron-complexed cations in a manner *anti* to the organometallic unit $Fe(CO)_3$, as deduced from the X-ray crystallographic analysis of some adducts: see, for example, the references 23,32,75,81d,83 (acyclic iron complexed cations) and 62e,84 (cyclic iron complexed cations).

Transoid Nonisolated η^5-Pentadienyl Iron-Complexed Cations

Despite the numerous results already obtained during the studies dealing with the reactivity of *isolated* η^5-pentadienyl iron-complexed cations with regard to the formation of new bonds, the general problem of the control of the regioselectivity of the nucleophilic attacks still remains only partially solved. An elegant solution was proposed recently in Uemura's group.[85] Transoid *nonisolated* η^5-iron-complexed cations generated under Lewis acid conditions react *in situ* with various nucleophiles present in the medium:

Me ... $Fe(CO)_3$ ... OAc, R^1, R^2 → (Nu; Lewis acid, CH_2Cl_2, - 78°C; 52 - 88 %) → Me ... $Fe(CO)_3$ ... Nu, R^1, R^2

Nu = Me_3Al, $CH_2 = CH - CH_2\,SiMe_3$, $CH_2 = C\,(OSiMe_3)\,Ph$

The best yielding Lewis acid is BF_3-OEt_2 and it is usable for allyltrimethylsilane and silylated enol ethers. Trialkylaluminiums do not need an additional Lewis acid. This high-yielding coupling procedure is totally regio- and stereoselective. The nucleophile is located on the carbon previously bearing the acetoxyl group with a net retention of configuration. This type of coupling was recently extended successfully to heteronucleophiles compatible with the presence of a strong Lewis acid such as phosphines or phosphites.[13,32]

n C_4H_9 ... $Fe(CO)_3$ ... OAc → (1. BF_3 - OEt_2 / CH_2Cl_2, PPh_3, 25°C, 7 h, 87 %; 2. aqueous $NaBF_4$) → n C_4H_9 ... $Fe(CO)_3$ ... $\overset{\oplus}{P}Ph_3$, $BF_4^{\ominus}$

MeO_2C ... $Fe(CO)_3$ ... OAc → (1. BF_3 - OEt_2 / CH_2Cl_2, P $(OMe)_3$, + 5°C, 14 h, 95 %) → MeO_2C ... $Fe(CO)_3$ ... P(O)(OMe)(OMe)

This last result is particularly noteworthy since the corresponding isolated η^5-pentadienyl iron-complexed cation, when reacted with $P(OMe)_3$, gives a mixture of η^4-dienyl complexes and a σ,η^3-π-allyl derivative as described before.[13a]

Remarks on the Mechanism

Although reactions of the iron-complexed cations are generally described as a two-electron concerted transfer reaction from the nucleophile to the unsaturated organometallic cationic ligand, the question of their mechanism arises. These nucleophilic additions could eventually be preceded by a discrete one-electron transfer step, as suggested by the isolation and characterization of dimeric products typical of transient organometallic radical species.[2,63,67,71a,e] These two processes can be described by the following equations:

Two-electron transfer reaction:

$$Fe(CO)_3L^{\oplus} + Nu^{\ominus} \rightarrow Fe(CO)_3LNu$$

One-electron transfer reaction:

$$Fe(CO)_3L^{\oplus} + Nu^{\ominus} \rightarrow \underset{\text{19 e radical species}}{Fe(CO)_3L\cdot} + Nu\cdot$$

$$2\,Fe(CO)_3L\cdot \rightarrow [Fe(CO)_3L]_2 \quad \text{dimeric product}$$

$$2\,Nu\cdot \rightarrow Nu_2 \quad \text{dimeric product}$$

Indeed, this problem was recently examined by J.K. Kochi's group using molybdenum organometallic anions.[63b] Depending upon the nature (cyclic or not) of the cations and temperature of the reaction, normal adducts and/or dimers were obtained:

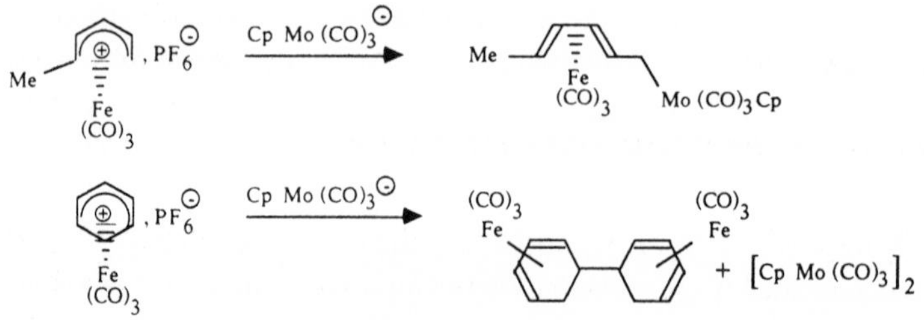

This electron transfer was proposed by Kochi to be a general mechanism during this type of iron-complexed cations—organometallic nucleophiles interactions providing that the additional homolytic scission of the nucleophilic adducts has been taken into consideration.[63b]

$$Fe(CO)_3LNu \rightarrow Fe(CO)_3L\cdot + Nu\cdot$$

More detailed experimental and theoretical studies would certainly be of much interest in order to gain a better understanding of the mechanistic details of these reactions.

C. Intramolecular Reactions of η^5-Pentadienyl Tricarbonyliron Complexed Cations

In view of the numerous results dealing with the reactivity of acyclic η^5-pentadienyl iron-complexed cations during bimolecular reactions, it was particularly astonishing to point out that their intramolecular trapping by appropriate nucleophiles was explored only very recently[87] (Scheme 4).

It is expected that this seminal concept, which takes a large benefit of the properties of η^4-acyclic iron complexes will be useful for the synthesis of natural products or analogues,[89] and the first results in this field appear very encouraging.

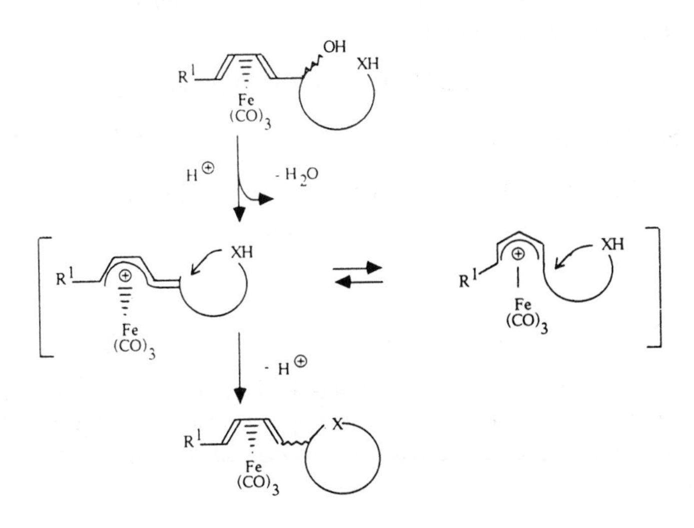

Scheme 4. Intramolecular trapping of iron complexed cations.

Preparation of Chiral Saturated Heterocycles

In Scheme 4 where the incoming nucleophile is an -OH group (X = O) separated from the potential η^5-pentadienyl iron-complexed cation by a variable tether, synthetic access to chiral tetrahydropyrans and tetrahydrofurans is described. This has been realized in the following schemes[90] beginning with the key corresponding η^4-dienyl iron-complexed diols **63, 64** (Scheme 5) and **71, 72** (see Scheme 8) in the Ψ-*exo* and Ψ-*endo* relative stereochemistries:

The Ψ-*exo* more-polar and Ψ-*endo* less-polar iron-complexed diols are easily separated by flash chromatography. The cyclization of the Ψ-*exo* (+)-**63** is high

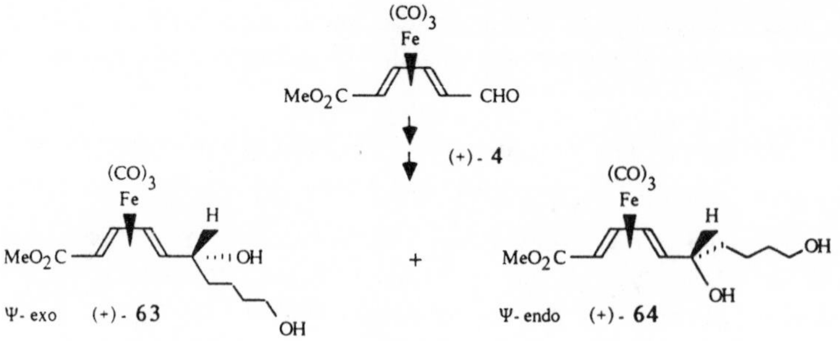

Scheme 5.

yielding when using the acidic complex $HBF_4 \cdot Et_2O$ leading to a (7/3) mixture of the two tetrahydropyrans (+)-**65** and (+)-**66** (Scheme 6) through the two respective depicted iron complexed cations.[20,60,72,82]

The absolute configuration of (+)-**65** has been unambiguously established by X-ray crystallographic analysis which indicates a (2*R*) absolute configuration for C2 of the heterocycle. The (2*S*) absolute configuration of (+)-**66** is established by chemical correlation through the sequence (+)-**66** → (–)-**68** → (–)-**67** providing the

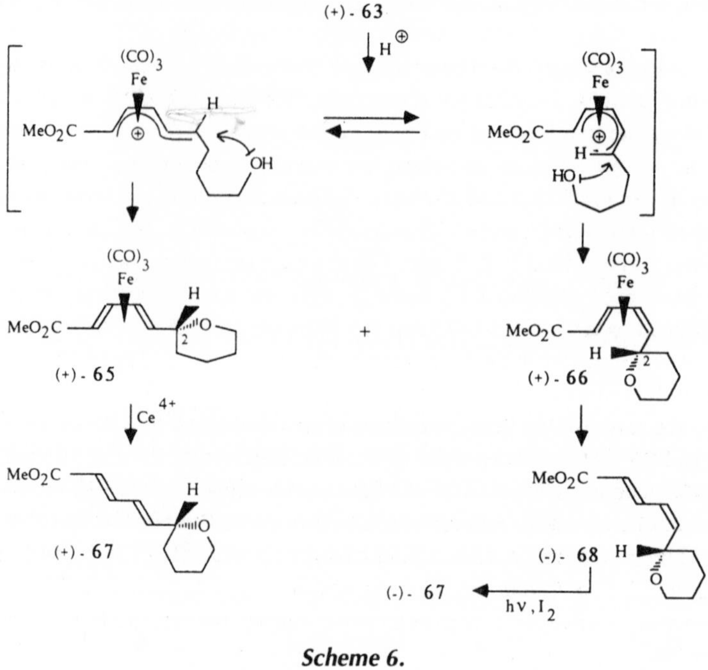

Scheme 6.

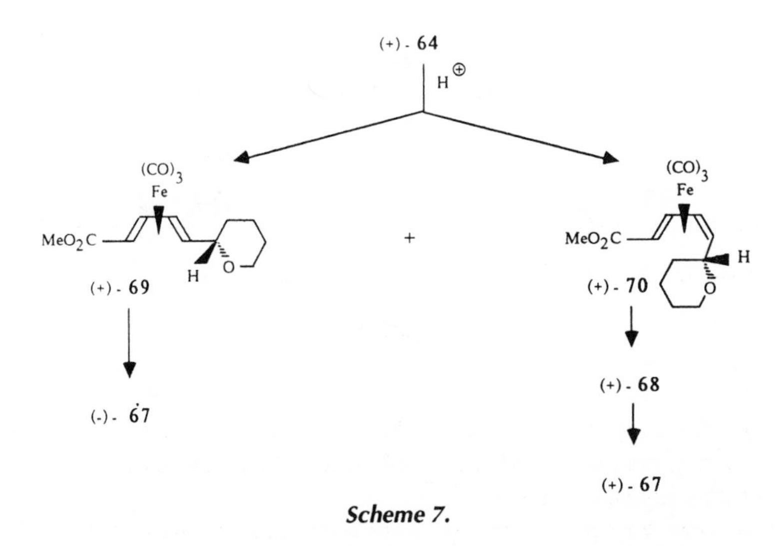

Scheme 7.

tetrahydropyran enantiomeric with (+)-**67**. These results can be rationalized assuming an attack of the –OH group *anti* to the $Fe(CO)_3$ unit on each transoid- or cisoid-shaped iron-complexed cation. These results agree the solvolytic studies of iron-complexed dinitrobenzoates which also occurred with retention of configuration.[57b]

In an analogous way, the diastereomeric Ψ-*exo*-diol (+)-**64** afforded the same tetrahydropyrans by a similar set of reactions (Scheme 7). This approach has been extended to the synthesis of optically active tetrahydrofurans[90,91] (Scheme 8). Under the same conditions as before, the two diastereoisomeric iron-complexed alcohols Ψ-*exo* (+)-**71a,b** and Ψ-*endo* (+)-**72a,b** afforded diastereoisomeric iron-complexed tetrahydrofuran (+)-**73** and (+)-**74**, respectively. No (*E,Z*)-complexed compound was obtained in that case. The absolute stereochemistries of (+)-**73** and (+)-**74** have been attributed by analogy with the tetrahydropyran series. Ce^{4+} salt-mediated oxidation of (+)-**73** and (+)-**74** easily gave the corresponding chiral dienyl tetrahydrofuran. In the case of the cyclization of Ψ-*exo* **71b** and Ψ-*endo* **72b**, the use of the milder acidic system (amberlyst 15 resin H ⊕ form, CH_2Cl_2, 20 °C) leads to the same set of iron-complexed tetrahydrofurans in a better yield of **73** (32%) and **74** (48%). Under these new acidic conditions, **73** and **74** equilibrate, raising the general question of the configurational stability of the C-2 vicinal to the organometallic unit. This epimerization, which depends upon the reaction conditions and the nature of the heterocycle, must be checked carefully in each case.

Integration of a *Z* double bond in appropriate position, such as in **75**, presented an interesting problem of regioselectivity during the nucleophilic cyclization (Scheme 9):[92]

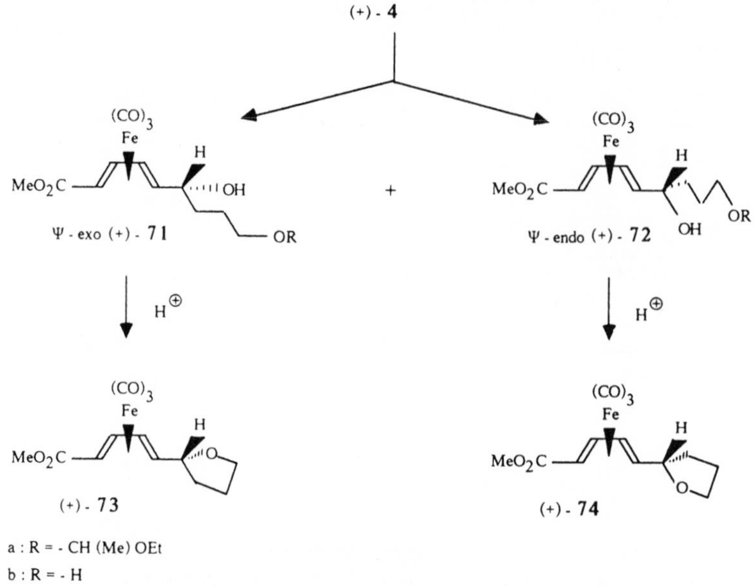

(+) - 4

$(CO)_3$ Fe H MeO_2C OH OR

Ψ - exo (+) - 71

+

$(CO)_3$ Fe H MeO_2C OR OH

Ψ - endo (+) - 72

$H^{\oplus}$ $H^{\oplus}$

$(CO)_3$ Fe H MeO_2C O

(+) - 73

$(CO)_3$ Fe H MeO_2C O

(+) - 74

a : R = - CH (Me) OEt

b : R = - H

Scheme 8.

MeO_2C CHO Fe $(CO)_3$

4

OH MeO_2C H OH Fe $(CO)_3$

75

+

MeO_2C H OH OH Fe $(CO)_3$

76

H O H MeO_2C ⊕ Fe $(CO)_3$

77

MeO_2C O H Fe $(CO)_3$

78

MeO_2C O H Fe $(CO)_3$

79

+

MeO_2C H O Fe $(CO)_3$

80

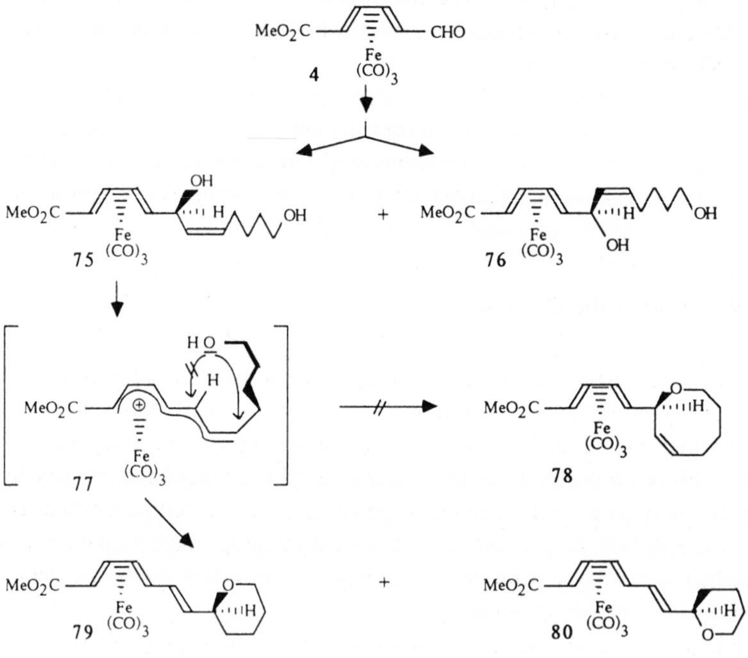

Scheme 9.

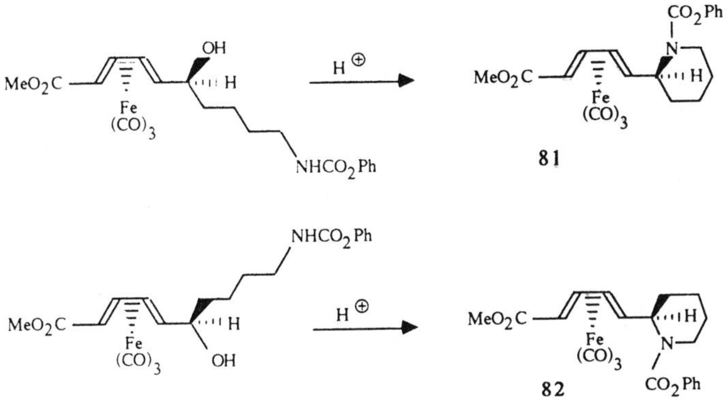

Scheme 10.

For instance, the Ψ-*exo* iron-complexed enediol **75**, when subjected to the usual acidic cyclization conditions, does not afford the iron-complexed oxocene **78**, but a mixture of the two (*E,E,E*)-vinylic tetrahydropyrans **79** (33%) and **80** (15%). The relative stereochemistry of the major adduct **79** has been tentatively assigned as depicted, assuming an *anti* nucleophilic attack of the –OH group on the delocalized iron-complexed cation **77**. This fact remains to be confirmed. Under the same conditions, its Ψ-*endo*-isomer **76** also furnished a mixture of the same compounds, **79** + **80**, but in a different ratio.

The same methodology using nitrogen[93] gave stereospecifically piperidines **81** and **82** (Scheme 10). The relative stereochemistry of each cyclized adduct has been assigned in agreement with the oxygenated series (retention of configuration for the incoming heteronucleophile). Preliminary experiments suggested a possible extension to sulfur nucleophiles.

Mechanism of the Cyclization

Recent interest in the mechanism of cyclization arose since Paquette studied the acid-catalyzed dehydration of diols **83** and **84** specifically labeled at the tertiary hydroxyl group by an ^{18}O label.[94] The formation of the two spirocyclic tetrahydrofurans **85** or **86** proceeds by the selective displacement of the primary hydroxyl functionality with a net retention of the label. The observed stereochemical retention exceeds 90% in this case. This striking reversal of conventional reactivity was ascribed to the destabilization of the tertiary carbocation by the vicinal electron-withdrawing methoxyl group.

When applied to the iron-complexed series[91] according to Scheme 11, no retention of the ^{18}O label was observed in the isolated iron-complexed tetrahydrofurans **73** and **74**, providing conclusive evidence for the heterolysis of the tertiary C–O bond vicinal to the iron-complexed moiety during the intramolecular trapping.

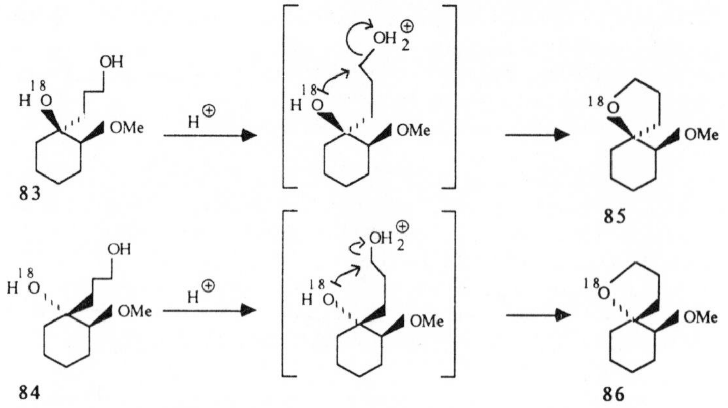

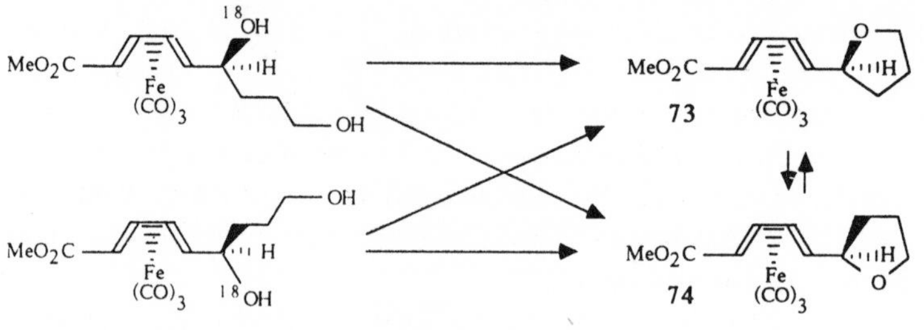

Scheme 11.

Preparation of Unsaturated Oxygenated Heterocycles

The intramolecular trapping of oxygenated nucleophiles by cationic iron complexed salts has been extended to the synthesis of unsaturated oxygenated heterocycles. Although these experiments are only preliminary, they clearly demonstrate the high synthetic potential of the underlying basic concept.

Iron-complexed (5,6)-dihydro 2*H*-pyrans[92] such as **88** and **89** are now easily accessible starting from diol **87** (Scheme 12). The stereochemistry of **88** and **89** has presently been attributed only by analogy with preceding results.

A complexed 2*H*-1-benzopyran **94** was also prepared via the new phosphorane **91** (Scheme 13).[95] The known Ψ-*exo* and Ψ-*endo* chlorhydrins **90**[96] are the starting materials for this synthesis. The reduction of the ketone **92** (*E* stereochemistry of the double bond) is totally stereoselective in favor of the Ψ-*endo*-diastereoisomer **93**, in agreement with previous results. The $E \rightarrow Z$ isomerization of the double bond during the cyclization is in good agreement with an allylic carbocation as interme-

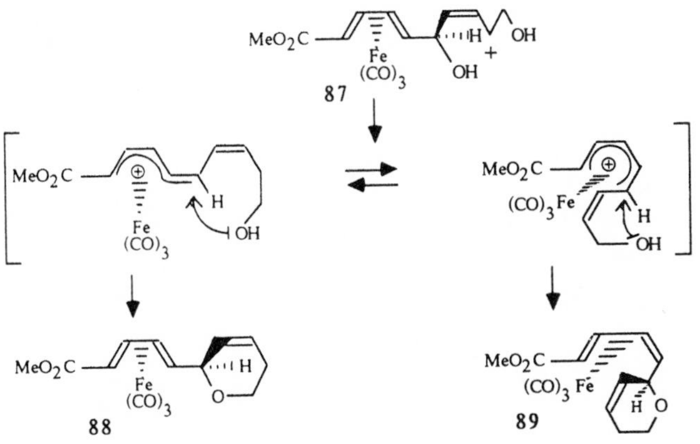

Scheme 12.

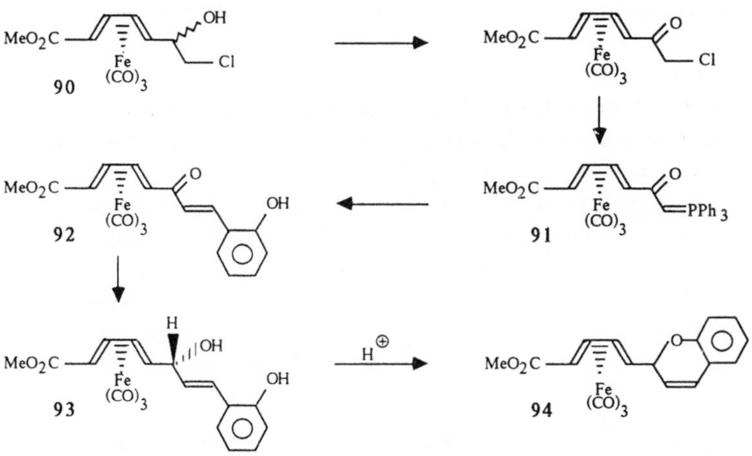

Scheme 13.

diate. The iron-complexed benzopyran **94** can be regarded as a potentially useful analogue of the anti-juvenile hormones, precocenes **95**.

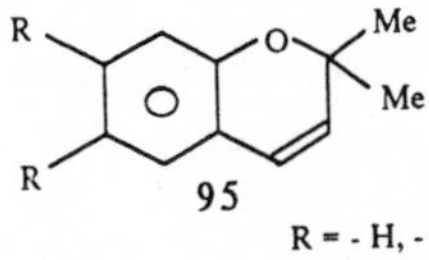

R = - H, - OMe

Further extension to medium ring-sized oxocenes has also been successfully realized following Scheme 14[95] and this reaction is stereospecific:

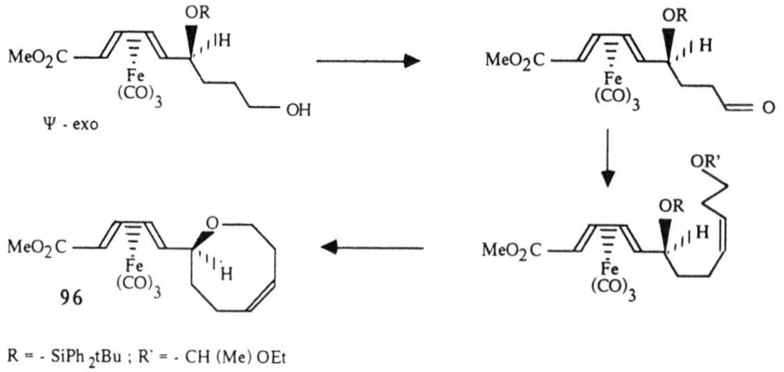

Ψ - exo

96

R = - $SiPh_2tBu$; R' = - CH (Me) OEt

Scheme 14.

Complexed heterocycle **96** (whose structure has been unambiguously established by X-ray analysis) is obtained starting from the Ψ-*exo*-diol, while its diastereoisomer has been isolated from the Ψ-*endo* diol. Preliminary results suggest also a possible extension to other medium-sized oxygenated heterocycles.

D. Conclusion

In conclusion, nucleophilic addition to iron-complexed pentadienyl cations can be regarded as one of the most efficient routes to new and useful polyfunctionalized diene–tricarbonyliron complexes. However, progress clearly has to be made in order to better understand the different factors governing the regioselectivities of addition, and thus to obtain reactions of great predictive value which can be integrated with confidence in the design of new total syntheses.

The intramolecular nucleophilic trapping of iron-complexed cations appears particularly attractive from a synthetic point of view. Even if much work remains to be done, we can expect for the future two possible lines of evolution in this peculiar field:

1. The extension of this seminal concept to nucleophiles other than heteronucleophiles with a special attention devoted to the formation of C–C bonds.
2. The integration of such a key step in synthetic schemes directed towards the synthesis of complex natural substances. In this case, combined strategies using the full palette of properties of the dienyl iron complexed moiety could offer efficient and elegant answers to new synthetic questions.

III. SYNTHESIS OF ACYCLIC POLYUNSATURATED NATURAL PRODUCTS

A. Arachidonic Acid Metabolites and Their Structural Analogues

The arachidonic acid cascade has been the subject of very intensive research during the last 30 years, both from the biological and the chemical point of

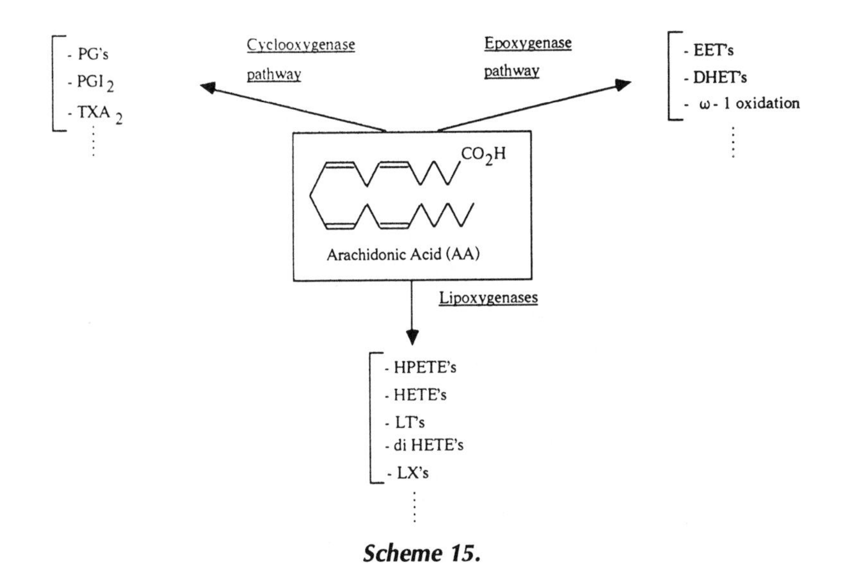

Scheme 15.

view.[97–99] In living systems, arachidonic acid is metabolized into a wide range of compounds and most of these natural derivatives have very potent physiological properties. The three main metabolic routes are indicated in Scheme 15. The first discovered was the cyclooxygenase pathway leading to the well-known protaglandins (PG's), prostacyclin (PGI_2) and thromboxanes (TXA_2)[97].

The epoxygenase pathway has been studied more recently; it leads, via cytochrome P_{450} type enzymes, mainly to epoxides (EET's) and corresponding diols (dHET's), or to oxidized derivatives in the ω or ω-1 position.[99] However, most attention has been paid during the last 15 years to the lipoxygenase pathway leading, *inter alia*, to the HPETE's, HETE's, leukotrienes (LT's), and lipoxins (LX's).[98] The structures of the most representative compounds are indicated in Scheme 16 (for the 5-lipoxygenase pathway) and Scheme 17 (in the case of 15-LO pathway).

Closely related and biologically active compounds could also be found in the 12-lipoxygenase pathway but less data are available presently for the other lipoxygenases (8-,9-, and 11-LO).[100] These lipoxygenase derived metabolites have, in many cases, very potent and selective biological properties. Thus, it is not surprising that many total syntheses were published in the recent literature, most starting with molecules from the chiral pool.

Due to their biosyntheses, all these LO metabolites have structures characterized by conjugated polyenic systems with lateral chains containing 1,2, or more stereogenic centers. Thus, it appeared very quickly that these derivatives were ideal target molecules to check the possibilities and limits for the use of chiral acyclic diene–tricarbonyliron complexes in synthesis.

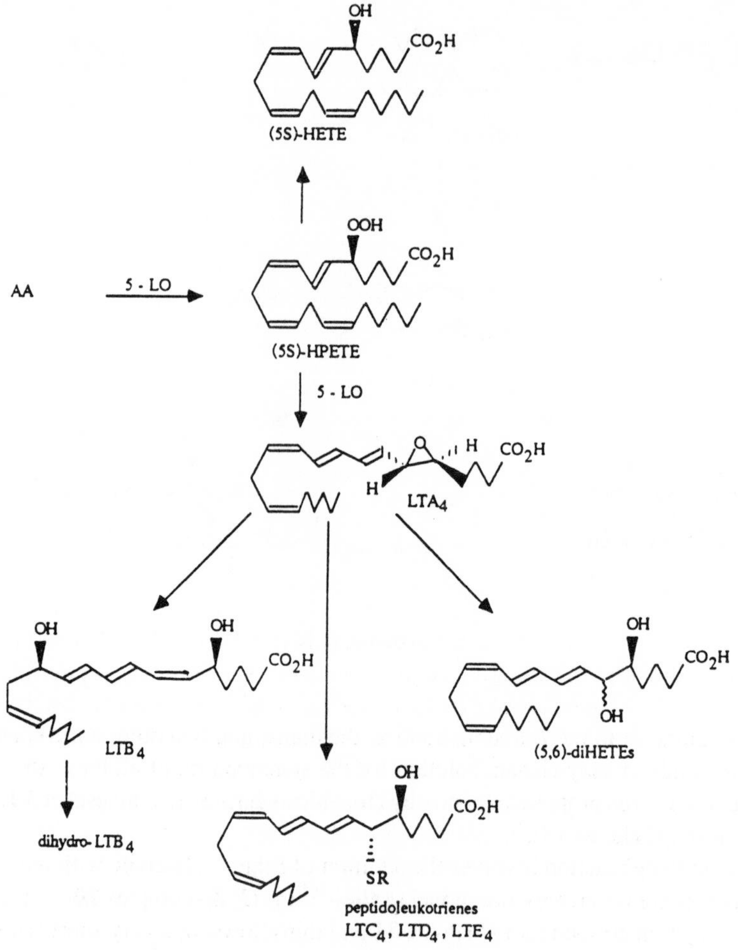

Scheme 16.

As will be described here, this new approach proved to be very fruitful since many of these metabolites have already been prepared in a stereoselective manner starting from these organometallic complexes. More importantly, *each of these successful syntheses gave the opportunity to test the compatibility of the organometallic moiety with new reactions as well as the stereoselectivity of these reactions.*

5-HETE Methyl Ester

The HETE's, which appear very early in the arachidonic cascade, share in common three characteristic features: an (*E*,*Z*)-diene unit, a chiral secondary

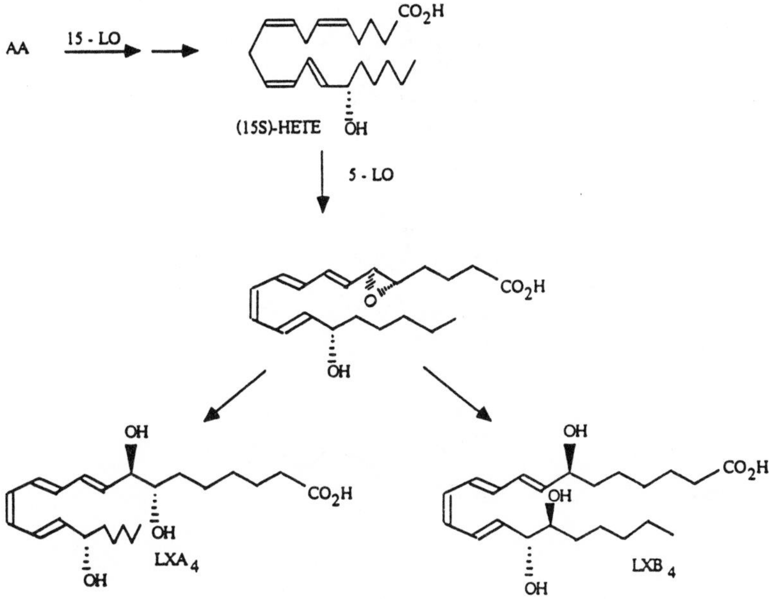

Scheme 17.

alcohol adjacent to the *E* double bond of the diene, and two other nonconjugated double bonds. A very elegant solution for the stereocontrol of all these structural elements has been proposed recently by Donaldson during his synthesis of 5-HETE methylester (Scheme 18).[21]

The first key reaction involves the addition of lithio (1,4)-diyne with the cation **97**. As expected from previous model studies,[101] the (*E,Z*)–complex **20** is obtained. It is easily transformed into aldehyde **21** and this allows, in a very short sequence of reactions, to control the stereochemistry of the four double bonds. The second key step involved a hetero Diels–Alder reaction with Danishefsky's diene. This gave a 2.1 : 1 mixture of desired dihydropyrone **98** with its stereoisomer. Reduction followed by Ferrier rearrangement led to the cyclic acetal which was transformed into the unsaturated lactone **99**. Although somewhat difficult, reduction of the electrophilic double bond to **100** could be done using either $Fe(CO)_5$/DABCO/DMF/H_2O (36%) or $[Ph_3PCuH]_6$/C_6H_6 (31%). Liberation of the ligand with concomitant transesterification gave racemic 5-HETE methyl ester. It is clear that this approach could be used for the synthesis of optically active 5-HETE since resolution of the starting complex has already been done.[21] It is also interesting to point out that, in agreement with other studies to be described later, the (*E,Z*)-geometry of the diene unit is maintained through the complete synthetic sequence.

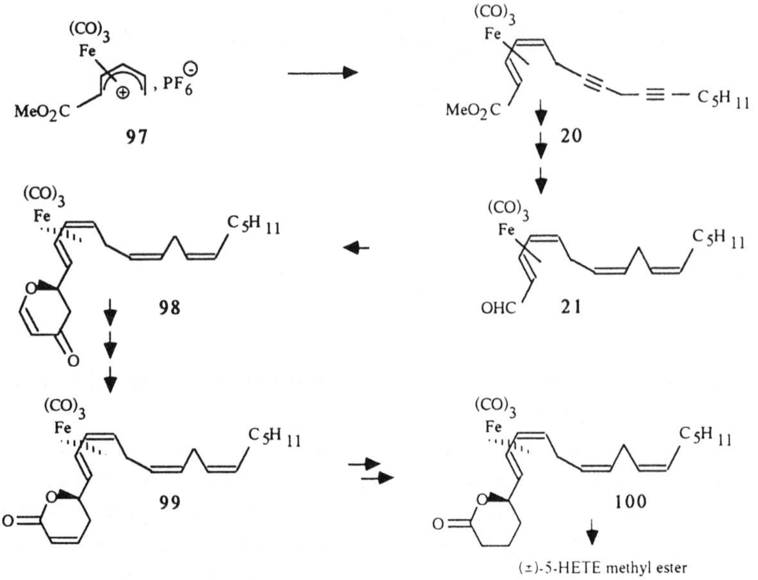

Scheme 18.

*LTA*₄

The second key compound in the 5-lipoxygenase pathway is the leukotriene A_4. This proved to be a very difficult molecule to prepare starting from these organometallic complexes, mainly due to the presence of the epoxide ring adjacent to the iron–carbonyl moiety. Indeed, all the preliminary studies performed on simple models were rather disappointing:

- Epoxidation of complexed trienes containing free double bonds adjacent to the complex by MCPBA were unsuccessful, giving only decomplexation products,[75] while no reaction was observed with TBHP and Mo $(CO)_6$.[102] However, it would certainly be interesting to test new reagents such as dioxiranes (under neutral conditions) for instance.
- All direct epoxidation methods failed in the case of **4** (Darzens reaction, sulfur or arsonium ylids);[75]

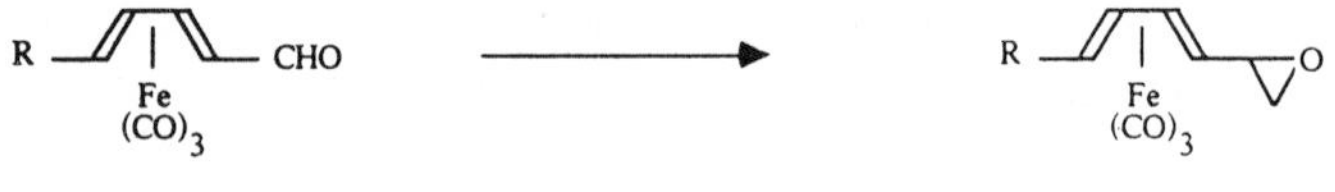

3 : R = - CH_3
4 : R = - CO_2CH_3

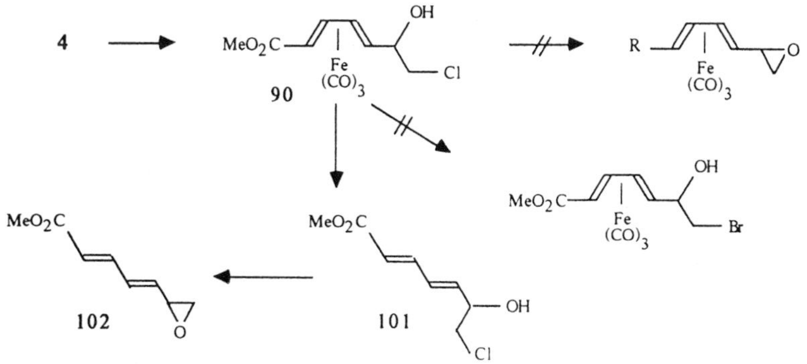

- Only in the case of **3** was it possible to isolate, in low yield and as a 1:1 mixture, the expected stereoisomeric epoxides.[103]
- The complexed chlorohydrins **90**, which are easily obtained from **4**, were very stable and poorly reactive: they could not be cyclized at this stage, under all standard reaction conditions tried; furthermore, no exchange with bromine could be accomplished. Although its role could not be clearly explained, the influence of the organometallic complex is established by the fact that the decomplexed chlorohydrin **101** was smoothly converted to epoxide **102**.[96]

These results were confirmed by Franck-Neumann who could prepare, in the following way, either (*E,Z*) or(*E,E*)-α-dienylepoxides:[104]

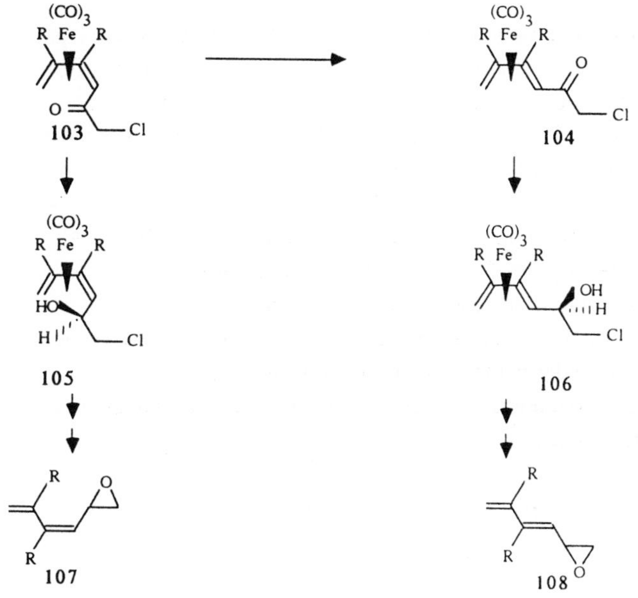

R = - H, - Me

The stereocontrol of this method is noteworthy: Friedel–Crafts acylation gives only *cis*-chlorodienone complexes **103** which can eventually be completely isomerized into **104**. Since reduction of ketones is, as usual, totally stereocontrolled, this synthesis gave after decomplexation and cyclization **107** or **108**. This method could be extended to disubstituted epoxides **109** and **110**:[104]

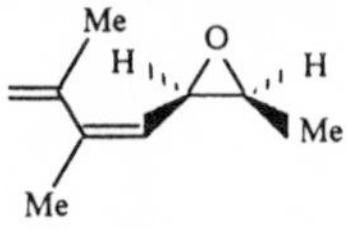

109

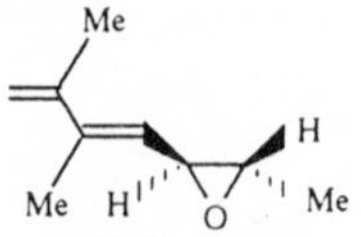

110

The key discovery towards epoxides vicinal to diene–tricarbonyliron complexes was done in Franck-Neumann's group when it was found that the halohydrins could be cyclized to epoxides by K_2CO_3 *but only when activated by 12-crown-6 ether*. This led first to a series of stable complexed epoxides **115a-g** according to Scheme 19.[105]

Starting from optically active diene–tricarbonyliron complexes, the corresponding epoxides were obtained without any racemization. The enoxysilanes vicinal to the complex moiety **112** (mixture of isomers) are also worthy of note: they are the first compounds of this type to be reported and they proved to be of interest in aldolization reactions.[106]

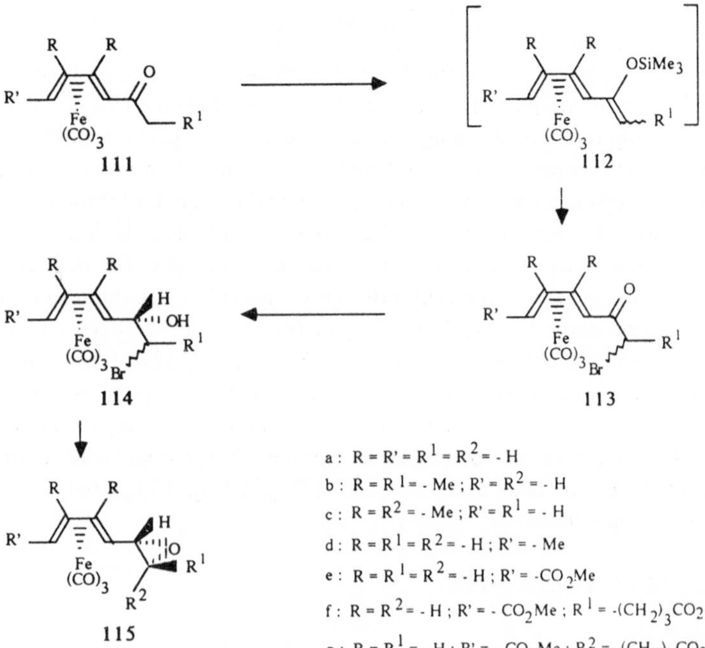

111 112 113 114 115

a : R = R' = R^1 = R^2 = - H
b : R = R^1 = - Me ; R' = R^2 = - H
c : R = R^2 = - Me ; R' = R^1 = - H
d : R = R^1 = R^2 = - H ; R' = - Me
e : R = R^1 = R^2 = - H ; R' = -CO_2Me
f : R = R^2 = - H ; R' = - CO_2Me ; R^1 = -$(CH_2)_3CO_2Me$
g : R = R^1 = - H ; R' = - CO_2Me ; R^2 = -$(CH_2)_3CO_2Me$

Scheme 19.

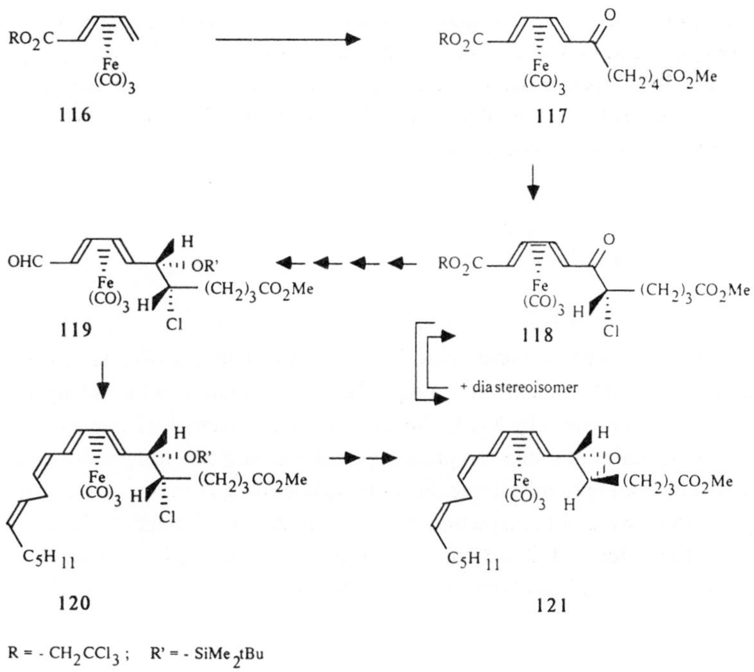

R = - CH_2CCl_3 ; R' = - $SiMe_2tBu$

Scheme 19A.

This approach was later applied to the enantioselective synthesis of the natural LTA_4, both in its complexed and decomplexed form (Scheme 19A).[105] This synthesis starts from the optically pure complex **116** with its special trichloroethyl ester group; this is compatible with the Friedel–Crafts acylation process and could be easily removed later. Chlorination via the enoxysilanes gave a 1:1 mixture of ketone **118** with its stereoisomer, but since they can be equilibrated, the final mass balance is good. Functional group transformations gave aldehyde **119** and, after Wittig reaction, protected chlorohydrin **120**. This was either decomplexed and transformed to the natural (–)-(5*S*,6*S*)-LTA_4 methyl ester, or cyclized under the previously described conditions to the complexed LTA_4 **121**. This compound, its stereoisomers, as well as the other preceding complexed epoxides would be very interesting models to study the influence of the $Fe(CO)_3$ moiety on the regioselectivity of the nucleophilic opening of the oxirane. Thus, it could lead not only to the biologically active peptidoleukotrienes (LTC_4, LTD_4, LTE_4) but also to some unnatural analogues.

LTB_4; (14,15)-Dehydro-LTB_4

Leukotriene B_4(LTB_4), produced by enzymatic hydrolysis of the LTA_4, is the major proinflammatory product of this 5-LO pathway and, as such, is implicated

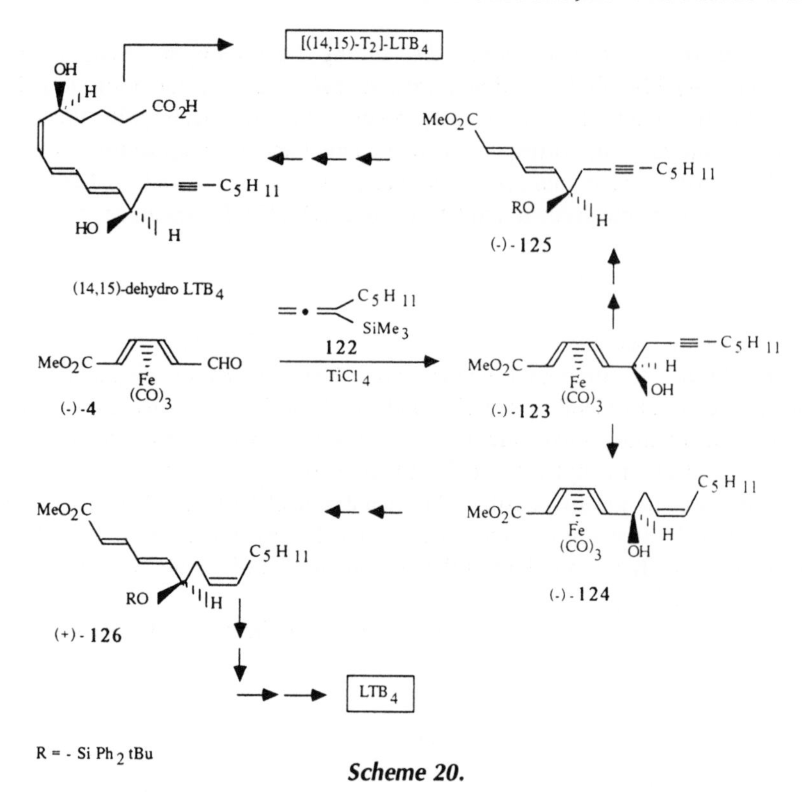

Scheme 20.

in numerous diseases. Hence, the preparation of LTB_4 and related compounds, especially labeled derivatives, was of much interest.

As shown on Scheme 20, a new short and highly enantioselective synthesis of key intermediates for the preparation of LTB_4 and its (14,15)-dehydro derivative was described.[107] The organometallic complex plays here key roles, at several stages of this synthesis, both as a protective and a stereoinducing group:

1. The reaction of allenylsilane with complex(–)-**4** is highly stereoselective giving only Ψ-*endo*-alcohol (–)-**123**. Thus, starting from an optically pure complex, it allows a complete control of the (*R*) absolute configuration at the secondary alcohol function.
2. The $Fe(CO)_3$ moiety efficiently protects the diene during this reaction: under the same conditions, the free diene corresponding to **4** gave, at best, 10% of **125**. This was in agreement with Danheiser's results indicating that this reaction with allenylsilanes did not work with conjugated enones.[108]

3. Protection of the diene occurs also during the semi-hydrogenation step giving (–)-**124**: if this reaction is applied, under the same conditions to **125**, not only is the triple bond semi-hydrogenated but the electrophilic diene is also reduced to the corresponding compound (and at a comparative rate).
4. These were formal total syntheses since polyenes (–)-**126** and (–)-**125** have already been transformed into LTB_4 and its [(14,15)-T_2] analogue.[109]

(6,7)-Dihydro-LTB_4 Methyl Ester

New metabolic pathways have been found recently for LTB_4; these include reductases leading to compounds characterized, *inter alia*, by a conjugated diene chromophore. If (10,11)dihydro LTB_4 and (10,11)-dihydro-12-oxo-LTB_4 have already been unambiguously identified,[110] it was also of interest to prepare (6,7)-dihydro-LTB_4-methyl ester **132** and its 5-oxo derivatives **131** since the corresponding acids are also possible LTB_4 metabolites.[111] This synthesis (Scheme 21) is a logical extension of preceding results. It was, however, an excellent opportunity to study the problem of the (1,4)-reduction of enones vicinal to the organometallic moiety.

Among several agents tried on a model compound, the most efficient was the Red-A1/CuBr/2-butanol mixture. Applied to enone **128**, it gave reduced complex

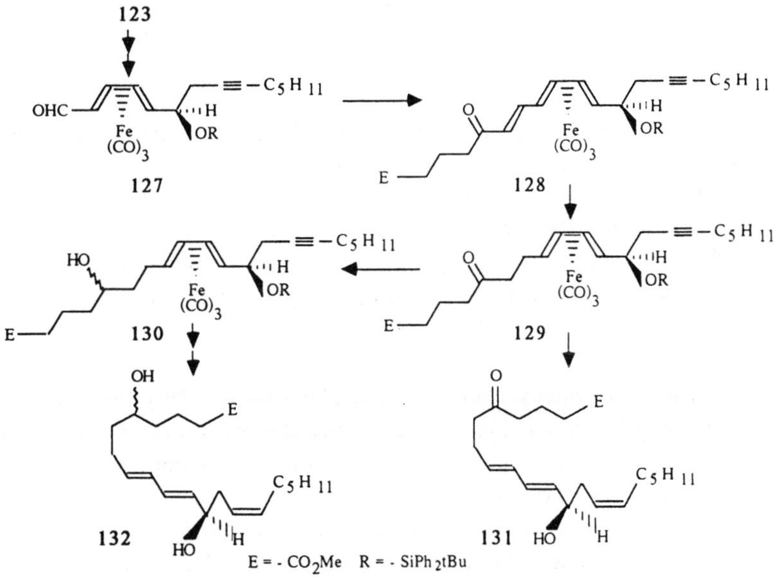

Scheme 21.

129 in 47% yield (together with 23% recovered **128**). This intermediate was then easily transformed into the target molecules **131** and **132**. This last compound was obtained as a mixture of stereoisomers at C-5 since the problem of the stereocontrol in the carbonyl reduction of **129** was not addressed at this stage. Comparison of these authentic standards with biological materials would clarify the question of their actual occurrence. Additionally, it would be interesting to study biological profile of these new derivatives.

(α,β)-DiHETEs and Analogues

The next attractive target molecules from these lipoxygenase pathways were polyhydroxylated compounds, including diHETEs and lipoxins. It is easy to recognize that all these derivatives share in common an α-diol structure vicinal to a conjugated (*E*,*E*,*Z*)-trienic system. A simple retrosynthetic analysis includes first a temporary protection of the (*E*,*E*)-diene and then, as a key step, the introduction of the two OH groups via a bishydroxylation process such as an osmylation reaction. The key points of these studies were then (1) the compatibility of the osmylation with $Fe(CO)_3$ moiety,(2) its chemoselectivity with regard to the other groups R^2 and R^3, and (3) its diastereoselectivity.

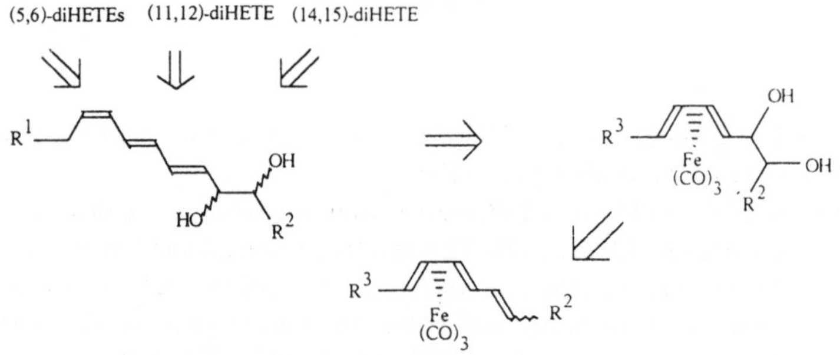

(5,6)-diHETEs (11,12)-diHETE (14,15)-diHETE

This original approach was very successful as established by the total syntheses of the (5*R*,6*S*)- and (5*S*,6*S*)-diHETEs, a formal synthesis of lipoxin A_4[112] and the first total synthesis of the (11*R*,12*S*)-diHETE.[113]

The key points regarding the osmylation reaction were studied first with complex(–)-**133** bearing a *Z*-free double bond (Scheme 22) and its (*E*)-isomer (–)-**138** (Scheme 23). In agreement with literature data,[114] this *fast* reaction is compatible with the $Fe(CO)_3$ moiety and gave in each case the expected diols in high yield. Its stereoselectivity was depending upon the stereochemistry of the free double bond.

The osmylation is highly stereoselective in the case of (–)- **133**, giving exclusively the erythro diol (–)-**134**. Its stereochemistry, corresponding to the addition of OsO_4 *anti* to the $Fe(CO)_3$ group was secured by an X-ray data analysis of

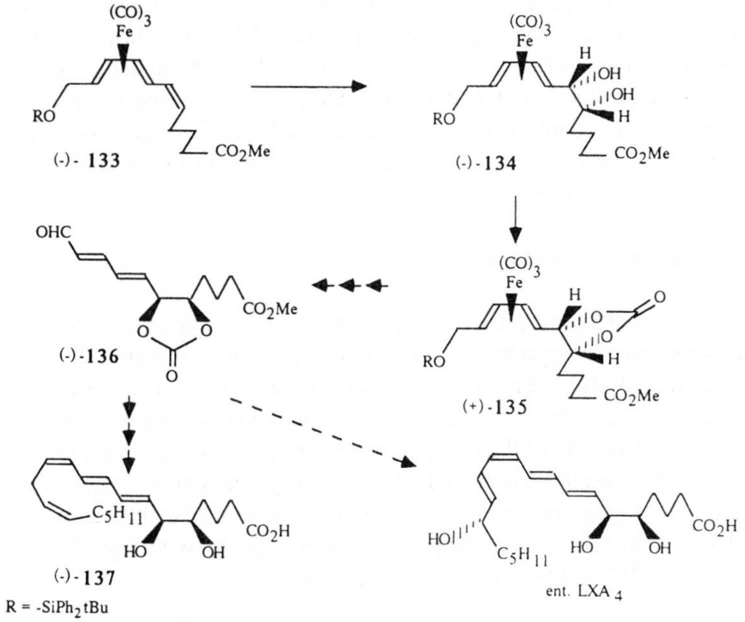

Scheme 22.

corresponding carbonate (+)-**135**.[115] This was easily transformed via the dienal (–)-**136** into the (5*R*,6*S*)-diHETE(–)-**137**.

In the case of the *E* olefin (–)-**138**, the osmylation was only stereoselective giving a 9/1 mixture of diols **139** and **140**. This result can be explained by the conformational equilibrium between the *s-cis* and *s-trans* form of the olefin with reaction of OsO_4, occurring on both forms, and *anti* to the $Fe(CO)_3$ group. The X-ray data analyses obtained on carbonates (–)-**141** and (+)-**142** are in full agreement with such an analysis.[115] It is also interesting to note that a similar explanation has been proposed for the stereoselective diazo-2-propane cycloaddition on some activated olefins with an *E* stereochemistry.[11,116]

Complex (–)-**141** was then transformed by the same sequence of reactions into the desired threo (5*S*,6*S*)-diHETE (–)-**144** . Furthermore, since the total synthesis of lipoxin A isomers starting from carbonates **136** and **143** has already been described,[117] this approach is a new formal synthesis of the enantiomer of lipoxin A_4.

The first total synthesis of the (11*R*,12*S*)-diHETE (+)-**146** and its (14,15)-didehydro analogue (+)-**145** (useful for the synthesis of labeled derivatives) was a natural extension of the preceding results (Scheme 24).[113] The interesting problem in that case was that the key osmylation reaction had to be realized in the presence of another unsaturated π-system corresponding to the (14,15)-double bond in the

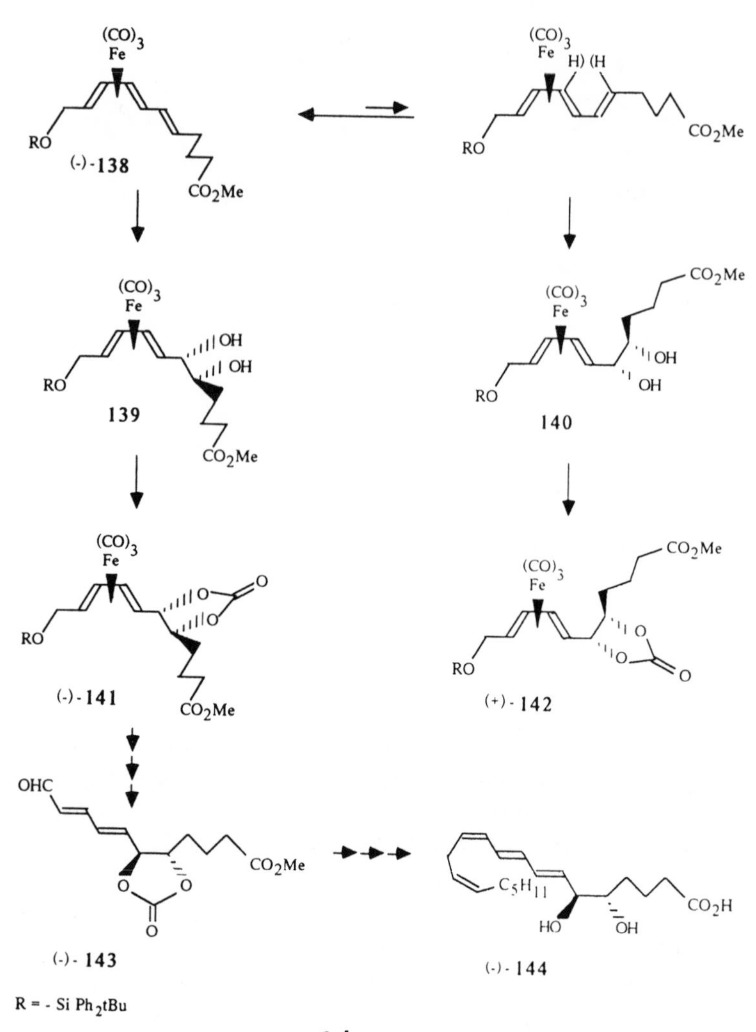

Scheme 23.

natural product. The triple bond proved to be the best choice since osmylation of complexed enyne (–)-**147** (1.5 equiv. of OsO_4) was not only completely stereoselective as before giving (+)-**148** (93% yield) but also fully chemioselective since

HO2C C5H11 (+)-145 HO OH

HO2C C5H11 (+)-146 HO OH

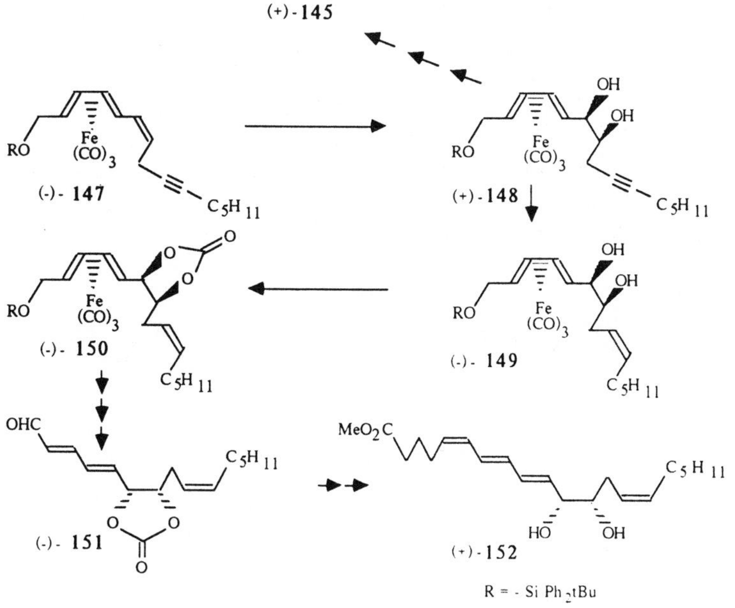

R = - Si Ph_2tBu

Scheme 24.

there was not evidence for reaction on the remote triple bond. Semireduction of the alcyne gave (–)-**149**, and this key intermediate was transformed by a sequence of reactions similar to the preceding ones into the methyl ester of (11*R*,12*S*)-diHETE (+)-**152**. The corresponding dehydro derivative (+)-**145**, which could be of use for the preparation of the labeled compounds, was also prepared similarly from (+)-**148**.

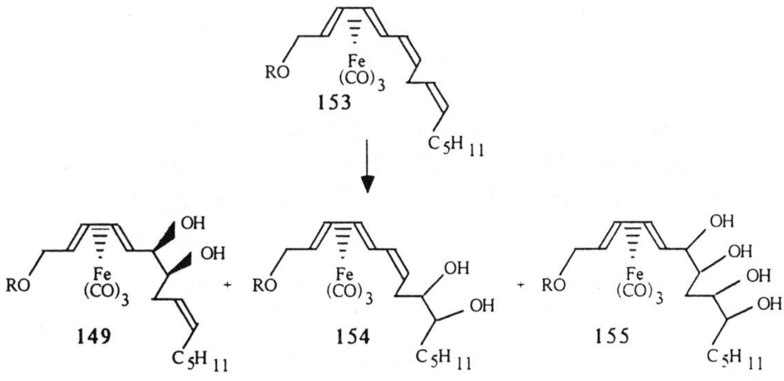

It is also of interest to note that the chemoselectivity of this osmylation appears exclusively due to the competition between the double bond and the alcyne and not to a strong influence of the organometallic moiety : the diolefin **153**, corresponding to **147**, gave under the same conditions a mixture of the two regioisomeric diols **149** and **154**, plus the tetrol **155** corresponding to a bis-addition.

Although all these osmylation reactions, performed only on small quantities, have been done under stoichiometric type conditions it would be also of interest to check the possible use of the catalytic version of this reaction.

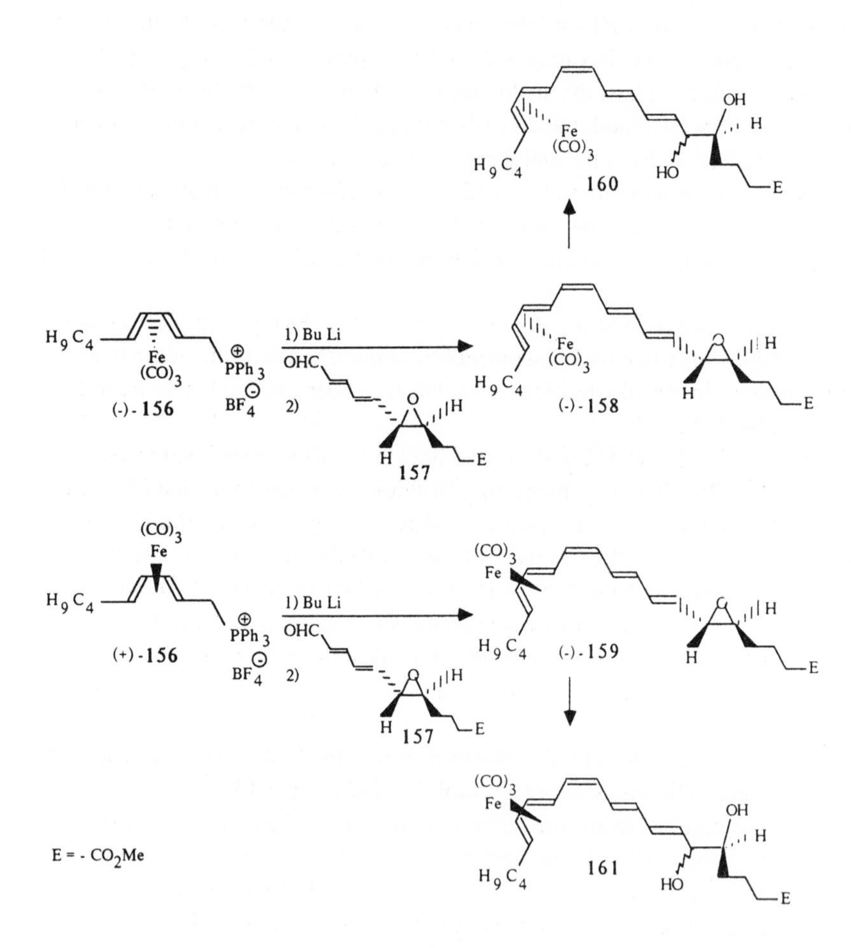

Scheme 25.

Iron-complexed Analogues of LTA_4 and (5,6)-DIHETEs

The next applications of these organometallic complexes in the icosanoid field were related to some selected structural analogues of the natural products. The first was dealing with the preparation of new iron-complexed analogues of LTA_4 and (5,6)-diHETEs designed as possible cold-labeled derivatives.[118] The precise location and assay of the binding sites of a biological effector circulating in fluids at picomolar level is of great biological importance and this is usually done using isotopically labeled molecules. However a new cold FT–IR assay technique, which relies on the strong IR carbonyl absorbance of organometallic species, has been developed recently ; it was used mostly in the field of organochromium complexes of steroidal hormones.[119] The η^4-dienyl $[Fe(CO)_3]$ moiety could be used, *a priori*, in the same way as a new FT–IR label. A crucial point in binding experiments with labeled analogues is that they must keep at least a part of the binding properties of the natural products. Thus, the analogues (–)-**158** and (–)-**159** (Scheme 25) were chosen since they have both the 20 carbon atom chain and the conjugated epoxy triene structure away from the bulky organometallic moiety.

The starting phosphonium salts (–)-**156** and (+)-**156** were obtained, in optically pure form, via the pentadienyl complexed cations as described previously.[13b] Their Wittig reaction with epoxide dienal **157** gave the desired LTA_4 analogues (–)-**158** and(–)-**159**.

Carefully controlled, hydrolysis gave the (5,6)-diHETEs type complexes **160** and **161**; it is noteworthy that the organometallic complex had no influence in that case on the reaction since analogous regioselectivity has been obtained in the hydrolysis of the natural LTA_4.

Shorter analogues (+)-**163** and (–)-**164** have been also synthesized in the same way (Scheme 26), but very interesting differences in reactivity have been observed.[13b] Reaction of (–)-**156** with the (5*S*,6*S*)-epoxy aldehyde **162** gave a low yield of (+)-**163** (18% yield; mismatched pair) while the enantiomeric ylid prepared from (+)-**156** gave a high yield of (–)-**164** (74% yield; matched pair). These original results could be rationalized using the two transition states indicated: if we assume that the reactions are occurring *anti* to the epoxidic moiety, then strong nonbonding interactions between the two chains should disfavor the formation of (+)-**163** as compared to (–)-**164**.

Very different results were also obtained during the hydrolysis step: under the same reaction conditions as before, the diols (+)-**165** and (–)-**166** (single diastereoisomer in each case) were obtained from (+)-**163** and (–)-**164**. This is probably due to the strong stabilization of complexed pentadienyl cations and the occurrence of an SN'_1 mechanism: after epoxide ring opening and allylic rearrangement the resulting carbocation reacts with water on the carbon vicinal to the complex and stereoselectively *anti* to the $Fe(CO)_3$ group.

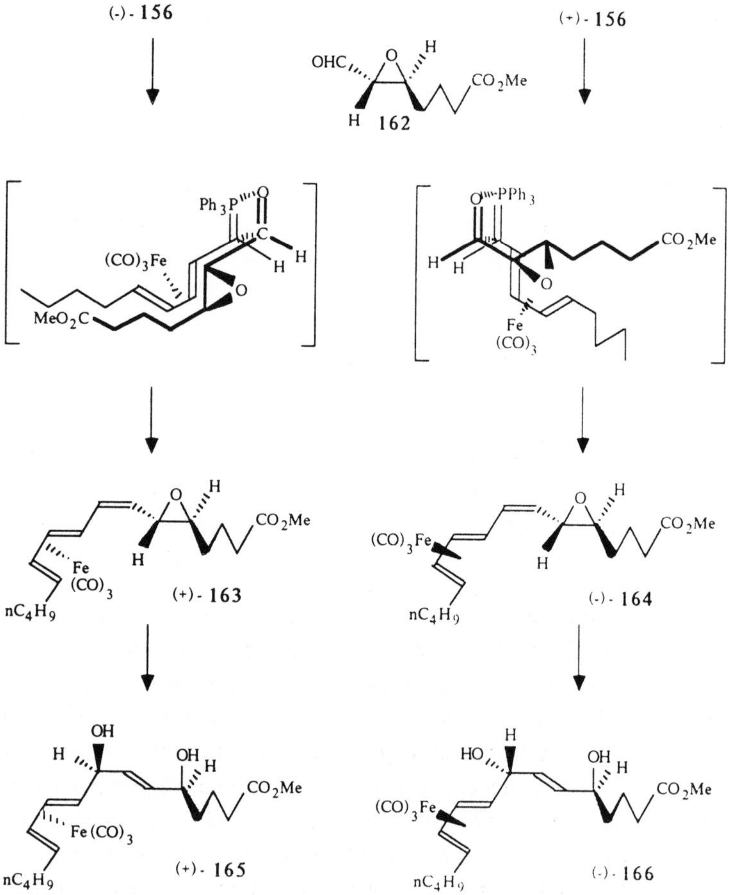

Scheme 26.

SM 9064

Another application of these complexes in the field of analogues of icosanoids was the synthesis of the diastereoisomers of SM 9064.[120] This compound was reported (as a racemic mixture of stereoisomers) to be a potent LTB_4 antagonist.[121] Thus, it was of interest to check the possible influence of the relative configuration of the two stereogenic centers on the biological activity. In that case, the organometallic complex was used as a "stereochemical relay" to control the relative configuration of the two secondary alcohols in positions 4 and 11.

The first stereogenic center was fixed by a (1,2)-addition of a Grignard reagent on aldehyde **167**, the more polar isomer having the usual Ψ-*exo*-structure **168**.

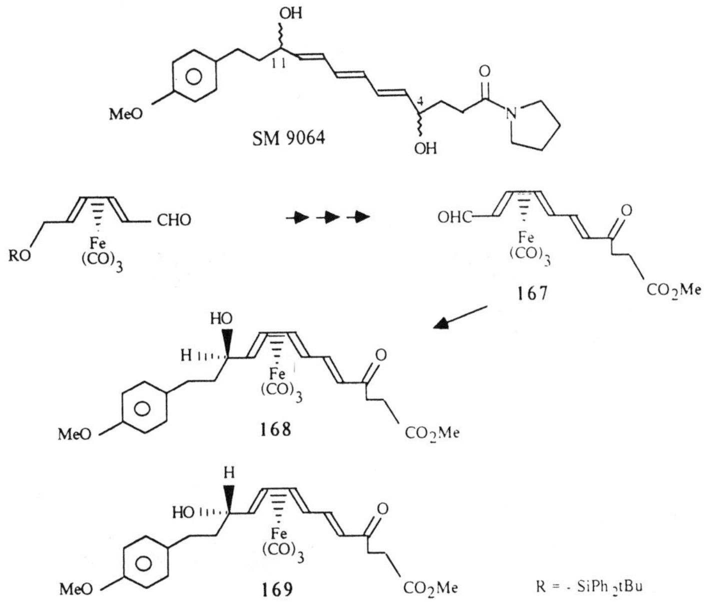

Reduction of the ketone **168** under Luche's conditions gave a mixture of diastereoisomers **170** and **171**, easily separated by chromatography (Scheme 27).

No attempts to improve the diastereoselectivity of this reaction were made in that case, since the different diastereoisomers were required for the biological tests. However, it was established on closely related molecules that high selectivities can be obtained using bulky reducing agents.[122] After transamidation and decomplexation the diastereoisomers **174** and **175** were obtained. Interestingly, neither the mixture nor the separated isomers displaced LTB_4 from its receptor in human polymorphonuclears; this raises questions about the mechanism of action of SM 9064. This principle of using $Fe(CO)_3$ as a relay to control the configuration of the stereogenic centers on the chains from both sides of the complex has already been used earlier[123] and certainly would be of wider applicability in synthesis.

B. Other Natural Products

These acyclic diene–tricarbonyliron complexes also find extensive synthetic use in other families of natural products including pheromones, terpenes, and toxins.

Pheromones are structurally simple natural products in which the biological action is strongly related to the stereochemical purity of the compounds. Thus, organoiron complexes have been used for the preparation of some pheromones bearing conjugated (*E,E*)-diene units.[124] Friedel–Crafts reactions on complexes of type **176** gave, after thermal isomerization of the initial (*E,Z*)-and (*E,E*)-mixture, the pure (*E,E*)-complexed dienes **177**. Mixed hydride reduction gave alcohols **178**, easily transformed later into stereochemically pure pheromones **179**.

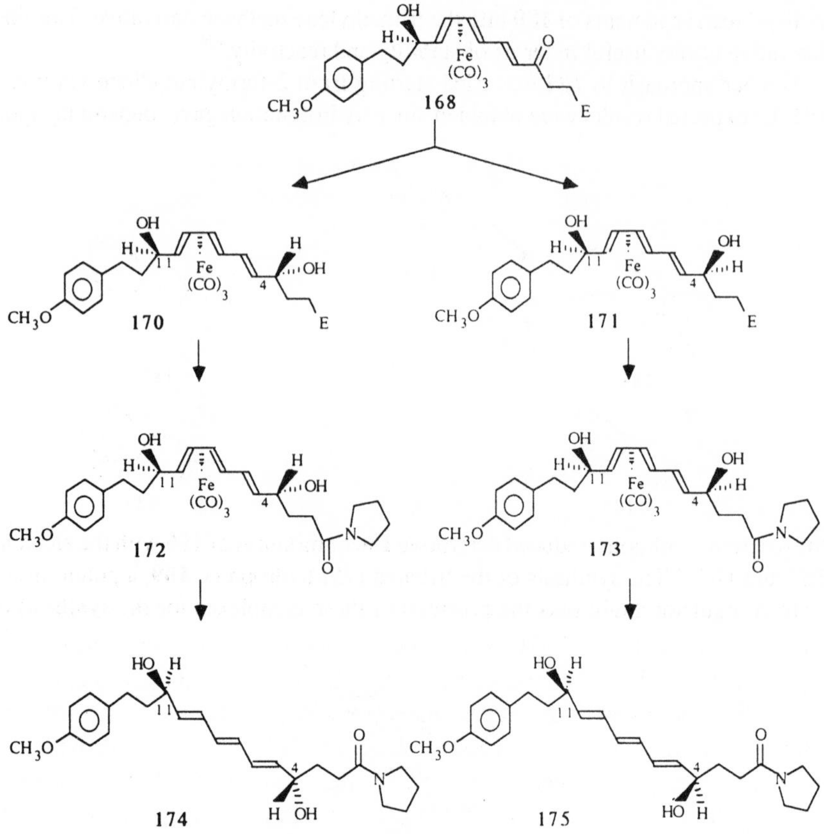

OH
H
Fe
$(CO)_3$
O
CH_3O
168
E

OH H 11 Fe $(CO)_3$ 4 H OH CH_3O 170 E

OH H 11 Fe $(CO)_3$ 4 OH H CH_3O 171 E

OH H 11 Fe $(CO)_3$ 4 H OH N O CH_3O 172

OH H 11 Fe $(CO)_3$ 4 OH H N O CH_3O 173

HO H 11 4 O N CH_3O 174 H OH

HO H 11 4 O N CH_3O 175 HO H

Scheme 27.

Interestingly, complexes of type **178** were also used recently as *propheromones*: they decompose very slowly at room temperature, releasing (over months!) minute amounts of the pure pheromones used to catch insects into designed traps.[125]

Other pheromones **183** and **184** have been prepared, in racemic form, from the very interesting isoprene anion equivalent **180**. The role of a metal M is important

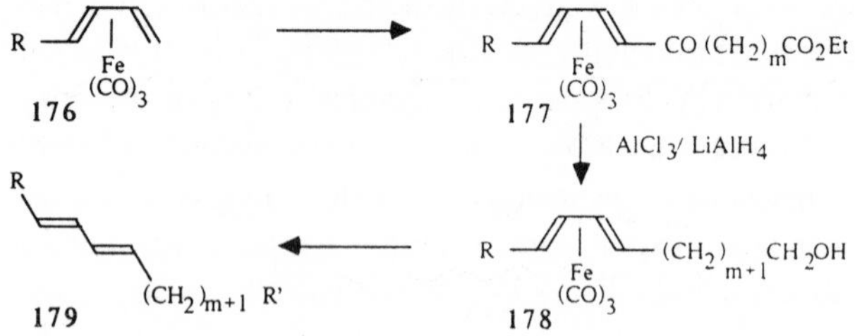

R Fe $(CO)_3$ 176

R Fe $(CO)_3$ CO $(CH_2)_m CO_2Et$ 177

$AlCl_3$/ $LiAlH_4$

R Fe $(CO)_3$ $(CH_2)_{m+1} CH_2OH$ 178

R $(CH_2)_{m+1}$ R' 179

a : R = - Et ; m = 6 : R' = - CH_2OAc ; b : R = - Me ; m = 5 ; R' = - CH_2OAc;
c : R = - Me ; m = 5 ; R' = - CH_2OH ; d : R = - nPr ; m = 7 ; R' = - CHO.

to avoid rearrangements of **180** into the trimethylene methane derivative. The zinc derivative is very useful in terms of stability and reactivity.[126]

Another approach to **184** was tried starting from 2-formylbutadiene via triene **185**. Unexpected results were obtained since hydroboration gave, depending upon

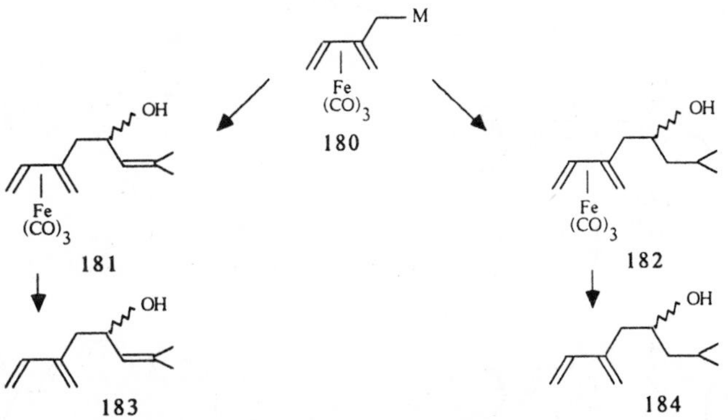

180 181 182 183 184

the reaction conditions, reduced derivative **186** or mixtures of **186** with the alcohols **187** and **182**.[11] The synthesis of the tritiated (7*S*)-hydroprene **189**, a potent insect growth regulator, highlights the potential of these complexes for the synthesis of

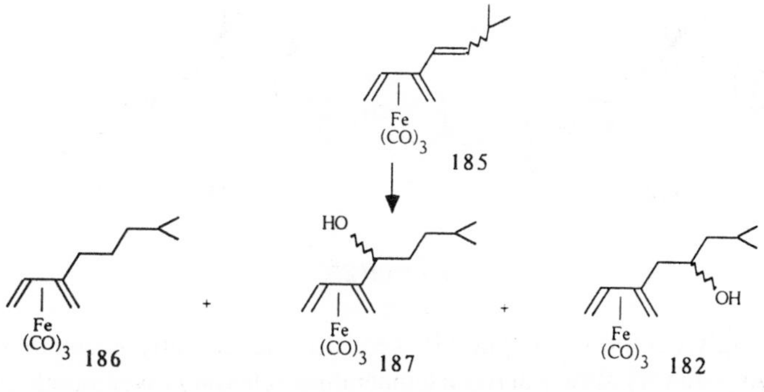

185 186 187 182

labeled compounds:[127] protection of the electrophilic diene by $Fe(CO)_3$ allowed a selective tritiation of the remote double bond giving, after decomplexation, the desired insecticide with a high specific activity.

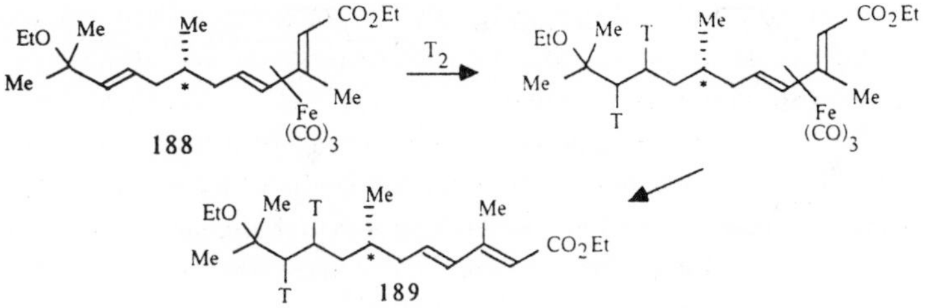

188 189

Retinoids and carotenoids are another important group of natural products which, obviously, appear relevant from the use of organoiron complexes. Interestingly, only a few examples are described yet: complex **190** was used for instance during a stereocontrolled synthesis of type **191** chromenes developed as new low-toxicity anticancer agents.[128]

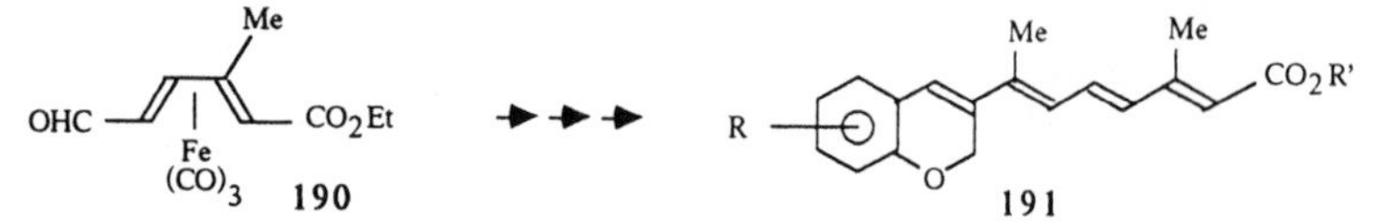

OHC Me CO_2Et Fe $(CO)_3$ **190** → R Me Me CO_2R' O **191**

The AF and AK host specific toxins **192** to **195** are another group of polyunsaturated natural products, very attractive both from the chemical and the biological

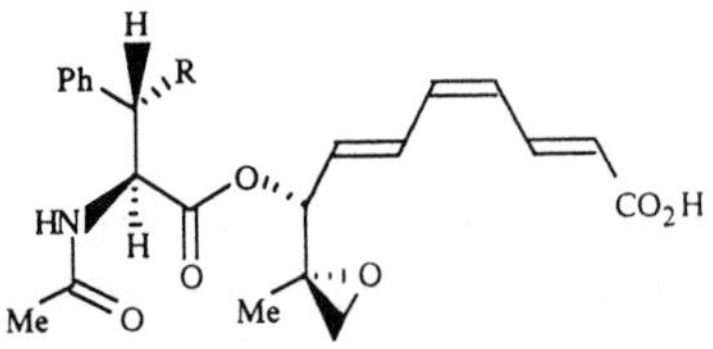

192a R = $-CH_3$ AK I Toxin
192b R = -H AK II Toxin

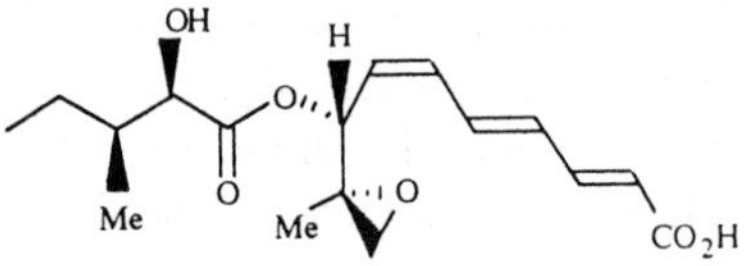

193 AF IIa Toxin

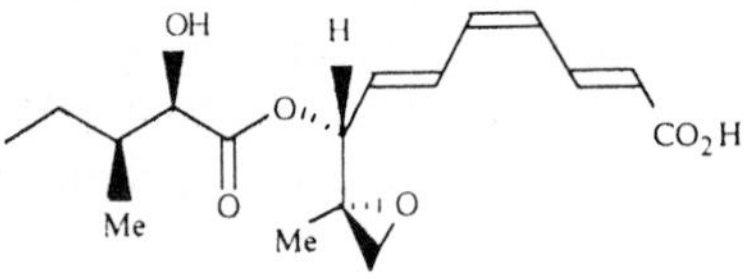

194 AF IIb Toxin

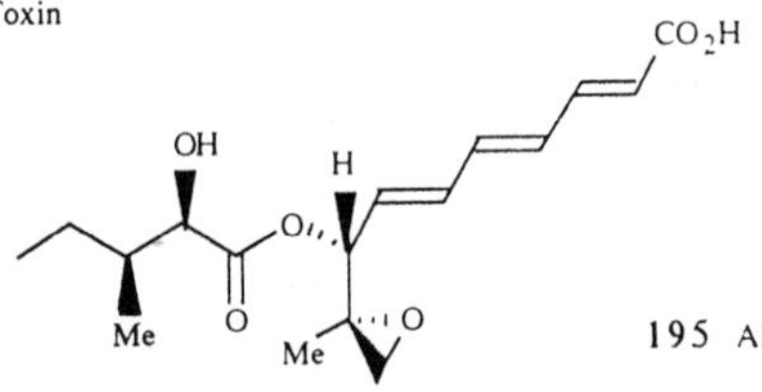

195 AF IIc Toxin

point of view.[129] These compounds were isolated several years ago from Japanese pears and strawberries infected by *Alternaria* species. These toxins are highly specific, not only for the fruit species but also for the fungus. Their mechanism of action has not been established, although some results indicated strong correlations between stereostructures and toxicities. Thus, it was of interest to design new stereoselective and versatile approaches to molecules of this type. In this regard, the design of diene-tricarbonyliron complexes were indeed very successful. Starting from either (–)-**196** or (+)-**197** it was possible to prepare, in optically pure form, the three key intermediates for the synthesis of these toxins and some of their analogues. In these syntheses the main difficulties were (1) the control of the

CO₂H R¹O OHC Fe (CO)₃ OR

(-)-196

R¹O CO₂H R¹O CO₂H

OHC Fe (CO)₃ OR

197

R = -SiMe₂tBu

absolute configuration at the stereogenic centers, (2) the possibility to realize direct epoxidations in the presence of the organometallic complex, and (3) the preparation and the use of (*E,Z*)-complexed dienes during multistep sequences. The preparation of the (*E,E,E*)-toxin **195** gave solutions to the first two problems: Reaction of the Grignard reagent from bromo-2-propene with the optically pure aldehyde (–)-**198** gave the more polar Ψ-*exo*-alcohol (–)-**199** and the less polar and minor Ψ-*endo*-isomer (–)-**200**, easily separated by chromatography. In agreement with preliminary

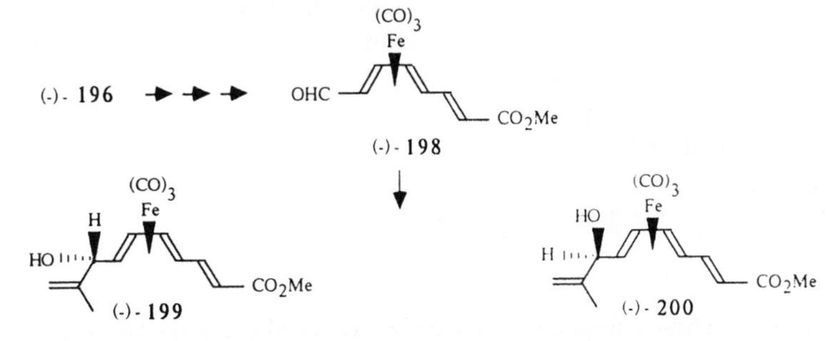

experiments on simpler models,[130] the Sharpless epoxidation procedure [VO(acac)$_2$ + tBuOOH] was compatible with the organometallic moiety. Furthermore, it was highly stereoselective, giving a 93/7 mixture of epoxides (–)-**201** and (–)-**202**. Decomplexation of the major isomer gave the optically pure intermediate (+)-**203** which had already been used for the synthesis of the natural AF IIc toxin. The unnatural (8*R*,9*R*)-diastereoisomer was also obtained by decomplexation of (–)-**202**.

The isomeric toxins were prepared by similar strategies from the (*E,Z*)-complex (+)-**205**. This compound was obtained in optically active form by hydrolysis of the complexed pentadienyl cation, as described before.[20] Several functional group

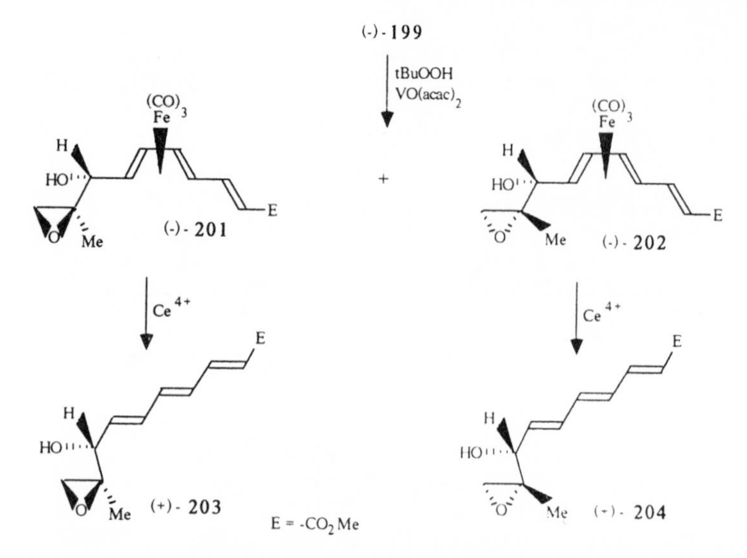

transformations led to triene aldehyde (+)-**207** via the complexes (–)-**197** and (+)-**206**. Grignard addition gave, as before, a mixture of Ψ-*exo*-alcohol (+)-**208** and its Ψ-*endo*-isomer (+)-**209**. Epoxidation of (+)-**209** was highly stereoselective, giving only (+)-**210** which after decomplexation gave (–)-**211**, the enantiomer (8*S*,9*R*)- of the intermediates for the AF IIa toxin. It is interesting to note that the complexes were stable during all these syntheses and *showed no tendency toward isomerization or toward bond shift.*

The synthesis of the third isomer also started from (–)-**197** but a different sequence of reactions had to be used. Preliminary experiments have established that Wittig-type reactions with a complex bearing the formyl group *on the Z double bond*, such as **212**, occurred with isomerization to an (*E,E,E*)-polyene **214**.[20] Thus, the triene system had to be prepared later in the synthesis (Scheme 28). After Grignard addition and separation of the two diastereoisomers, alcohol (–)-**215** is decomplexed giving the (*E,Z*)-diene with the (*R*) required absolute configuration. Elaboration of the (*E,Z,E*)-triene followed by the epoxidation gave the desired

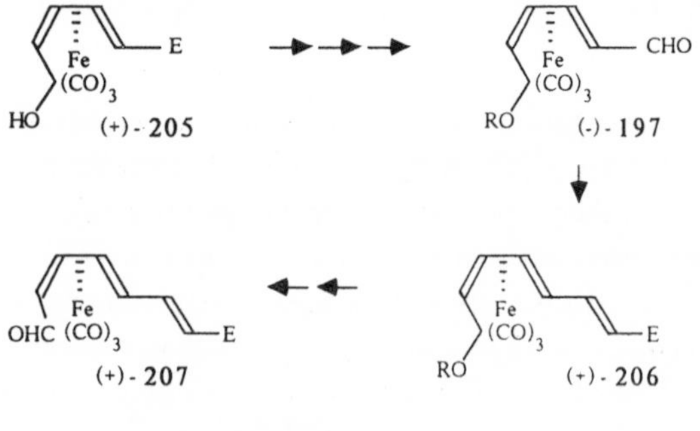

R = $-SiMe_2tBu$, E = $-CO_2Me$

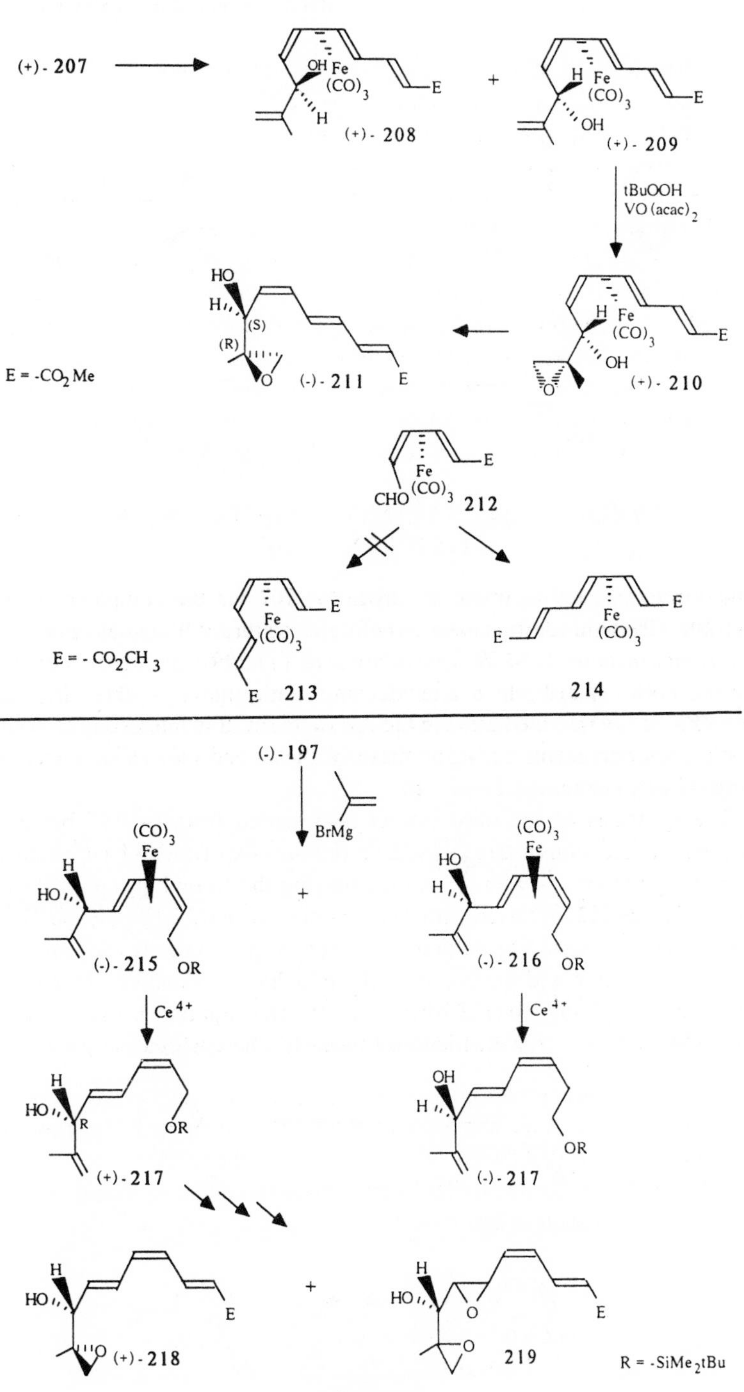

(+)-207 → OH Fe (CO)$_3$ H E (+)-208 + H Fe (CO)$_3$ OH E (+)-209

tBuOOH
VO(acac)$_2$

HO H (S) (R) O (-)-211 E ← H Fe (CO)$_3$ OH O E (+)-210

E = $-CO_2Me$

Fe (CO)$_3$ CHO E 212

Fe (CO)$_3$ E E 213   E Fe (CO)$_3$ E 214

E = $-CO_2CH_3$

(-)-197

BrMg

(CO)$_3$ Fe H HO OR (-)-215 + (CO)$_3$ Fe HO H OR (-)-216

Ce^{4+} Ce^{4+}

H HO R OR (+)-217   OH H OR (-)-217

H HO E O (+)-218 + H HO O O E 219

R = $-SiMe_2tBu$

Scheme 28.

intermediate (+)-**218** together with a small amount of bisepoxide **219**. Again, this is a formal total synthesis since preparation of AK and AF IIb toxins starting from (+)-**218** have already been described.

In conclusion, these diene–tricarbonyliron complexes are highly versatile intermediates for the synthesis in chiral, non-racemic, form of various polyunsaturated natural products. In most cases, these new routes are flexible enough to offer a good potential for the preparation of (1) the corresponding labeled compounds, (2) the stereoisomers of these natural derivatives, or (3) their structural analogues. This is obviously of much importance for the studies in relation with biological problems.

Probably, the most original aspect in the use of these complexes is that they allow a *simultaneous control of the stereogenic center(s)* **and** *the geometry of the polyenic system*. This "semi-immolative" asymmetric synthesis appears then as an interesting and useful alternative to the natural chiral pool approach.

IV. STEREOCONTROLLED SYNTHESES OF CARBO- AND HETEROCYCLES

In the preceding part, the organometallic complex was used as a chiral protective group for a diene unit *which, at the end of the synthesis, was fully integrated into the polyenic skeleton of the desired compound.* Of course, it is possible to consider numerous other potential applications of these complexes and corresponding dienes. Thus, this part will be dealing with new syntheses of carbo- and heterocycles and special attention will be paid to three different aspects which have been studied during the last years:

- use of diene–tricarbonyliron complexes as stereodirecting groups during intermolecular cycloadditions,
- syntheses via intramolecular reactions of the free dienes obtained after decomplexation,
- possible use of the $Fe(CO)_3$ group to induce a very long distance remote stereocontrol in macrolide type molecules.

A. Pyrethroids and Other Cyclopropanes

Cyclopropanes are interesting compounds, both as natural products or as synthetic intermediates. Thus, it is not surprising that all their methods of preparation have been studied within the context of organoiron chemistry.

Dichlorocarbenes can be used under phase transfer conditions giving corresponding cyclopropanes such as **220**[131] or **221**.[10b]

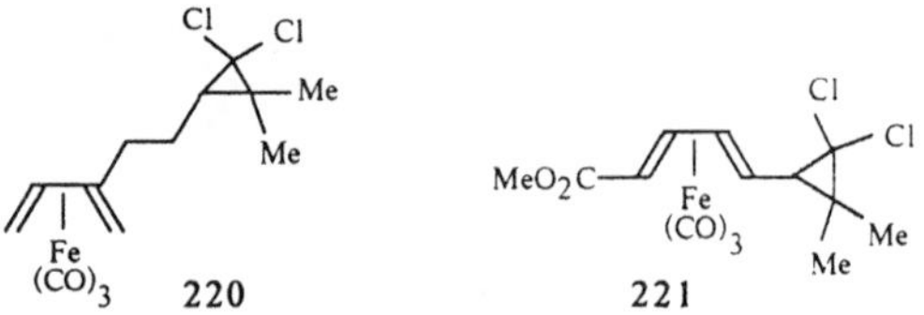

Cl Cl Me Me Fe $(CO)_3$ 220

MeO_2C Fe $(CO)_3$ Cl Cl Me Me 221

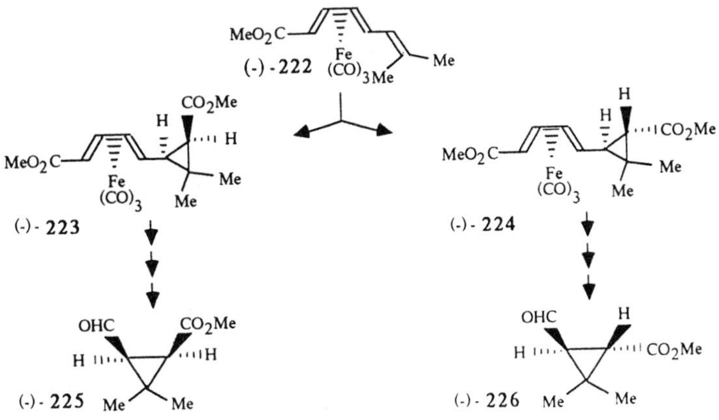

Scheme 29.

More interestingly, the carbene from methyl diazoacetate reacts with optically pure complex (–)-**222** giving a 1:1 mixture of cyclopropanes (–)-**223** and (–)-**224** (Scheme 29). After chromatographic separation, decomplexation and ozonolysis, the optically pure hemicaronaldehydes (–)-**225** and (–)-**226** were obtained.[132]

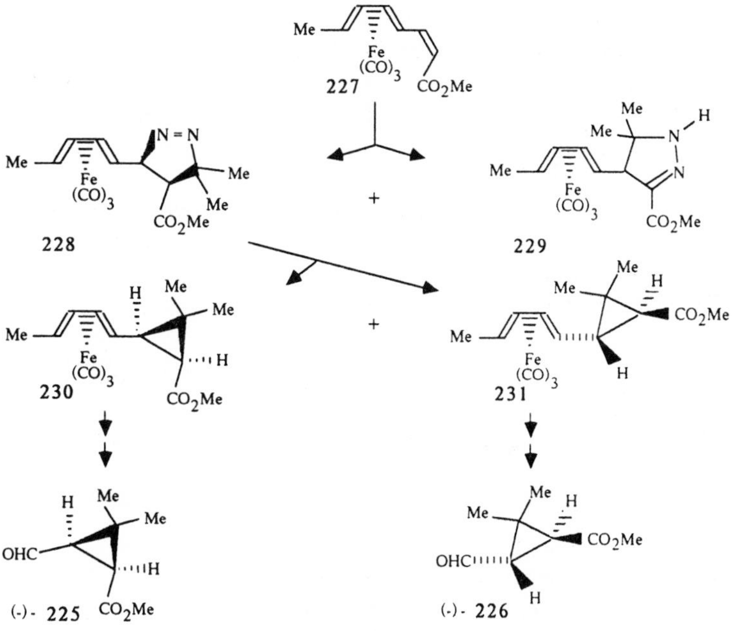

Scheme 30.

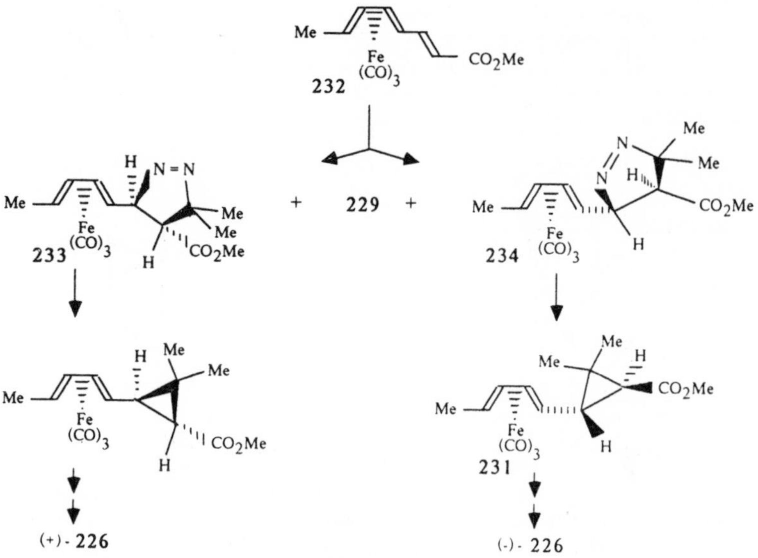

Scheme 31.

These important intermediates for the synthesis of pyrethroids, were also prepared by (1,3)-dipolar cycloadditions of 2-diazopropane[133] (Scheme 30). Reaction on the (*Z*)-olefin **227** gave adduct **228** together with Δ^2-pyrazoline **229**. Thermolysis of **228** yielded cyclopropanes **230** and **231**. After separation by chromatography, decomplexation and ozonolysis, the pure hemicaronaldehydes (–)-**225** and (–)-**226** were obtained.

In the case of the (*E*)-olefin **232** (Scheme 31), the cycloaddition was only stereoselective giving a mixture of pyrazolines **233** and **234**, together with **229**. The same reactions as before gave also pure hemicaronaldehydes. Similar results were also obtained in the case of complex **235** substituted in position 2.[11]

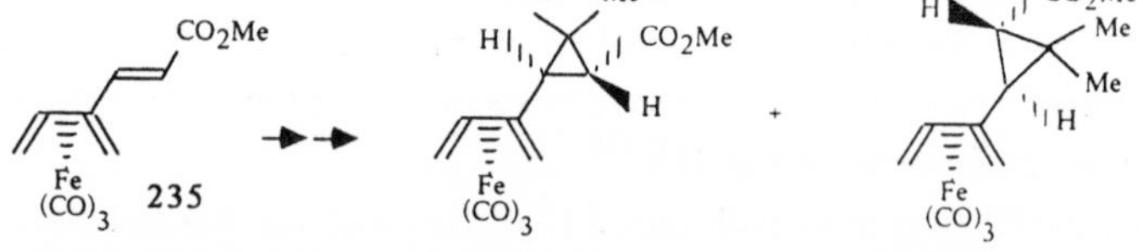

Optically pure electrophilic cyclopropanes, which are of interest in synthesis as "homo Michael" reagents, have been prepared from complex (–)-**236** by two routes (Scheme 32).[10] Diazomethane cycloaddition gave a single pyrazoline **237** which, by treatment with $Ce(NH_4)_2(NO_3)_6$, lead directly to the cyclopropane **238**. Ozonolysis gave then the (2*S*)-formyl derivative (+)-**239**. Sulfur ylid addition on

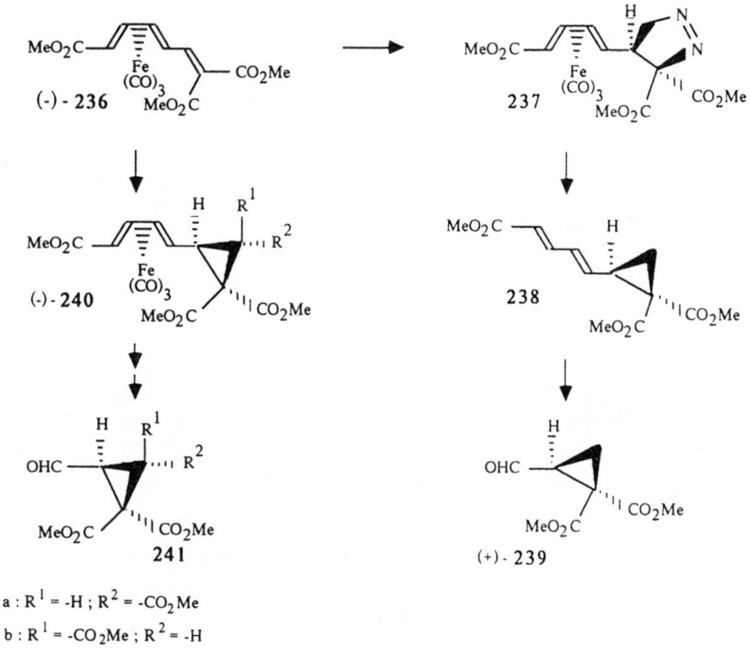

a : R^1 = -H ; R^2 = -CO_2Me

b : R^1 = -CO_2Me ; R^2 = -H

Scheme 32.

(–)-**236** also gave, via (–)-**240a** and (–)-**240b**, the optically pure cyclopropanes (–)-**241a** and (–)-**241b**.

It is important to note here that, through the sequence decomplexation plus ozonolysis, these diene–tricarbonyliron complexes appear as *synthetic equivalents of a chiral formyl group.*

B. Heterocycles and Cyclohexenes

Various type of heterocycles could also be prepared in the presence of the tricarbonyliron unit: the β-lactam **242** was obtained by a [2 + 2] cycloaddition and the pyrrolidines **243a,b** by reaction of an azomethine ylid.[10b] Nitrone **244** was the first example of a (1,3)-dipole directly linked to a diene–tricarbonyliron complex; it led to isoxazolidines such as **245**.[134]

Nitrile oxides also react with type **247** olefins and this was used in an original synthesis of (+)-(*S*)-6-Gingerol (Scheme 33).[135] In agreement with preliminary results obtained on simple models,[35] cycloaddition of the *in situ* generated nitrile oxide **246** on (+)-**247** gave a 89:11 mixture of (+)-**248** and **249**.

After decomplexation the major diastereoisomer was hydrogenated giving gingerol directly in optically active form (96% ee). Here, it is interesting to point out

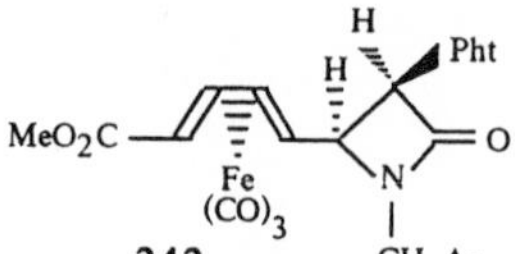

242

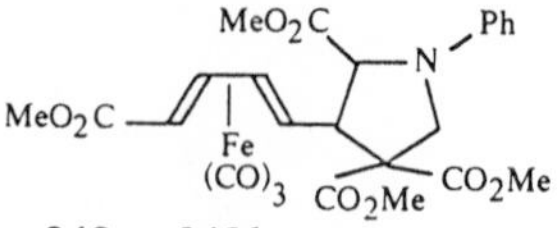

243a , 243b

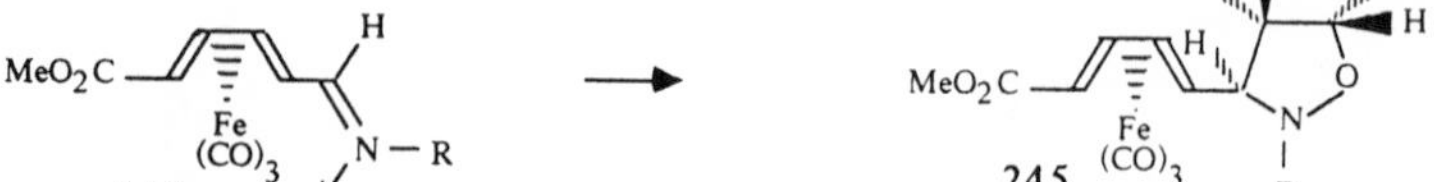

244 R = -Me, -Ph

245

a *new use of the organometallic complex as a precursor of saturated hydrocarbon chains with possible applications for the synthesis of labeled compounds.*

Functionalized cyclohexenes were also prepared by Diels–Alder reactions on linear polyenes **250**, selectively protected by $Fe(CO)_3$.[37] These highly stereoselective reactions appear presently limited to strongly electrophilic olefins, although an efficient catalysis by $TiCl_4$ was observed in the case of **250c**. Selectively complexed tetraene **252** also reacted with *N*-methylmaleimide by an *endo* approach and *anti* with regard to the $Fe(CO)_3$ to give **253** as a single adduct.[37]

Also, worthy of note was the successful hetero Diels–Alder reaction involved in a key step of the 5-HETE synthesis.[21] Thus, it appears that these diene-tricarbonyliron complexes are compatible with the main classes of cycloadditions commonly used in synthesis. Although the preceding results were obtained on a relatively small number of examples, several points appear particularly worthy of note:

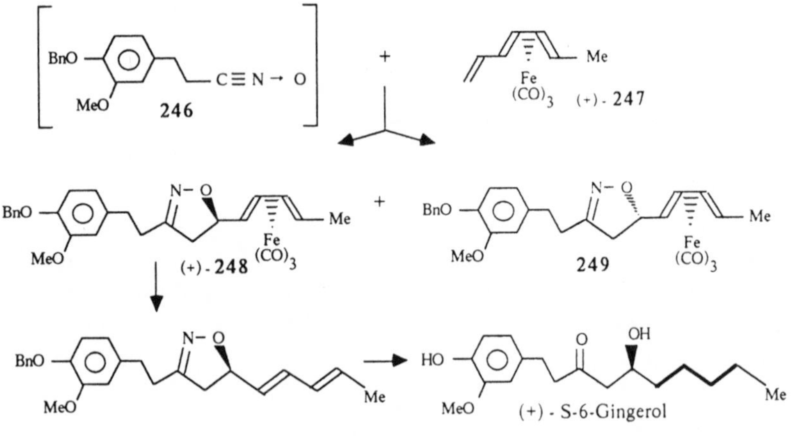

Scheme 33.

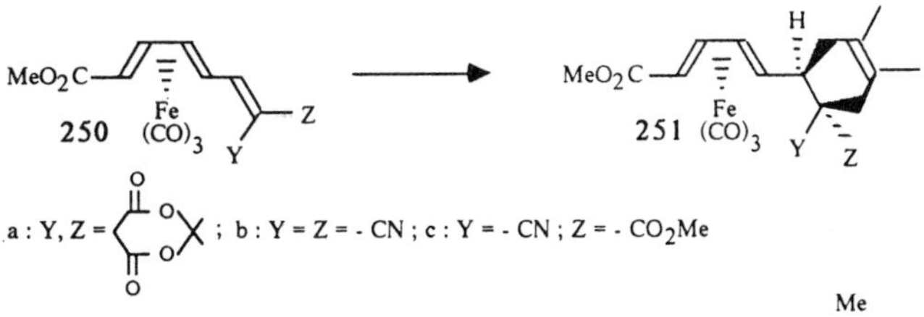

a : Y, Z = ; b : Y = Z = - CN ; c : Y = - CN ; Z = - CO_2Me

236 : Y = Z = - CO_2Me

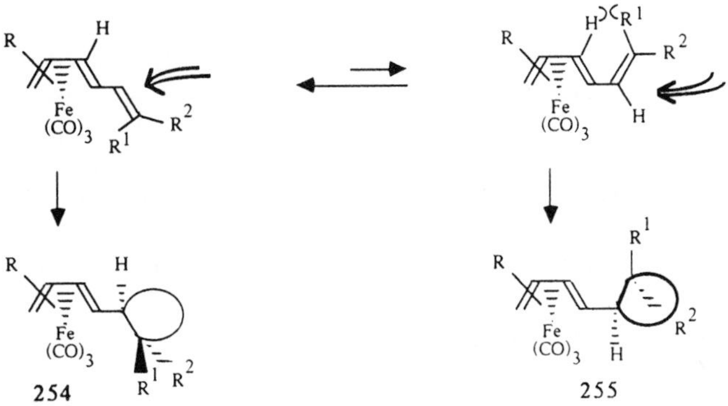

1. No isomerization of the complexes by $Fe(CO)_3$ bond-shift was observed, even during reactions occurring at higher temperature (≈ 80 °C).
2. No racemization of the optically active complexed polyenes could be detected under the same reaction conditions.
3. The stereoselectivity of the cycloadditions has been established unambiguously in most of the preceding reactions, either by X-ray data or by chemical correlations. It can be easily rationalized taking into account two key factors: (a) the cycloadditions occur always *anti* to the bulky $Fe(CO)_3$ group, (b) the ground state structure, and especially the *s-cis* ↔ *s-trans* equilibrium had to be considered.

If $R^1 \neq H$, this equilibrium is strongly shifted to the left and only reactions on the *s-trans* conformer are detected. If $R^1 = H$, the difference is not large enough and reactions occur on *both* conformers giving mixtures (usually around 9:1) of type **254** and type **255** adducts.

C. Syntheses Using the Free Diene Units

Another very attractive aspect of these complexes has to be considered next: after stereocontrolled reactions on the lateral chains, followed by the decomplexation process, free diene are recovered; *these dienes offer interesting opportunities to build more elaborated molecules via different type of inter- or intramolecular reactions.*

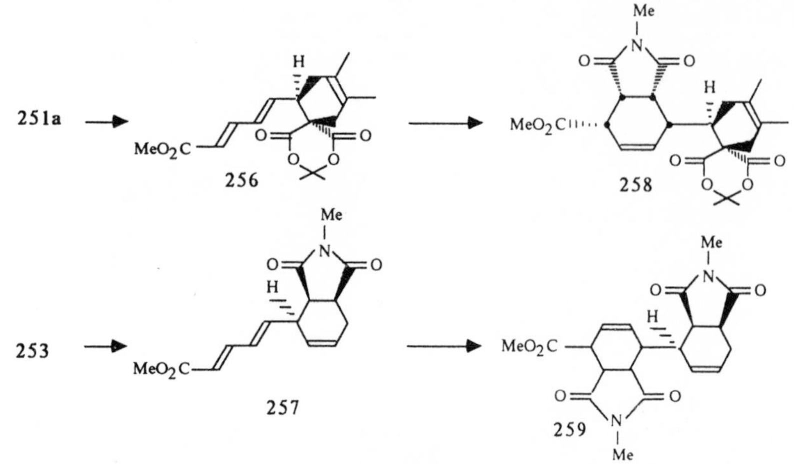

Diels–Alder reactions, for instance, have been described:[37] dienes **256** or **257** (obtained by decomplexation of **251a** and **253**) add to *N*-methylmaleimide giving new polycylic systems **258** and **259**. During these *tandem* Diels–Alder reactions the

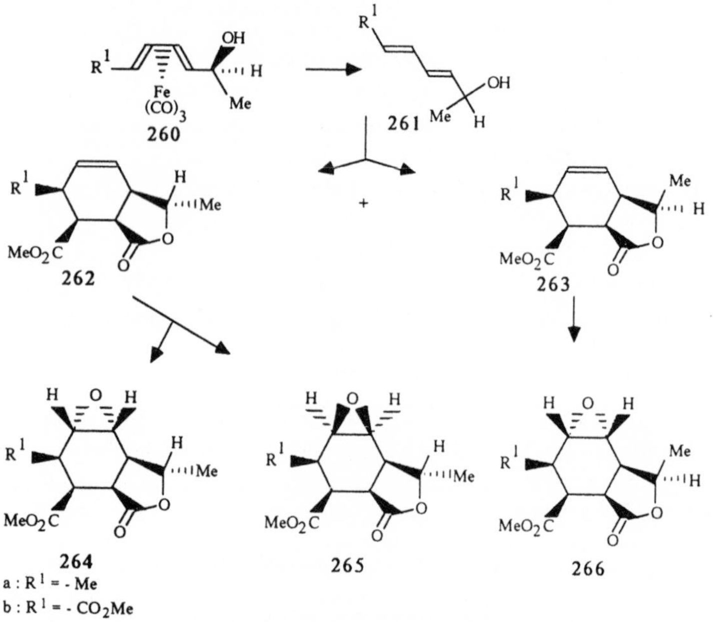

a : R^1 = - Me
b : R^1 = - CO_2Me

Scheme 34.

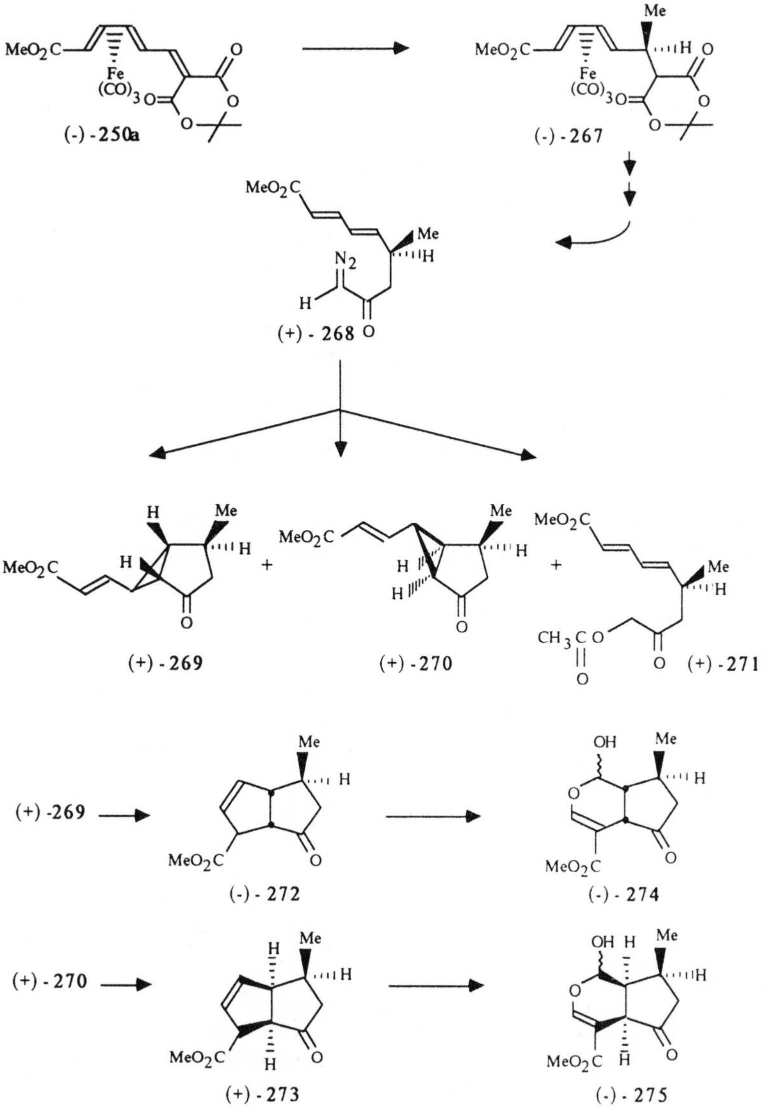

Scheme 35.

complex is used to control not only the two successive reaction sites but also the stereochemistry of the cycloadducts, with possible extension to optically active compounds.

Intramolecular Diels–Alder cycloadditions have been also considered (Scheme 34).[136] Reaction of dienol **261** with maleic anhydride, followed by an esterification, gave a 1:1 mixture of adducts **262** and **263**. Epoxidation of **263** was highly stereoselective giving only **266** while its isomer **262** gave mixtures of **264** and **265** (95:5) in the case of **262a** and (80:20) in the case of **262b**. Thus, it appears possible to prepare, by a very short sequence of reactions, polyfunctionalized lactones in controlling the configuration of the seven stereogenic centers.

An intramolecular carbene-type addition has also been used during the first total synthesis of (–)-verbenalol (Scheme 35).[36] The (1,4)-addition of methyl Grignard on (–)-**250a** gave by an exclusive *exo* attack the complex (–)-**267** with the desired (*S*) absolute configuration at the newly created stereocenter. Decomplexation and functional group transformation gave the key diazoketone (+)-**268**. Intramolecular reaction of the corresponding carbene gave a 1:1 mixture of cyclopropanes (+)-**269** and (+)-**270**, with a small amount of (+)-**271**. The cyclopropane (+)-**269** was transformed first into the diquinane (–)-**272** and then to the desired verbenalol (–)-**274**. Epiverbenalol (–)-**275**, which is closely related to the rare iridoids with the unusual (5*R*,8*S*,9*S*)-absolute configuration,[137] was prepared in the same way from (+)-**270**.

If the free diene has an electrophilic character, due for instance to the presence of an ester group, it becomes possible to use also intramolecular (1,6)-Michael type reactions. This was applied for instance to the synthesis of γ-lactones (–)-**277** and (–)-**278** by intramolecular addition of the anion derived from malonate (–)-**276**:[138]

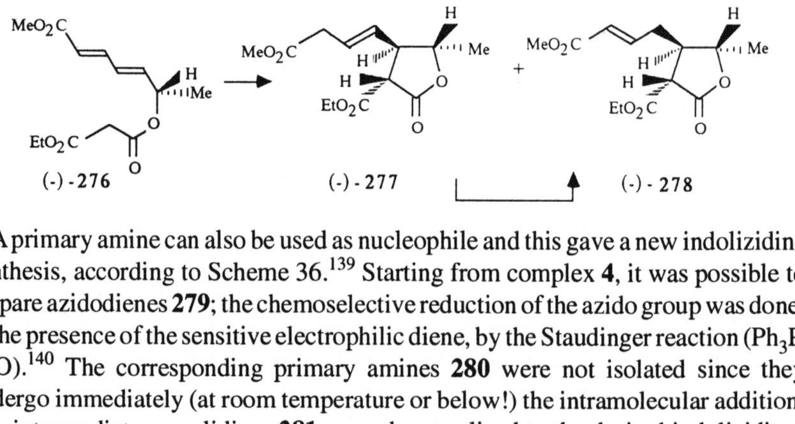

(-)-276 (-)-277 (-)-278

A primary amine can also be used as nucleophile and this gave a new indolizidine synthesis, according to Scheme 36.[139] Starting from complex **4**, it was possible to prepare azidodienes **279**; the chemoselective reduction of the azido group was done, in the presence of the sensitive electrophilic diene, by the Staudinger reaction (Ph_3P, H_2O).[140] The corresponding primary amines **280** were not isolated since they undergo immediately (at room temperature or below!) the intramolecular addition. The intermediate pyrrolidines **281** were then cyclized to the desired indolizidines **282** and **283**.

Although more examples are needed to discuss in detail the stereoselectivity of the (1,6)-Michael addition, it clearly appears to depend upon the nature of the R^1 and R^2 groups. Preliminary experiments have shown that this methodology could be extended to quinolizidine synthesis.[139b] Furthermore, type **279** azidodienes have also been used recently by Hudlicky to prepare some pyrrolizidines.[141] Thus, it is obvious that these organometallic complexes have some interesting potential in alkaloid synthesis.

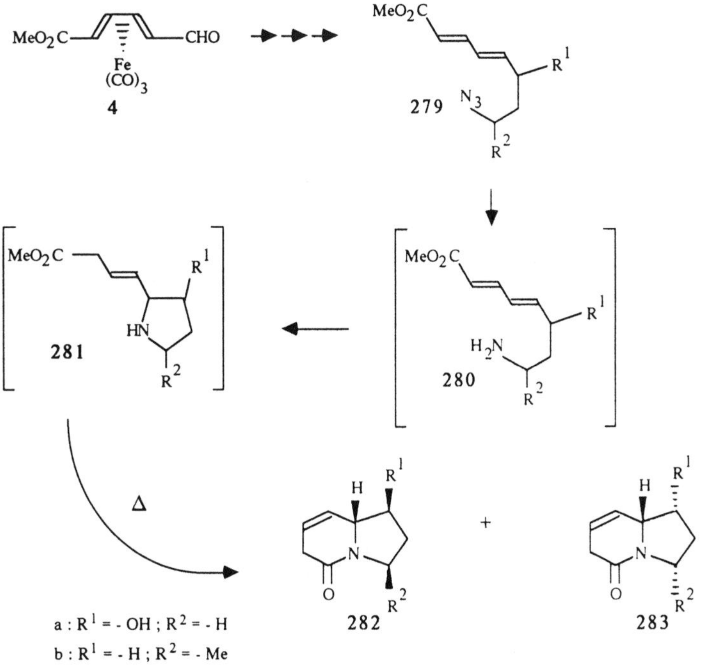

a : R^1 = - OH ; R^2 = - H
b : R^1 = - H ; R^2 = - Me

Scheme 36.

D. Macrocycles

Until now, the $Fe(CO)_3$ group has been utilized to control the configuration of new stereogenic centers *only* in close vicinity to the organometallic complex. The next step then was to check the possible use of the bulky $Fe(CO)_3$ moiety to induce a *remote stereochemistry control at a very long distance (5 bonds or more).* This is a fundamental problem in chemistry which has rarely been addressed until now.

In the case of medium-sized cycles, remote stereocontrol (up to distances of 3 to 4 bonds) has been experimentally established by Still;[142] furthermore, molecular mechanics calculations successfully rationalized most of these results.

The problem is considerably more difficult in the macrocyclic area. Several examples,[143] including an elegant synthesis of 3-deoxy rosaranolide,[144] have demonstrated the possibility to also use a remote stereocontrol. However it appears particularly difficult, at least presently, to rationalize these results[144,145] and to *use them in a predictive fashion* for stereocontrolled syntheses of macrocycles. Thus, it appeared that organoiron complexes of macrolides (of general type **284**) would be ideal models to study such a problem for several reasons:

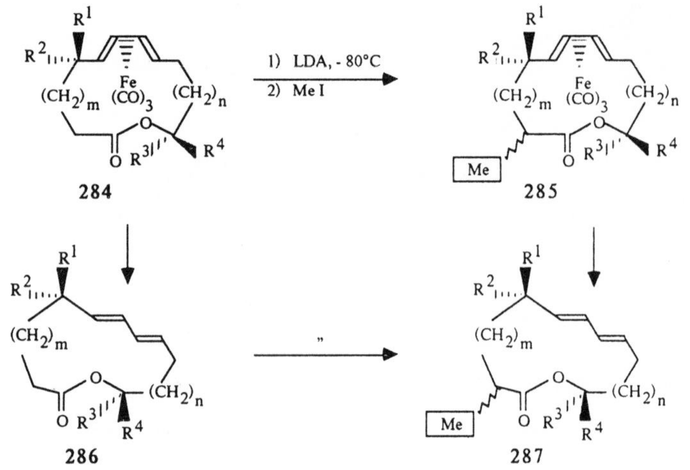

a : R^1 = - OMe ; R^2 = R^3 = R^4 = - H ; m = n = 5
b : R^1 = R^3 = R^4 = - H ; R^2 = - OMe ; m = n = 5
c : R^1 = - OMe ; R^2 = R^4 = - H ; R^3 = - Me ; m = n = 5
d : R^1 = - OMe ; R^2 = R^3 = - H ; R^4 = - Me ; m = n = 5
e : R^1 = - OMe ; R^2 = R^3 = R^4 = - H ; m = n = 3

Scheme 37.

- the diene-tricarbonyliron moiety, with its *s-cis*-conformation of the diene, will afford more rigidity to the macrocycle skeleton;
- the $Fe(CO)_3$ group is very bulky and then should be very useful in order to test the limits of the method; and
- the organometallic moiety is chiral and could lead, ultimately, to a synthesis of macrolactones in optically active form. This is of particular interest since polyenic macrolides are difficult target molecules from the synthetic point of view and useful antibiotics in medicine.[146]

During the first exploratory part of this program only the alkylation reactions of enolates derived from type **284** complexes were studied, according to general Scheme 37.[27] The synthesis of these complexed macrolactones was not an easy problem and the preparation of **284a** is a representative example (Scheme 38). We chose to start with complex **4**, accessible in both optically active forms, and to add the two lateral chains via successive (1,2)-additions of functionalized nucleophiles. One of the main difficulties was the formation of the $-CH_2CH_2-$ bond vicinal to the complex. After several unsuccessful attempts, including catalytic reduction of double bonds, it was solved via a reductive deoxygenation (Et_3SiH, CF_3CO_2H)[27] of an alcohol function α to the organometallic complex (**288** → **289**). Functional group transformations gave **292** and then hydroxy acid **293**. Macrolactonization was efficient (78% yield) under high-dilution conditions.[147] The other models were

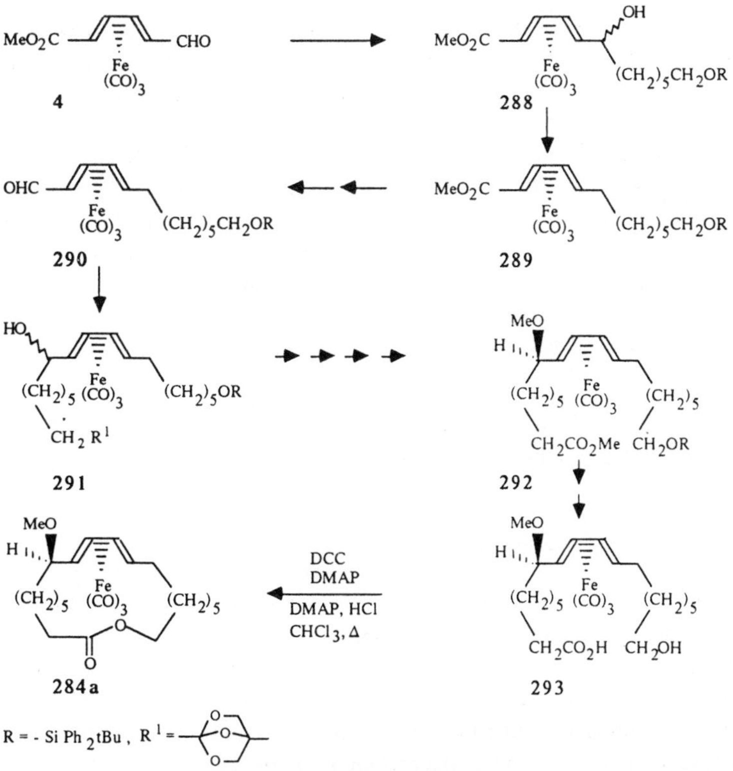

Scheme 38.

prepared using closely related sequences and corresponding free dienes were easily obtained by decomplexation; these last derivatives are important standards to estimate the intrinsic role of $Fe(CO)_3$ on the remote stereocontrol.

Alkylations were done under reaction conditions reported to be kinetically controlled: formation of the enolate using LDA at –80 °C and reaction with an excess of CH_3I at –80 °C in the presence of HMPA. The configuration of the adducts were unambiguously established using combination of X-ray data and chemical correlations.

The results depended mainly upon the nature of the R^3 and R^4 substituents α to the lactone oxygen and these macrocycles could be divided into two classes:

1. Macrolactones without substituent in that position ($R^3 = R^4 = H$): (a) as expected, no diastereoselectivities were observed during the alkylation of the uncomplexed derivatives **286a**, **286b**, **286e**, giving exactly 1:1 mixtures of stereoisomers of **287a**, **287b**, **287e**; and (b) *very low* diastereoselectivities were obtained in the case of the 20-member complexed lactones; a 58:42

mixture of stereoisomeric **285a** was observed starting from **284a**. A similar result (55:45) was obtained with **284b**, and this was a confirmation that the stereochemistry of the remote methoxy group has no influence in that case. The ring size also does not appear to be a significant factor since smaller macrolactone **284e** (16 membered) yielded a 42:58 mixture of **285** stereoisomers.

2. Macrolactones with a methyl substituent α to the oxygen (R^3 or R^4 = –Me): (a) a *low* selectivity was observed for the free diene **286c**, giving a 65:35 mixture of lactones **287c**; and (b) a *complete stereoselectivity* was obtained for its complexed counterpart **284c**, giving exclusively the macrolactone **285c**, with the two methyl in *cis* position. This result appears very important since it established *that a complete remote stereocontrol can exist over a seven bonds distance*!

The structure of **285c** has been established by X-ray analysis and gave the following selected distances in the solid state (Scheme 39):

$$\text{Fe–C}_1 : 8.260\ \text{Å} \qquad \text{Fe–Me}_2 = 9.394\ \text{Å}$$

$$\text{O}_1\text{–C}_1 : 5.389\ \text{Å} \qquad \text{O}_1\text{–Me}_2 : 6.839\ \text{Å}$$

Thus, it appears quiet impressive that such a complete stereocontrol can be induced by a group so remote from the reacting center!

The stereochemistry of the methyl group α to the oxygen appears of less importance since **284d**, the diasteroisomer of the preceding lactone **284c**, also gave a single alkylated lactone **285d**.[27]

A tentative rationalization for the preceding results and especially the very high selectivities observed in the last examples can be proposed. It takes into account the two key factors: *the local conformation of the enolates and the remote stereocontrol induced by the bulky* $Fe(CO)_3$ *group.*

For these enolates, two conformers, called *endo* and *exo*,* have to be considered. In the case of the enolates bearing the γ methyl substituent, the conformational equilibrium is shifted to the right due to the strong (1,3)-interaction between the oxygen and the methyl group. Peripheral attack of the electrophile then occurs exclusively from the face opposite to $Fe(CO)_3$ in the *exo*-enolate leading to **285c**, as the only isolated product. This is in full agreement with the results obtained in the case of medium sized cycles.[142]

The lower selectivity observed in the case of the corresponding decomplexed lactone clearly demonstrates the key role of the bulky $Fe(CO)_3$ group in this remote stereocontrol. Reactions on *both* faces of the *exo*-enolate in the case of **286c** would appear as an attractive possible explanation for the 65:35 mixture obtained; however, is not possible to exclude in that case a different position in the conformational equilibrium and reaction on both the *endo*- and the *exo*-enolates.

---

* *endo* means that the oxygen enolate is on the same face of the complex as the $Fe(CO)_3$ group; *exo* means that the oxygen enolate is on the opposite face of the complex as the $Fe(CO)_3$ group.

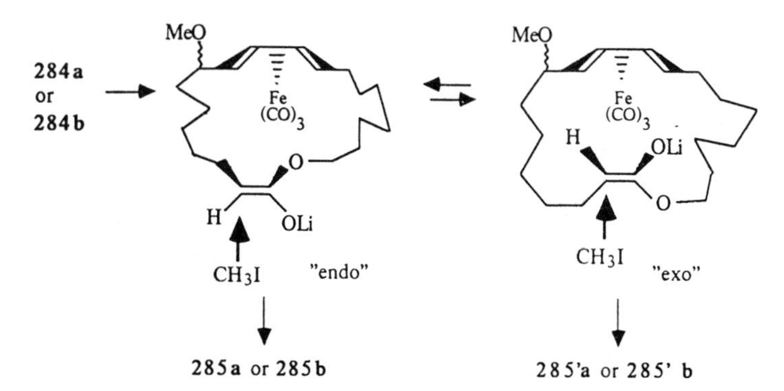

In the case of the lactones unsubstituted on the carbon α to the oxygen, the corresponding enolates should be in rapid equilibrium between the *endo* and the *exo* forms. Then, peripheral attack on both conformers should lead to the mixture of diastereoisomers experimentally observed.

Another important point has to be noted here: these last results exclude an eventual interaction between the enolate and the $Fe(CO)_3$ group as a key factor for the remote stereocontrol. Such an interaction should stabilize the *endo*-enolates,

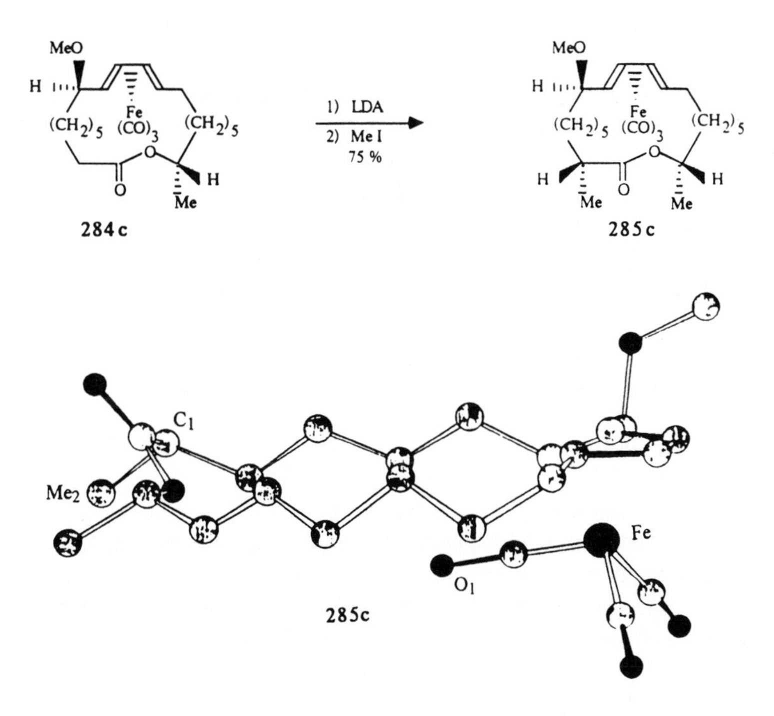

Scheme 39.

whatever the substituents of the lactones, and then induce a preferential attack of CH_3I, leading to **285a** or **285b** contrary to the results which have been obtained.

A great amount of work remains to be done in this area including shorter, more efficient syntheses of the complexed macrolactones and extension to other types of reactions such as (1,4)-additions or cycloadditions on (α,β)-unsaturated lactones. However, these preliminary results appear very encouraging and indicate that there is good potential for the use of these diene–tricarbonyliron complexes for very long distance remote stereocontrol. In that case, it should be of *predictive value*, provided that the local conformational problems at the reacting center can be dealt with.

V. APPENDIX: SYNTHETIC OPERATIONS THAT ARE COMPATIBLE WITH THE η^4-DIENYL TRICARBONYLIRON MOIETY

The synthetic operations that follow show the diversity of the chemistry and the great synthetic potentialities of the iron-complexed moiety, providing these two essential factors are taken into consideration:

1. The organometallic unit must be used during the considered synthetic step without any structural modification.
2. Because of its planar chirality, accent will be put on the reactions that use π-facial discrimination of an unsaturated unit induced by the organometallic unit.

Indeed, these reactions can occur in close vicinity of the complexed unit or, in some cases, far from it. Therefore, the numerous synthetic examples involving the iron-complexed moiety are registered and classified in order to give a ready access and detailed information about the reaction, the experimental conditions, the yields and stereoselectivity whenever possible, and the essential literature data. Key information is thus available for the synthetic chemist who would be eager to integrate this type of organoiron chemistry in his own synthetic strategies.

A. Friedel-Crafts Acylations

TYPE OF REACTION	FRIEDEL-CRAFTS ACYLATION OF IRON COMPLEXED DIENES 1/5	References
1.	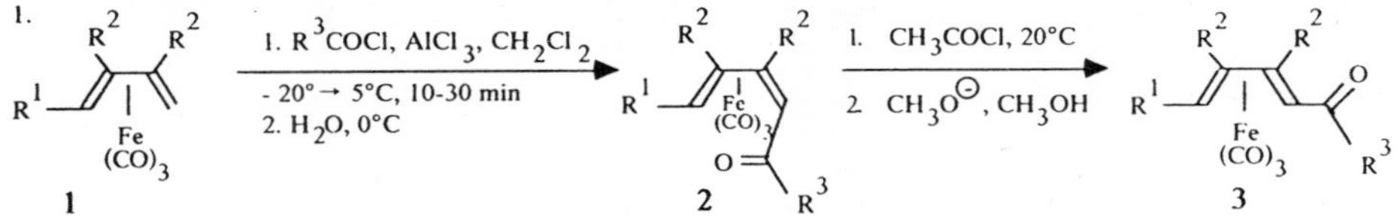 1 → 2: 1. R^3COCl, $AlCl_3$, CH_2Cl_2, - 20° → 5°C, 10-30 min; 2. H_2O, 0°C. 2 → 3: 1. CH_3COCl, 20°C; 2. $CH_3O^{\ominus}$, CH_3OH	41, 104. 148

R^1	R^2	R^3	Yield of the two steps 1 → 3
-H	-H	- CH_2Cl	43 %
-H	- CH_3	- CH_2Cl	62 %
-H	- CH_3	- $CHClCH_3$	50 %
-H	- CH_3	- CH_2CH_3	88 %
-H	-H	- CH_3	60 %
- CH_3	-H	- CH_3	93 %
- CO_2CH_3	-H	- CH_3	83 %
- CO_2CH_3	-H	- $(CH_2)_4CO_2CH_3$	68 %
- CH_3	-H	-H	93 %
- CH_3	-H	- CH_2CH_3	90 %
- CH_3	-H	- Cyc C_6H_{11}	88 %

COMMENTS

The acylation is regiospecific towards the *less substituted carbon* of the iron complexed diene. The mechanism of this reaction has been discussed [149] and involved a π-allyl complex in which the iron atom is bonded to the oxygen of the acyl group. Therefore the *cis* iron complexed dienone 2 is always formed at *the first time*. Isomerization to the thermodynamically more stable *trans* iron complexed dienone can be performed using various sets of acidic or basic conditions [28,149,150].

TYPE OF REACTION	FRIEDEL-CRAFTS ACYLATION OF IRON COMPLEXED DIENES 2/5	References
2.	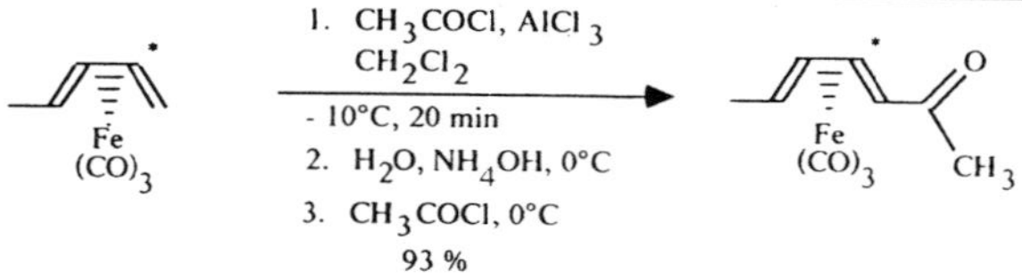 Fe $(CO)_3$ → 1. CH_3COCl, $AlCl_3$, CH_2Cl_2, - 10°C, 20 min; 2. H_2O, NH_4OH, 0°C; 3. CH_3COCl, 0°C; 93 % → O, Fe $(CO)_3$, CH_3	19
COMMENTS	The acylation reaction preserves the optical purity of the starting iron complexed diene.	
3.	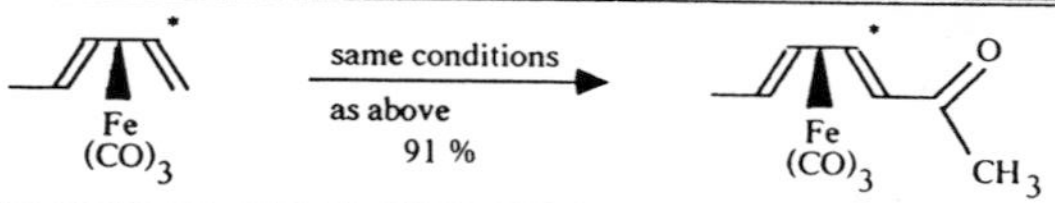 Fe $(CO)_3$ → same conditions as above; 91 % → O, Fe $(CO)_3$, CH_3	19
4.	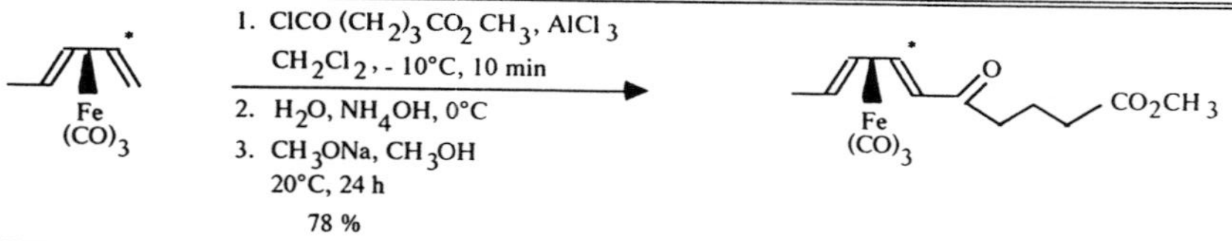 Fe $(CO)_3$ → 1. $ClCO(CH_2)_3CO_2CH_3$, $AlCl_3$, CH_2Cl_2, - 10°C, 10 min; 2. H_2O, NH_4OH, 0°C; 3. CH_3ONa, CH_3OH, 20°C, 24 h; 78 % → O, Fe $(CO)_3$, CO_2CH_3	19
5.	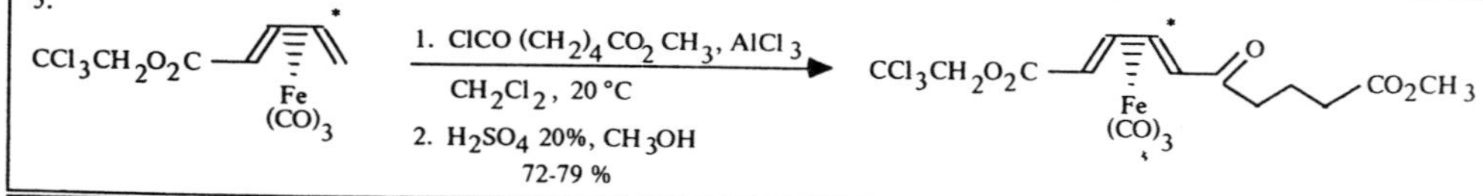 $CCl_3CH_2O_2C$, Fe $(CO)_3$ → 1. $ClCO(CH_2)_4CO_2CH_3$, $AlCl_3$, CH_2Cl_2, 20 °C; 2. H_2SO_4 20%, CH_3OH; 72-79 % → $CCl_3CH_2O_2C$, O, Fe $(CO)_3$, CO_2CH_3	105

TYPE OF REACTION	FRIEDEL-CRAFTS ACYLATION OF IRON COMPLEXED DIENES 3/5	References
6. 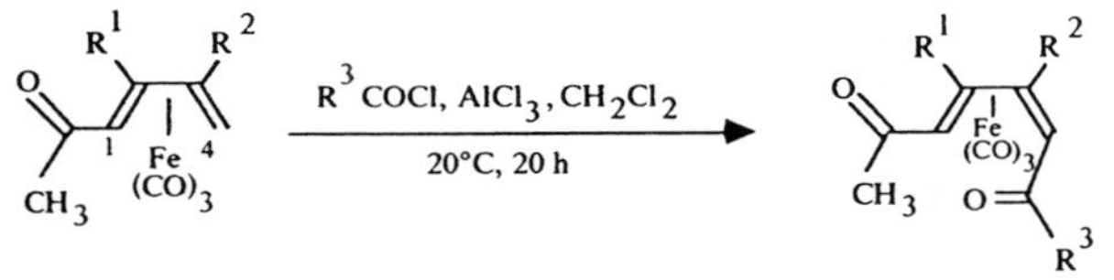	R^3COCl, $AlCl_3$, CH_2Cl_2 → 20°C, 20 h	41, 149
COMMENTS	In contrary of previous data [28,151], the second acylation, althought slow, is possible at the 4-position of the iron complexed dienones. This double acylation works well only with the substituted complexed dienones. Apparently, the isomerization process of the cis-trans complexed diene diones has not been described or studied.	

R^1	R^2	R^3	Yield
- H	- CH_3	- CH_3	86 %
- CH_3	- CH_3	- CH_3	98 %
- CH_3	- CH_3	- CH_2CH_3	93 %
- CH_3	- CH_3	- $CH(CH_3)_2$	75 %

TYPE OF REACTION	FRIEDEL-CRAFTS ACYLATION OF IRON COMPLEXED DIENES 4/5	References
7.		39

SiR^1_3 — 1 — 4 — $Fe(CO)_3$ — **1**

1. R^2COCl, $AlCl_3$, CH_2Cl_2, 0-20°C, 1 h
2. CH_3COCl, 20°C, 30 min

→ **2** + **3**

R^1	R^2	Yield of 2	Yield of 3
- $CH(CH_3)_2$	- CH_3	91 %	-
- CH_2CH_3	- CH_3	86 %	12 %
- CH_2CH_3	- CH_3	86 %	12 %
- CH_2CH_3	- $CH(CH_3)_2$	85 %	12 %
- CH_2CH_3	- $C(CH_3)_3$	67 %	8 %
- CH_2CH_3	- Ph	62 %	17 %
- CH_2CH_3	- $(CH_2)_7CO_2Me$	83 %	9 %

COMMENTS

As encountered in the previous example, the *cis* iron complexed dienones can be isolated if desired before the isomerization step. The acylation is *regiospecific* in the case of the bulkier - $Si(iPr)_3$ moiety. No ipso substitution of the trialkylsilyl group is ever observed. The *major* or *total* acylation at the position 4 of the complex **1** is best explained by steric interactions and by the β effect of the silicium atom.

8.

SiEt 3, Fe (CO)3, O, R 1, **1** — $R^2COCl, AlCl_3, CH_2Cl_2$; 20°C, 20 h → SiEt 3, Fe (CO)3, O, R 1, O, R 2, **2**

R^1	R^2	Yield of 2
- Me	- Me	88 %
- Me	- iPr	50 %
- Me	- $(CH_2)_4CO_2Me$	76 %
- iPr	- Me	40 %
- tBu	- Me	32 %
- Ph	- Me	55 %
- $(CH_2)_7CO_2Me$	- Et	90 %

References: 39

COMMENTS

As observed previously, the second acylation at the second terminal carbon of 1 is slower again without *any ipso* substitution of the triethylsilyl residu. The dienediones **2** are isolated. In one case (R^2 = - CH_3, R^1 = - CH_3), **2** has been isomerized to the corresponding trans-trans dienedione under basic conditions (CH_3ONa/CH_3OH, 20°C, 20 h : 92 %).

B. Palladium Mediated Acylations

TYPE OF REACTION	PALLADIUM MEDIATED ACYLATION OF IRON COMPLEXED DIENES 1/1	References
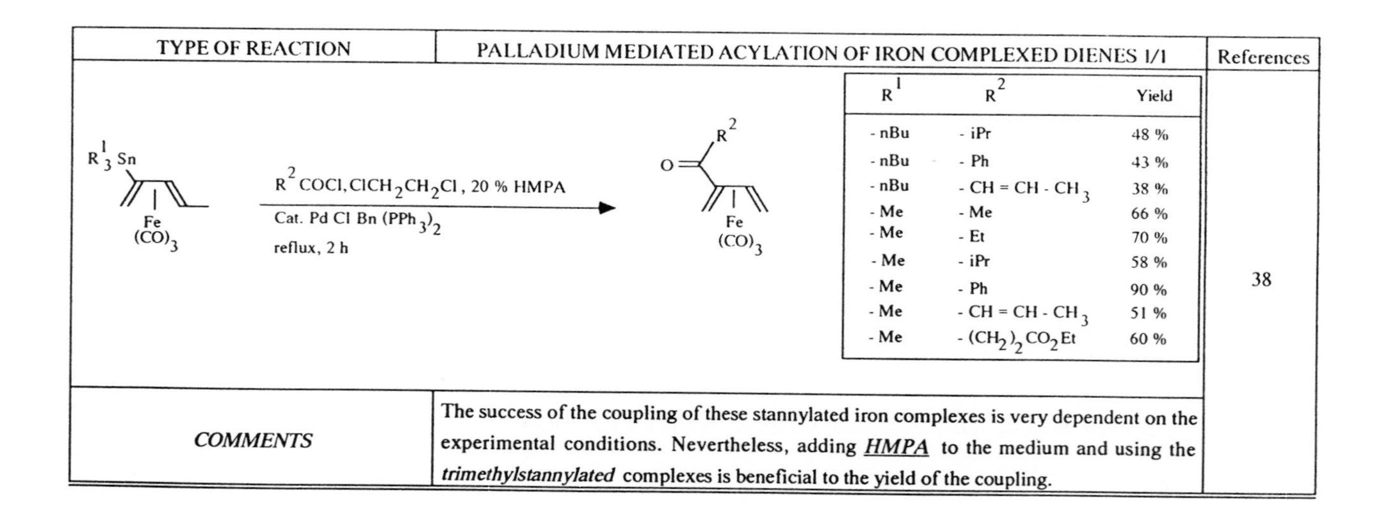		38

R^1	R^2	Yield
- nBu	- iPr	48 %
- nBu	- Ph	43 %
- nBu	- CH = CH - CH_3	38 %
- Me	- Me	66 %
- Me	- Et	70 %
- Me	- iPr	58 %
- Me	- Ph	90 %
- Me	- CH = CH - CH_3	51 %
- Me	- $(CH_2)_2CO_2Et$	60 %

COMMENTS	The success of the coupling of these stannylated iron complexes is very dependent on the experimental conditions. Nevertheless, adding ***HMPA*** to the medium and using the ***trimethylstannylated*** complexes is beneficial to the yield of the coupling.

TYPE OF REACTION	WITTIG REACTION USING IRON COMPLEXED PHOSPHONIUM SALTS 1/3	References
1. CH_3 ... Fe $(CO)_3$ **1** $\overset{\oplus}{PPh_3}$, $\overset{\ominus}{BF_4}$ 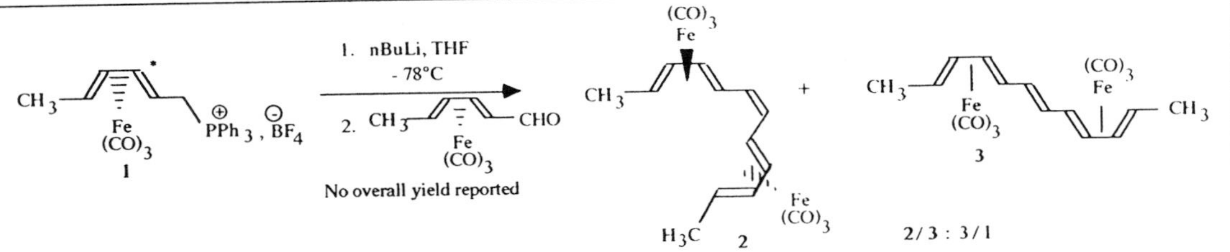	1. nBuLi, THF, - 78°C; 2. CH_3 ... Fe $(CO)_3$ ... CHO. No overall yield reported. Products: **2** + **3** (Fe $(CO)_3$ complexes). 2/3 : 3/1	65
COMMENTS	The organoiron complexed phosphonium salt 1, when deprotonated by nBuLi, forms a stable ylide. The two isomeric alkenes 2 and 3 are chromatographically separated. The *transoid* arrangement of the two organometallic units has been determined by an X-ray cristallographic analysis. This same arrangement is undetermined for 3. 1 does not show, in these experimental conditions, a high Z selectivity.	

TYPE OF REACTION	WITTIG REACTION USING IRON COMPLEXED PHOSPHONIUM SALTS 2/3	References
2. 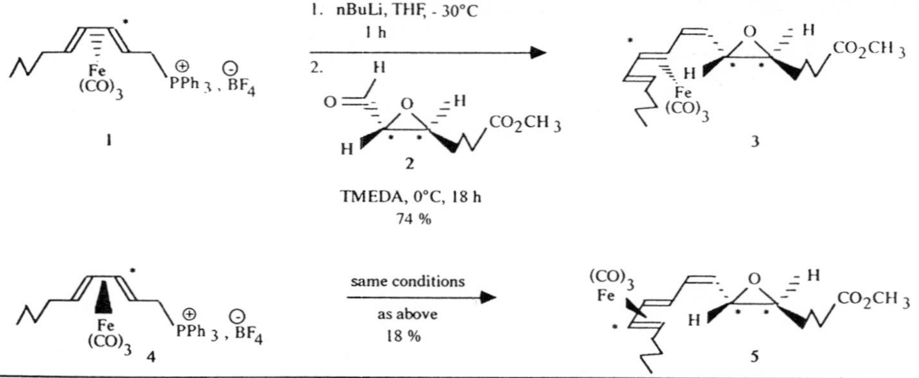 Fe $(CO)_3$ PPh_3, BF_4 1 Fe $(CO)_3$ PPh_3, BF_4 4	1. nBuLi, THF, - 30°C 1 h 2. O H O H CO_2CH_3 H 2 TMEDA, 0°C, 18 h 74 % → H Fe $(CO)_3$ O H CO_2CH_3 3 same conditions as above 18 % → $(CO)_3$ Fe H O H CO_2CH_3 5	13
COMMENTS	These two Wittig reactions are *unique* examples of the use of *chiral* iron complexed phosphonium salts. 1 and 2 are the matched pair (74 % yield in 3), 4 and 2 is the mismatched pair (18 % yield in 5). Having in mind the formation of the *erythro* oxaphosphetane in each case (Z alkene formed), these results are rationalized by an attack of a *transoid* ylid *anti* to the $Fe(CO)_3$ group and antiperiplanar to the epoxidic moiety. It is interesting to note that the Z alkene is obtained stereoselectively in these experimental conditions.	

TYPE OF REACTION	WITTIG REACTION USING IRON COMPLEXED PHOSPHONIUM SALTS 3/3	References
3. 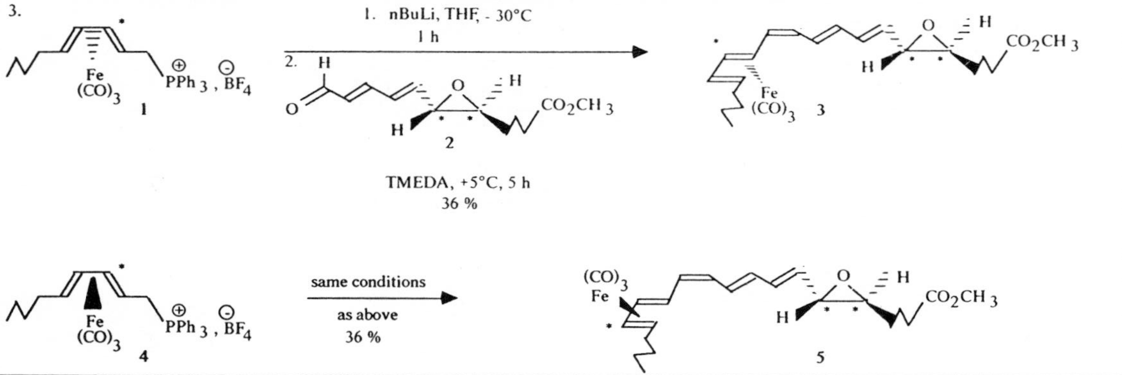 Fe $(CO)_3$ $\oplus$ PPh_3 , $\ominus$ BF_4 **1**	1. nBuLi, THF, - 30°C, 1 h 2. H, O, H, CO_2CH_3, H, **2** TMEDA, +5°C, 5 h 36 % → Fe $(CO)_3$ **3**, O, H, CO_2CH_3, H	118
Fe $(CO)_3$ $\oplus$ PPh_3 , $\ominus$ BF_4 **4**	same conditions as above 36 % → $(CO)_3$ Fe, O, H, CO_2CH_3, H, **5**	
COMMENTS	The formation of the Z alkenes **3** and **5** is again observed. In agreement with the reactivity model of **1** and **4** which is discussed in the previous example, the two chiral phosphonium salts show similar reactivity with the chiral epoxide **2**.	

D. Wittig-Horner Reactions

TYPE OF REACTION	WITTIG-HORNER REACTION USING IRON COMPLEXED PHOSPHONATES 1/2	References
1. EtO_2C, Fe $(CO)_3$, P, OEt, OEt, O	1. nBuLi, THF, - 78°C; 2. OHC, Fe $(CO)_3$, CO_2Et; 85 % → EtO_2C, Fe $(CO)_3$, $(CO)_3$ Fe, CO_2Et	152
COMMENTS	The olefination reaction is highly stereoselective since the only alkene formed is the E-geometry one. The relative arrangement of the two $Fe(CO)_3$ groups has not been determined.	
2. CH_3O_2C, Fe $(CO)_3$, P = O, OEt, OEt	1. NaH, DME/HMPT 0°C, 1 h; 2. $OHC-CH(OEt)_2$; 75 % → CH_3O_2C, Fe $(CO)_3$, $CH(OEt)_2$	32
COMMENTS	The *E* isomer is the only isolated alkene. No bond shif of the Fe $(CO)_3$ group is observed in these experimental conditions.	

TYPE OF REACTION	WITTIG-HORNER REACTION USING IRON COMPLEXED PHOSPHONATES 2/2	References
3.	1. NaH, DME/HMPT 0°C, 40 min; 2. RCHO 15°C, 2-5 h (1 → 2)	13a, 66
COMMENTS	The iron complexed phosphonate 1 forms a stable sodio anion when treated with NaH. The combination of NaH in a mixture DME/HMPT is the best experimental one. This sodio anion reacts chemoselectively with aldehydes (the ketones do not react at all) with a high stereoselectivity towards the formation of the E-alkene. This phosphonate anion is highly sensitive to steric effects (see the case of the ortho methoxy benzaldehyde). It is important to point out that there is no bond shift of the organometallic group during the olefination. This fact has been established by a combination of deuterium labelling and high-field ^{1}H-NMR.	

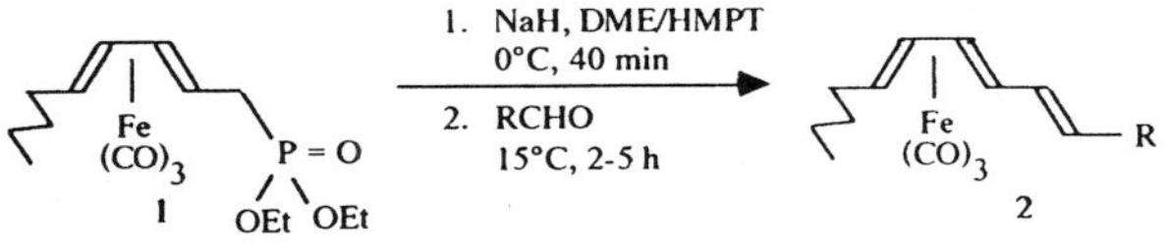

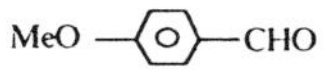

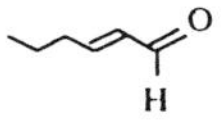

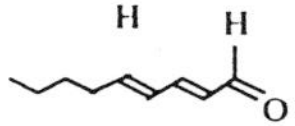

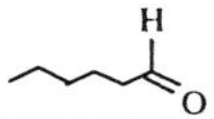

R CHO	Yield	% of E-alkene
Ph CHO	65 %	≥ 98 %
o-MeO-C$_6$H$_4$-CHO	0 %	
MeO-C$_6$H$_4$-CHO	50 %	95 %
(E)-2-hexenal	55 %	≥ 98 %
(E,E)-2,4-decadienal	45 %	≥ 98 %
hexanal	90 %	≥ 98 %

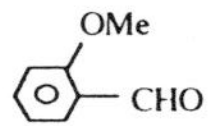

E. Oxidation Reactions

TYPE OF REACTION	OXIDATION REACTION OF IRON COMPLEXED DIENOLS 1/3	References
1.	oxidizing conditions (see the table)	19, 75

R^1	R^2	R^3	R^4	oxidizing reagent	R^5	Yield
$-CH(OCH_3)_2$	-H	-H	-OH	$Py\text{-}SO_3$, DMSO, Et_3N, 20°C	-H	67 %
$-CH_3$	$-CH_3$	-H	-OH	MnO_2, pentane, 20°C	$-CH_3$	30 %
$-CH_3$	$-CH_3$	-H	-OH	$Py\text{-}SO_3$, DMSO, Et_3N, 20°C	$-CH_3$	0 %
$-CH_3$	$-CH_3$	-H	-OH	$(COCl)_2$, DMSO, Et_3N, CH_2Cl_2	$-CH_3$	8 %
$-CH_3$	$-CH_3$	-H	-OH	Ac_2O, DMSO, 20°C	$-CH_3$	40 %
$-CH_3$	-OH	-H	$-CH_3$	DMSO, DCC, TFA - Py 20°C and 60°C	$-CH_3$	0 %
$-CH_3$	$-CH_3$	-H	-OH	PDC, Py, Bn, 20°C	$-CH_3$	60 %
$-CH_3$	-OH	-H	$-CH_3$	PDC, Py, Bn, 20°C	$-CH_3$	35 %
$-CH_3$	- ≡ - Ph	-H	-OH	MnO_2, pentane, 20°C	- ≡ - Ph	87 %
$-CH_3$	-OH	-H	- ≡ - Ph	MnO_2, pentane, 20°C	- ≡ - Ph	87 %
$-CH_3$	-Ph	-H	-OH	PDC, CH_2Cl_2	-Ph	75 %

TYPE OF REACTION	OXIDATION REACTION OF IRON COMPLEXED DIENOLS 2/3	References
COMMENTS	These oxidations steps are *always* delicate reactions since the organometallic unit is sensitive to oxidizing reagents. Decomplexation is the major competitive reaction. Nevertheless, major conclusions could be drawn from these results : the *Ψ-exo* alcohols are more stable and react more slowly, increasing the decomplexation process. The oxidation of secondary alcohols is more difficult than that of the primary ones. Yields rarely exceed 75 %. Good results are observed with activated alcohols (case of alkynyl and phenyl alcohols). So, any oxidation step in the iron complexed serie, needs to be carefully studied at the laboratory level.	
2. HO ... Fe $(CO)_3$ ... Fe $(CO)_3$ ... OH $\xrightarrow[20°C,\ 56\ \%]{Ag_2CO_3,\ Bn}$ OHC ... Fe $(CO)_3$ ... Fe $(CO)_3$ ... CHO		152
3. R^1 ... Fe $(CO)_3$ ... OH $\xrightarrow[-30°C \rightarrow -40°C,\ 3hours - 4 hours]{PDC,\ 4Å\ mol.\ sieves}$ R^1 ... Fe $(CO)_3$ ... CHO R^1 / Yield: - CO_2CH_3 60 %; - $CO(CH_2)_2CO_2CH_3$ 55 %		120, 129
COMMENTS	The organometallic unit do not shift during the oxidation steps.	

TYPE OF REACTION	OXIDATION REACTION OF IRON COMPLEXED DIENOLS 3/3	References
4. CH_3O_2C … $Fe(CO)_3$ … OH	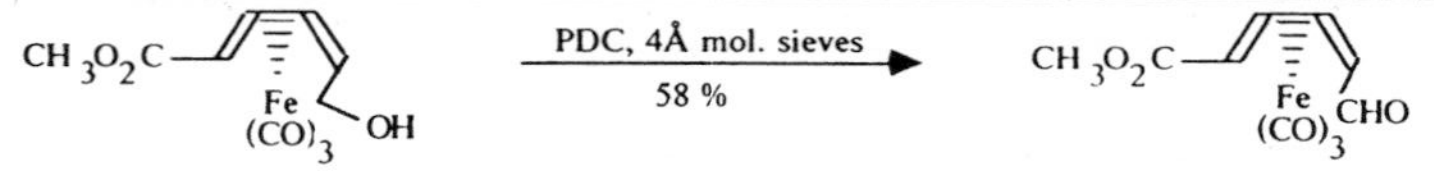 PDC, 4Å mol. sieves; 58 % → CH_3O_2C … $Fe(CO)_3$ … CHO	20
COMMENTS	The (E,Z)-geometry of the iron complexed dienol is retained during the oxidation.	
5. HO … $Fe(CO)_3$ … CO_2CH_3	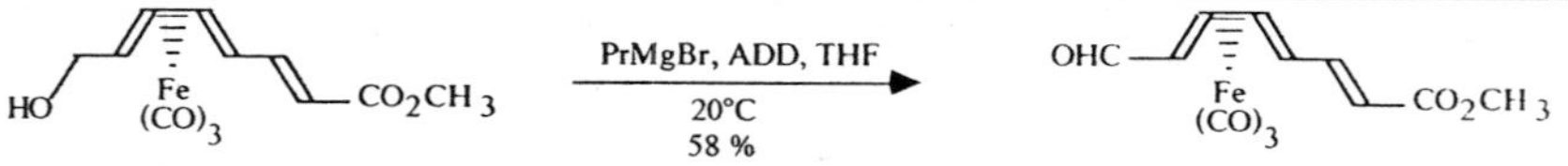 PrMgBr, ADD, THF; 20°C; 58 % → OHC … $Fe(CO)_3$ … CO_2CH_3	129
6. HO … $Fe(CO)_3$ … H, $OSiMe_2tBu$	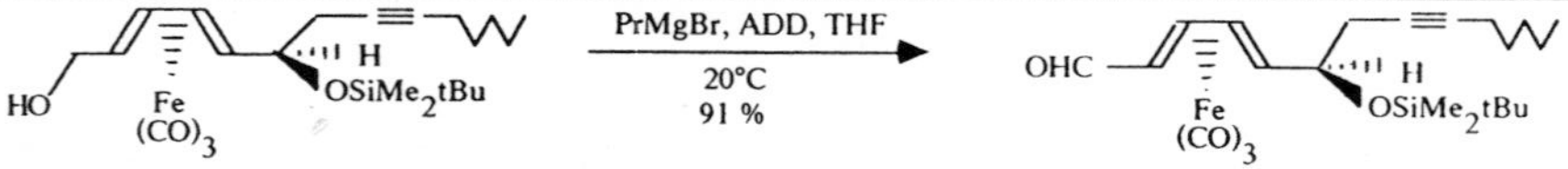 PrMgBr, ADD, THF; 20°C; 91 % → OHC … $Fe(CO)_3$ … H, $OSiMe_2tBu$	111
7. HO … $Fe(CO)_3$ … CO_2CH_3	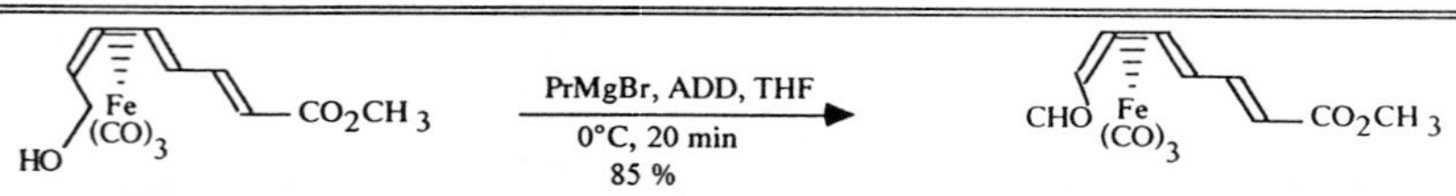 PrMgBr, ADD, THF; 0°C, 20 min; 85 % → CHO … $Fe(CO)_3$ … CO_2CH_3	129
COMMENTS	The (Z,E)- geometry of the iron complexed dienol is retained.	
8. $tBuMe_2SiO$ … $Fe(CO)_3$ … OH	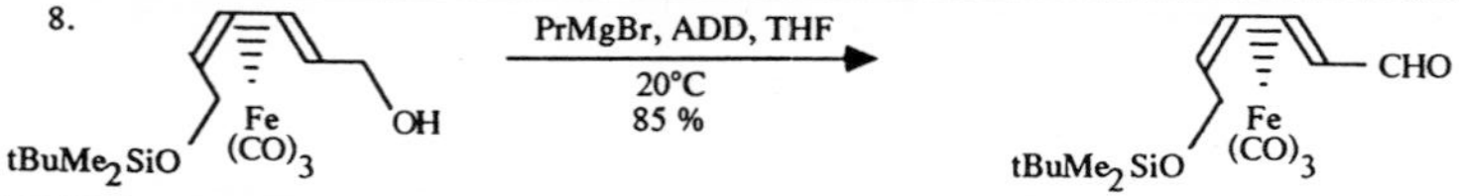 PrMgBr, ADD, THF; 20°C; 85 % → $tBuMe_2SiO$ … $Fe(CO)_3$ … CHO	129
COMMENTS	The (Z,E)- geometry of the iron complexed dienol is retained.	
9. $Fe(CO)_3$ … OH	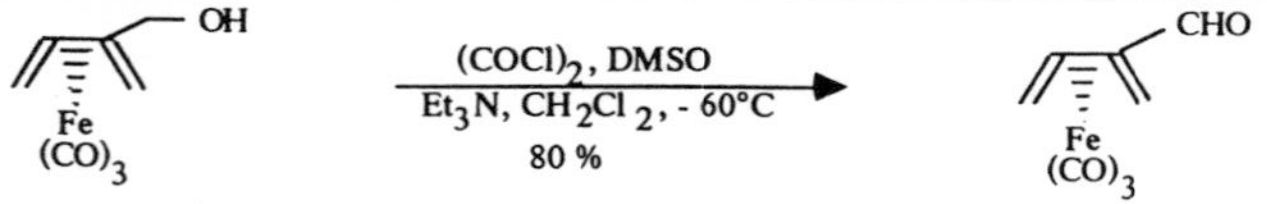 $(COCl)_2$, DMSO; Et_3N, CH_2Cl_2, - 60°C; 80 % → $Fe(CO)_3$ … CHO	11, 116

TYPE OF REACTION	REDUCTION OF IRON COMPLEXED DIENONES 1/4	References

1. R^1, R^2, R^3, Fe$(CO)_3$, R^4 — **1** —reducing agents (see the table)→ **2** : Ψ- exo more polar + **3** : Ψ- endo less polar

R^1	R^2	R^3	R^4	Experimental Conditions	Yield of 2	Yield of 3
- CH_3	- H	- H	- CH_3	$NaBH_4$/ CH_3OH-H_2O 20°C	5 %	95 %
- Ph	- H	- H	- CH_3	" "	2 %	98 %
- CH_3	- CH_3	- H	- CH_3	" "	5 %	95 %
- CH_3	- H	- CH_3	- CH_3	" "	1 %	99 %
- CH_3	- H	- H	- CH_3	$NaBH_4$/ CH_3OH, O°C 2 h 30 min		100 %
- CH_3	- H	- H	- Ph	" "	2 %	98 %
- CH_3	- H	- H	- ≡ - Ph	" "	10 %	90 %
- CH_3	- H	- H	- ≡ - Ph	$LiAlH_4$/ Ether, -78°C 0.5 - 4 hours	2 %	98 %
- CH_3	- H	- H	- ≡ - CH $(OEt)_2$	" "	2 %	98 %

References: 19, 26, 153

COMMENTS

This reduction is highly stereoselective and, depending on the steric requirements of the R^4 group, sometimes stereospecific. The major alcool which is *always* the *Ψendo* less polar one, arising from an approach of the reducing agent *anti* to the organometallic unit on the major *s-cis* conformation of the iron complexed dienone. Improvements of the diastereoselectivity are observed with slight modifications of the experimental conditions.

TYPE OF REACTION	REDUCTION OF IRON COMPLEXED DIENONES 2/4	References
2. 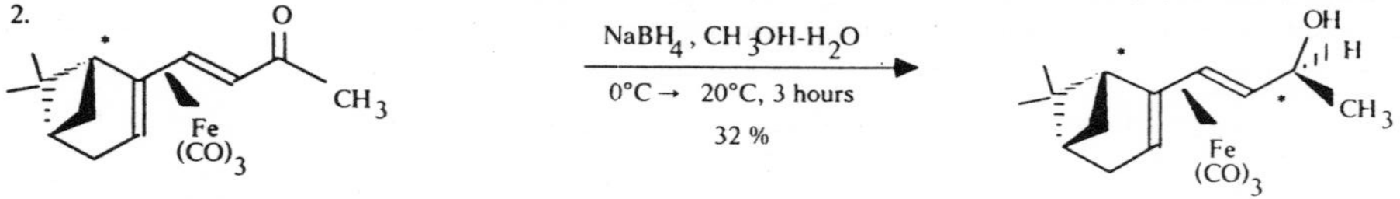	$NaBH_4$, CH_3OH-H_2O; 0°C → 20°C, 3 hours; 32 %	31b
COMMENTS	*One* Ψ-endo alcohol is obtained.	
3. 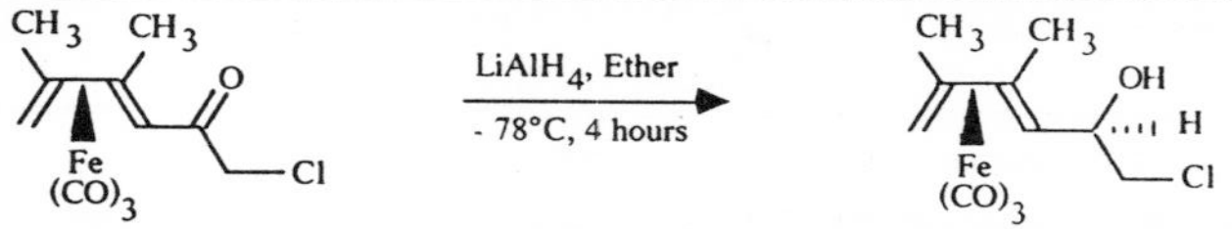	$LiAlH_4$, Ether; - 78°C, 4 hours	104
COMMENTS	*One* Ψ-endo alcohol is isolated. The reducing conditions are compatible with the chloro group.	
4. 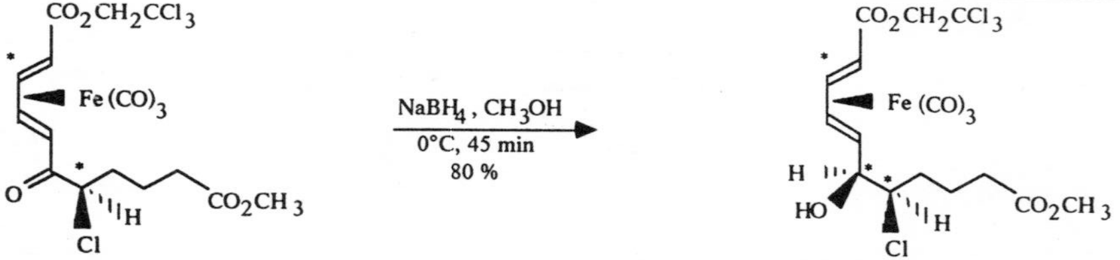	$NaBH_4$, CH_3OH; 0°C, 45 min; 80 %	105
COMMENTS	*One* Ψ-endo chlorohydrine is obtained.	

TYPE OF REACTION	REDUCTION OF IRON COMPLEXED DIENONES 3/4	References
5.	(reaction scheme: 1 → 2 + 3; reducing agents (see the table))	154
COMMENTS	The stereochemical outcome of the reduction is now sensitive to conformationnal effects of the cyclic iron complexed dienones. Boat conformations of **1** (R^1 = - H ; - CH_3) explain the observed stereoselectivities.	
6.	(reaction scheme: $LiAlH_4$, Ether, - 78°C, 45 min)	105
COMMENTS	The (E,Z)-geometry of the complex is retained during the reduction. No epoxide is formed. The relative stereochemistry of the unique alcohol which is depicted is deduced from ^{1}H-NMR coupling data of the corresponding uncomplexed epoxide. This highly stereoselective reduction is best explained by an *anti* attack of the reducing agent and by the highly strained nature of the iron complexed dienones.	

Fe $(CO)_2$ $P(OPh)_3$ — reducing agents (see the table) → **2** + **3**

R^1	Experimental Conditions	Yield of 2	Yield of 3
- H	$NaBH_4$, CH_3OH, Et_2O, 0°C, 2 h	37.5 %	62.5 %
- H	L- Selectride		100 %
- CH_3	$NaBH_4$, CH_3OH, Et_2O, 0°C, 2 h	100 %	

$LiAlH_4$, Ether, - 78°C, 45 min

R^1	R^2	R^3	R^4	Yield
- H	- H	- Cl	- H	80 %
- CH_3	- H	- Cl	- H	80 %
- CH_3	- CH_3	- Cl	- H	92 %
- CH_3	- CH_3	- H	- Cl	90 %

TYPE OF REACTION	REDUCTION OF IRON COMPLEXED DIENONES 4/4	References
7. 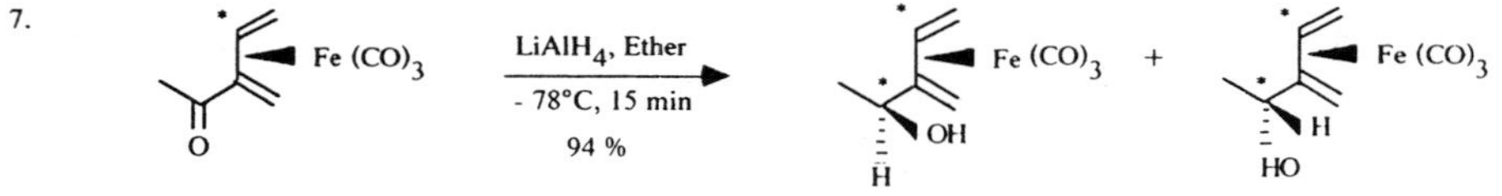 Fe (CO)$_3$ → LiAlH$_4$, Ether, - 78°C, 15 min, 94 % → Fe (CO)$_3$ (OH, H) + Fe (CO)$_3$ (H, HO); 1 : less polar 71 / 29 2 : more polar		24, 105
COMMENTS	In the case of this acyl-2 iron complexed dienone, *two* diastereoisomers are obtained. The absolute configuration of 2 has been previously determined by X-ray cristallographic analysis of its camphor derivative. The addition of $LiAlH_4$ occurs *anti* to the organometallic unit on the slightly thermodynamically prefered *s-cis* conformation of the iron complexed dienone. The optical purity of 1 (better than 94 %) has been measured by ^{1}H-NMR in the presence of Eu(hfc)$_3$.	
8. 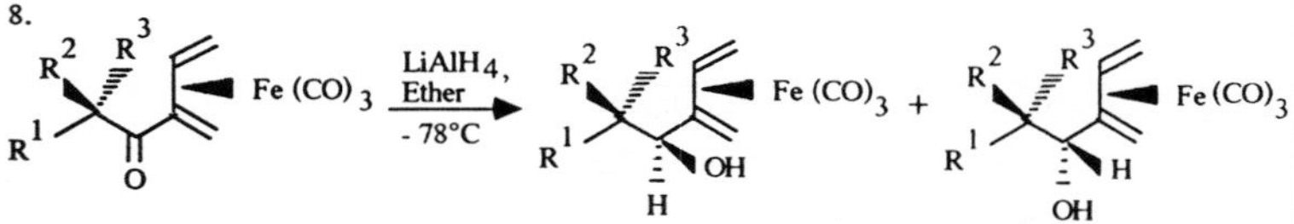 R^1, R^2, R^3, Fe (CO)$_3$ → LiAlH$_4$, Ether, - 78°C → Fe (CO)$_3$ (OH, H) + Fe (CO)$_3$ (H, OH); 1 : less polar 2 : more polar		105
COMMENTS	The relative stereochemistries of the two diastereoisomeric alcohols 1 and 2 have been attributed based on previous data [24,105] and ^{1}H-NMR correlations. In contrary to the reduction of acyl-1 iron complexed dienones, the reduction is <u>not</u> stereospecific. This lack of selectivity is best explained by the lack of a highly preferred conformation of the iron complexed dienone.	

R^1	R^2	R^3	Yield 1 + 2	ratio 1/2
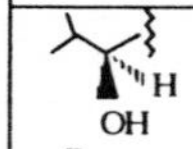 (structure, H, OH)	- H	- H	75%	77/23
- Br	- H	- H	91 %	92/8
- CH_3	- Br	- H	95 %	100/0
- CH_3	- H	- Br	85 %	55/45

TYPE OF REACTION	REDUCTION REACTION OF IRON COMPLEXED DIENYL CARBONYL COMPOUNDS OR DIENOLS 1/4	References
1. Fe $(CO)_3$ O O CH_3 H CO_2Et **1**	$LiAlH_4/AlCl_3$, Ether, - 10°C, 30 min, 85 % → Fe $(CO)_3$ **2**	15, 19
COMMENTS	The optical purity of **2** (better than 96 %) has been controlled by ^{1}H-NMR of the Friedel-crafts acetylated product in the presence of the chiral shift reagent $Eu(tfc)_3$. Therefore, **2** is obtained with the same optical purity than the starting complex **1**.	
2. $(CO)_3$ Fe R R O O CH_3 H CO_2Et	$LiAlH_4/AlCl_3$, Ether, - 10°C, 30 min → $(CO)_3$ Fe R R R: - H, Yield 85 %; R: - CH_3, Yield 44 %	15, 19
3. Fe $(CO)_3$ CHO **1**	$LiAlH_4/AlCl_3$, Ether, - 5°C, 30 min, > 95 % → Fe $(CO)_3$ **2**	19, 28, 29, 30
COMMENTS	The optical purity of the iron complex **2** has been controlled as above.	
4. $(CO)_3$ Fe CHO	$LiAlH_4/AlCl_3$, Ether, - 5°C, 30 min, > 95 % → $(CO)_3$ Fe	19, 28, 29, 30

TYPE OF REACTION	REDUCTION REACTION OF IRON COMPLEXED DIENYL CARBONYL COMPOUNDS OR DIENOLS 2/4	References
5. $LiAlH_4/AlCl_3$, Ether, - 78°C, 30 min		19

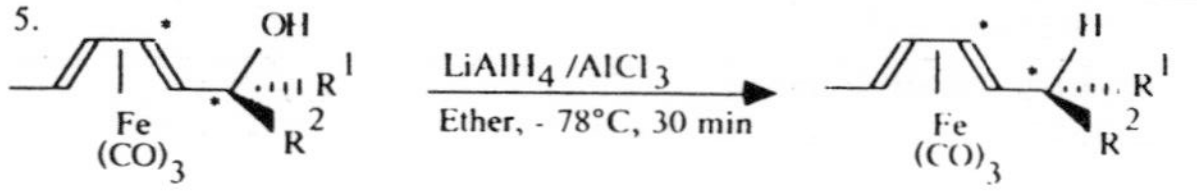

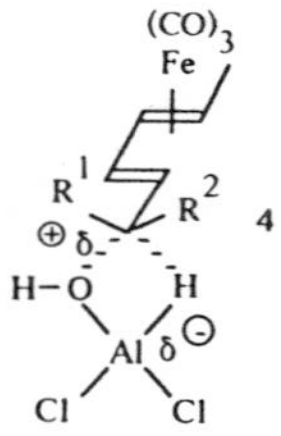

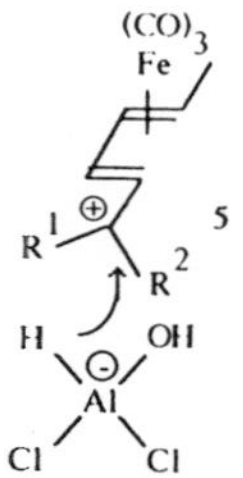

Metal	R^1	R^2		Yield
$Fe(CO)_3$	- CH_3	- Ph	1	: 85 %
$Fe(CO)_3$	- Et	- CH_3	2	: 86 %
$Fe(CO)_3$	- CH_3	- Et	3	: 86 %

COMMENTS

In each case, only *one* diastereoisomer is detected using high-field ^1H-NMR analysis. Nevertheless, chemical correlation experiments on 1 showed that the reduction proceeds with *retention* of configuration and modest optical purity (about 70%). Racemization problems arising during the chemical correlation on 1 let this question opened. The mechanism of the reduction involving the Lewis acid $HAlCl_2$ [155] is discussed : it can go through a four centers intermediate (see structure 4) with an intramolecular transfer of hydride or through the nucleophilic hydride attack of an iron complexed transoid cation [85] (see the structure 5).

TYPE OF REACTION	REDUCTION REACTION OF IRON COMPLEXED DIENYL CARBONYL COMPOUNDS OR DIENOLS 3/4	References
6. CH_3O_2C–diene–$Fe(CO)_3$–OH	CF_3CO_2H, Et_3SiH / CH_2Cl_2, 20°C, 1h 45 min, 89 % → CH_3O_2C–diene–$Fe(CO)_3$–CH_3	27
COMMENTS	The (E,E)-structure of the iron complexed dienol is retained in these highly acidic conditions.	
7. CH_3O_2C–diene–$Fe(CO)_3$–$C(R^1)(R^2)$–$(\ )_{n^1}$–R^3	CF_3CO_2H, Et_3SiH / CH_2Cl_2, 20°C → CH_3O_2C–diene–$Fe(CO)_3$–$(\ )_{n^2}$–R^3	27
COMMENTS	The (E,E)-structures of the iron complexed dienols are retained. The Ψ-exo alcohols are ***more reactive*** than the Ψ-endo ones. Consistently, this high-yielding reduction gives a best yield for the Ψ-exo alcohols. Among various acids which have been tested, CF_3CO_2H gives the best results. More likely, transoid cations [85] generated in the medium are the reactive species. Limitations are encountered with iron complexed allylic alcohols and monosubstituted alkynes.	

R^1	R^2	R^3	n^1	n^2	time	yield
- OH	- H	- H	1	2	2 h	99 %
- H	- OH	- H	1	2	1 h 15 min	88 %
- OH	- H	- H	3	4	30 min	93 %
- H	- OH	- H	3	4	1 h 15 min	83 %
- OH	- H	- $OSiPh_2tBu$	6	7	20 min	92 %
- H	- OH	- $OSiPh_2tBu$	6	7	1 h 45 min	59 %

TYPE OF REACTION	REDUCTION REACTION OF IRON COMPLEXED DIENYL CARBONYL COMPOUNDS OR DIENOLS 4/4	References
8. CH_3O_2C–[Fe(CO)$_3$ diene]–C(R^1)(R^2)–C≡C–Ph	CF_3CO_2H, Et_3SiH / CH_2Cl_2, 20°C → CH_3O_2C–[Fe(CO)$_3$ diene]–CH_2–C≡C–Ph	27
COMMENTS	The reaction conditions are compatible with a bisubstituted alkyne.	
9. HO– macrocyclic lactone, Fe(CO)$_3$, ( )$_6$, ( )$_5$	CF_3CO_2H, Et_3SiH / CH_2Cl_2, 20°C, 1 h, 38 % → macrocyclic lactone, Fe(CO)$_3$, ( )$_6$, ( )$_5$	27
10. CH_3O_2C–[Fe(CO)$_3$ diene]–CH(OH)–… OSiPh$_2$tBu	CF_3CO_2H, Et_3SiH / CH_2Cl_2, 20°C, 1h 45 min → CH_3O_2C–[Fe(CO)$_3$ diene]–… OSiPh$_2$tBu **1**	27
COMMENTS	The optical purities of the two diastereoisomeric complexes **1** have been measured as high as 95 % by high-field ^{13}C-NMR analysis.	

Entry 8:

R^1	R^2	time	yield
- OH	- H	15 min	77 %
- H	- OH	20 min	73 %

Entry 10:

Metal	Yield
Fe(CO)$_3$ (hashed)	71 %
Fe(CO)$_3$ (wedge)	75 %

TYPE OF REACTION	REDUCTION REACTION OF AN IRON COMPLEXED DIENE OXIME 1/1	Reference
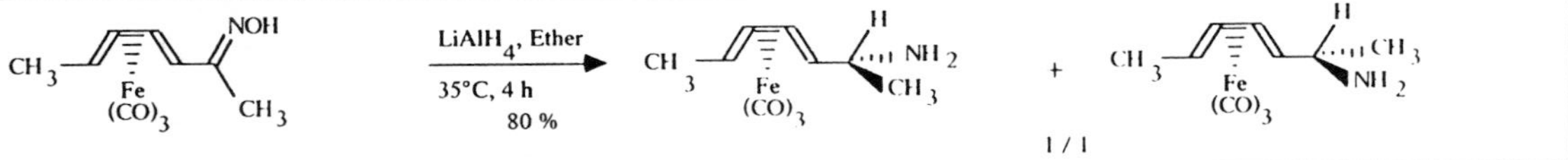 CH_3, Fe $(CO)_3$, NOH, CH_3	LiAlH_4, Ether; 35°C, 4 h; 80 % → CH_3, Fe $(CO)_3$, H, NH_2, CH_3 + CH_3, Fe $(CO)_3$, H, CH_3, NH_2; 1 / 1	156
COMMENTS	This reduction is *not* stereoselective. Most likely, there is no thermodynamic preference for the s-cis or s-trans form of the iron complexed oxime.	

TYPE OF REACTION	REDUCTION / SEMI-REDUCTION REACTION OF DOUBLE / TRIPLE BONDS LINKED OR FAR FROM THE IRON COMPLEXED DIENES 1/4	References
1. 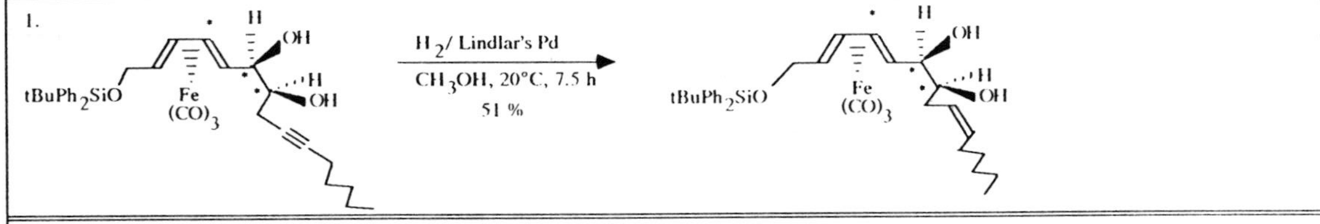 tBuPh_2SiO, Fe $(CO)_3$, H, OH, H, OH	H_2/ Lindlar's Pd; CH_3OH, 20°C, 7.5 h; 51 % → tBuPh_2SiO, Fe $(CO)_3$, H, OH, H, OH	113
2. 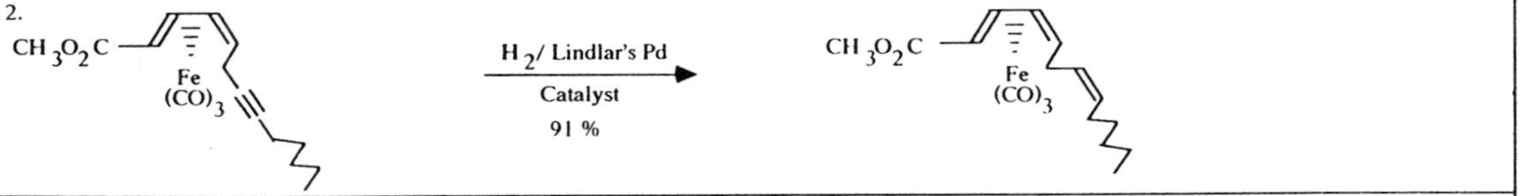 CH_3O_2C, Fe $(CO)_3$	H_2/ Lindlar's Pd; Catalyst; 91 % → CH_3O_2C, Fe $(CO)_3$	82
COMMENTS	The (E,Z)-geometry of the complexed diene is retained.	

TYPE OF REACTION	REDUCTION / SEMI-REDUCTION REACTION OF DOUBLE / TRIPLE BONDS LINKED OR FAR FROM THE IRON COMPLEXED DIENES 2/4	References
3. CH_3O_2C, Fe, $(CO)_3$, R^1, R^2, n, R^3	H_2, 10 % Pd / C; CH_3OH, 4-30 h → CH_3O_2C, Fe, $(CO)_3$, R^1, R^2, n, R^3	75, 92, 130
COMMENTS	It is interesting to note that, in each case, the reduction stops at olefinic stage without producing any over-reduction alkane by-products. These results are independent of the relative configuration of the alcohol and of the organometallic unit. The Ψ-endo alcohols are always reduced more rapidly. Most likely, steric factors due to the bulky iron complex play a great role.	
4. R^1, R^2, EtO_2C, Fe, $(CO)_3$, H, Me, $OSiMe_2tBu$	H_2, 10 % Pd / C; CH_3OH, 20°C, 1 h → R^1, R^2, Fe, $(CO)_3$, CO_2Et, H, Me, $OSiMe_2tBu$	77
COMMENTS	As described previously [75,92,130], no over-reduction by-products are isolated. The (E,Z)-stereochemistry of the complexed dienes is retained during the semi-reduction.	

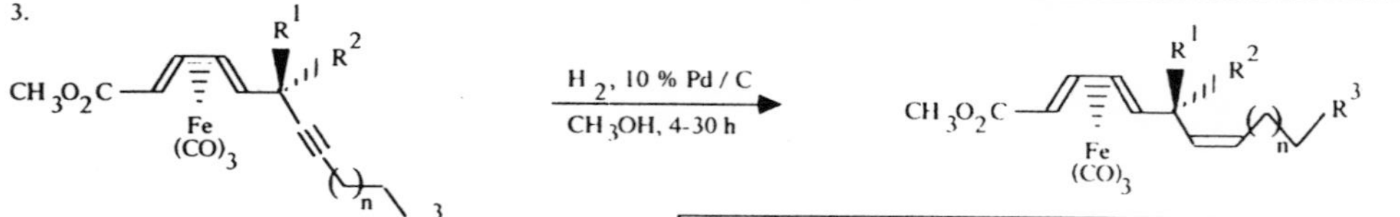

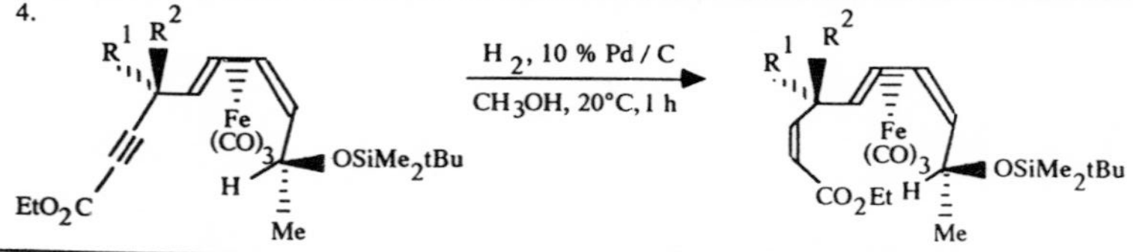

Reaction 3:

R^1	R^2	R^3	n	yield
- OH	- H	- OH	1	> 95 %
- H	- OH	- OH	1	> 95 %
- OH	- H	- OH	3	> 95 %
- H	- OH	- OH	3	> 95 %
- OH	- H	- H	2	98 %
- H	- OH	- H	2	85 %

Reaction 4:

R^1	R^2	yield
- OH	- H	64 %
- H	- OH	89 %
- O	- O	69 %

TYPE OF REACTION	REDUCTION / SEMI-REDUCTION REACTION OF DOUBLE / TRIPLE BONDS LINKED OR FAR FROM THE IRON COMPLEXED DIENES 3/4	References
5. 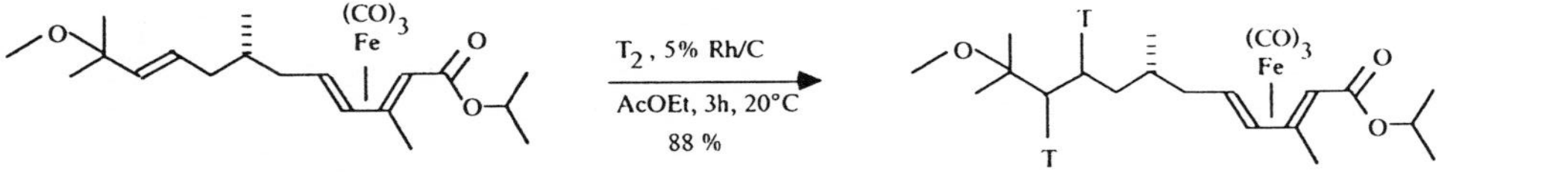 $(CO)_3$ Fe	T_2, 5% Rh/C; AcOEt, 3h, 20°C; 88 %; T, T, $(CO)_3$ Fe	127
COMMENTS	The organometallic unit can be considered as a protecting group for the dienoate fonction during the reduction step.	
6. 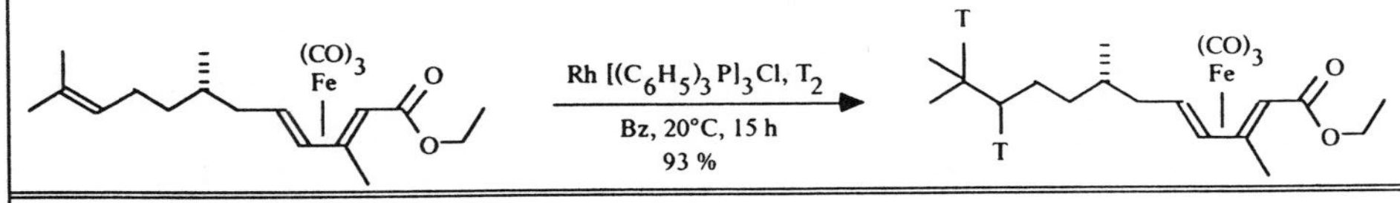 $(CO)_3$ Fe	Rh $[(C_6H_5)_3P]_3Cl$, T_2; Bz, 20°C, 15 h; 93 %; T, T, $(CO)_3$ Fe	127
7. 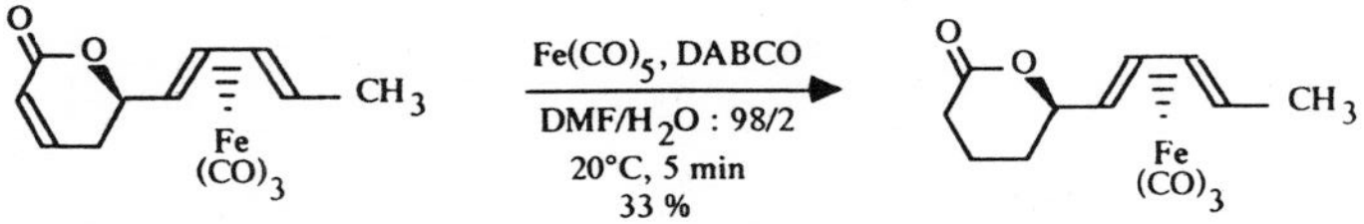 CH_3, Fe $(CO)_3$	$Fe(CO)_5$, DABCO; DMF/H_2O : 98/2; 20°C, 5 min; 33 %; CH_3, Fe $(CO)_3$	101

TYPE OF REACTION	REDUCTION / SEMI-REDUCTION REACTION OF DOUBLE / TRIPLE BONDS LINKED OR FAR FROM THE IRON COMPLEXED DIENES 4/4	References
8. 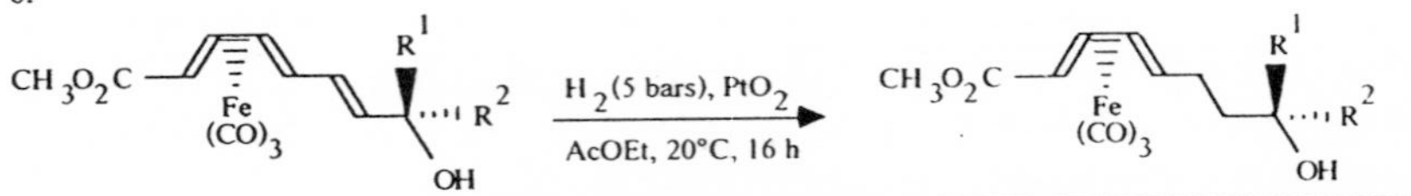 CH_3O_2C — Fe $(CO)_3$ — R^1, R^2, OH	H_2(5 bars), PtO_2 / AcOEt, 20°C, 16 h → CH_3O_2C — Fe $(CO)_3$ — R^1, R^2, OH	139
	R^1: - H, R^2: - CH_3, yield: 98 % R^1: - CH_3, R^2: - H, yield: 98 %	
COMMENTS	This reaction is a *rare* exemple of the catalytic reduction of a double bond directly linked to the organometallic unit. A medium pressure of H_2 and an hydroxyl as activating group seem necessary for the success of the reduction. The relative stereochemistry of the complex versus the hydroxyl group have no influence on the speed and yield of the reduction.	
9. 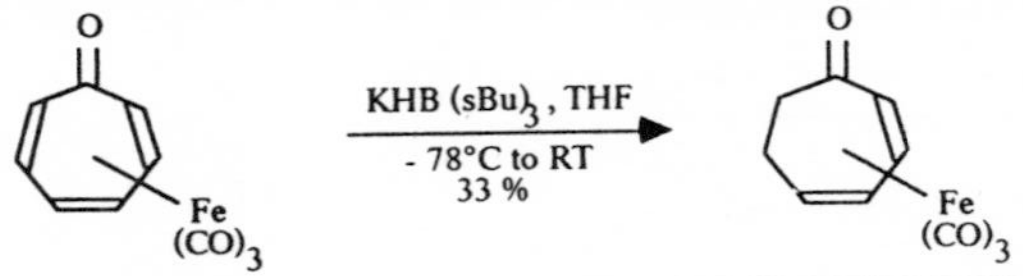 O, Fe $(CO)_3$	KHB $(sBu)_3$, THF / - 78°C to RT, 33 % → O, Fe $(CO)_3$	157
COMMENTS	Activation of the double bond linked to the organometallic unit is provided by the carbonyl group allowing a true *enone* character for the iron complexed tropone. No stereochemical data have been described.	

G. Hydroboration Reactions

TYPE OF REACTION	HYDROBORATION REACTION OF IRON COMPLEXED TRIENES 1/3	References
1. RO, Fe $(CO)_3$	1. BH_3, THF; 2. H_2O_2, OH^-; 80-90 % → RO, OH, Fe $(CO)_3$ + RO, OH, Fe $(CO)_3$	158
COMMENTS	Interestingly, the oxidative work-up of the hydroboration medium is compatible with the organometallic unit. The complexation of the conjugated diene protects it from the attack of BH_3.	
2. Fe $(CO)_3$	1. BF_3, $NaBH_4$, THF, 20°C, 12 h; 2. CH_3OH, alumina, 25°C, 48 h → Fe $(CO)_3$ **1** + Fe $(CO)_3$ **2**	159
COMMENTS	The ratio **1/2** and the yield have not been described. Isomerisation of the (Z,E)-complex seems to occur during the hydroboration.	
3. Fe $(CO)_3$	1. BF_3, $NaBH_4$, THF, 20°C, 12 h; 2. CH_3OH, alumina, 25°C, 48 h; 73 % → Fe $(CO)_3$	159

TYPE OF REACTION	HYDROBORATION REACTION OF IRON COMPLEXED TRIENES 2/3	Références
4.	1. BF_3, $NaBH_4$, THF, 20°C, 22 h; 2. H_2O_2, KOH, H_2O, 20°C, 30 min	160
COMMENTS	The sequence hydroboration of a double bond close to the complex-oxidative work-up is compatible with the organometallic unit. The stabilization of a neighbouring positive charge by the complex as required for borane reduction and steric factors, likely, explain the observed regioselectivity.	
5.	1. BH_3, THF ; THF, 0°C; 2. H_2O_2, KOH 1N, 1 min	91
COMMENTS	The relative Ψ-exo or Ψ-endo stereochemistry of the complexed trienols has no effect on the results of the hydroboration. Noteful is the *short* time of the oxidative work-up.	
6.	1. BH_3, THF ; THF, 0°C; 2. H_2O_2	91

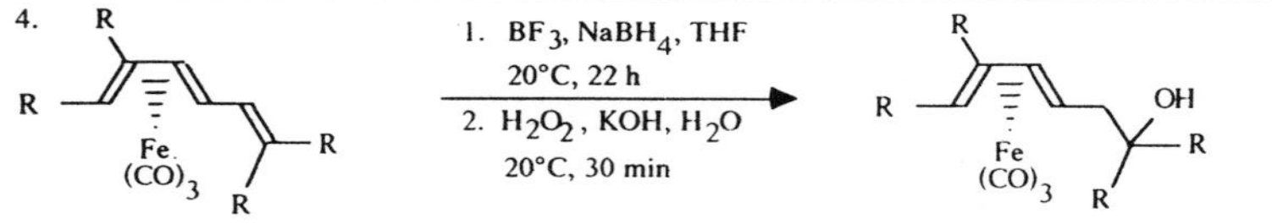

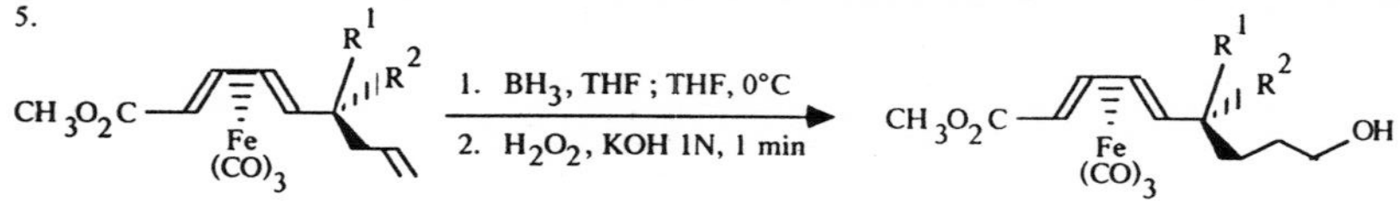

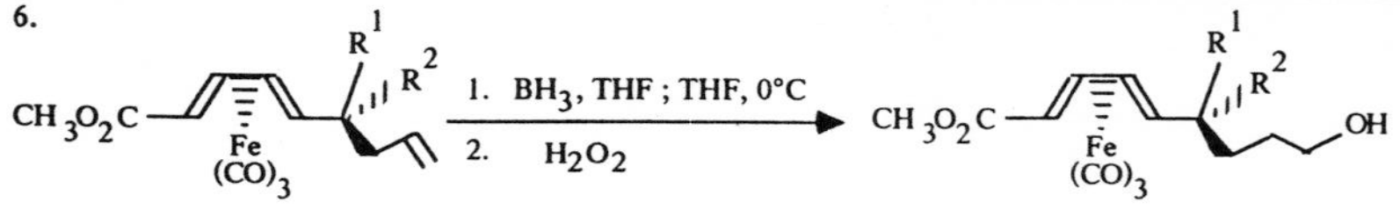

Reaction 4:

R	Yield
- H	42 %
- CH_3	63 %

Reaction 5:

R^1	R^2	Yield
- OH	- H	95-98 %
- H	- OH	95-98 %

Reaction 6:

R^1	R^2	Yield
- H	- $OSiMe_2tBu$	75 %
- $OSiMe_2tBu$	- H	75 %

TYPE OF REACTION	HYDROBORATION REACTION OF IRON COMPLEXED TRIENES 3/3	References
7. 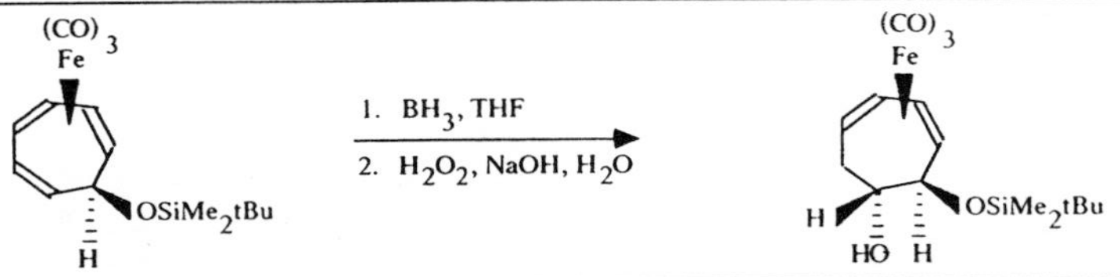 $(CO)_3Fe$ complex, OSiMe$_2$tBu, H	1. BH_3, THF; 2. H_2O_2, NaOH, H_2O → $(CO)_3Fe$ complex, H, HO, H, OSiMe$_2$tBu	161
COMMENTS	The hydroboration is totally regio- and stereoselective. The organometallic unit and also the other function control the regiochemistry of the hydroboration *anti* to the $Fe(CO)_3$ unit. No yield was reported.	
8. 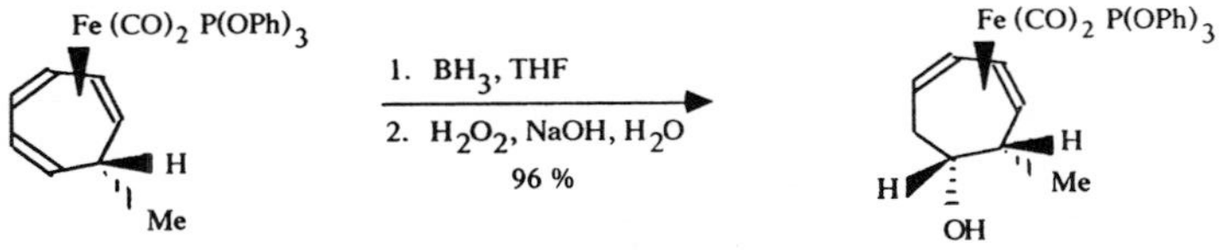 $Fe(CO)_2P(OPh)_3$ complex, H, Me	1. BH_3, THF; 2. H_2O_2, NaOH, H_2O; 96 % → $Fe(CO)_2P(OPh)_3$ complex, H, OH, H, Me	161
9. 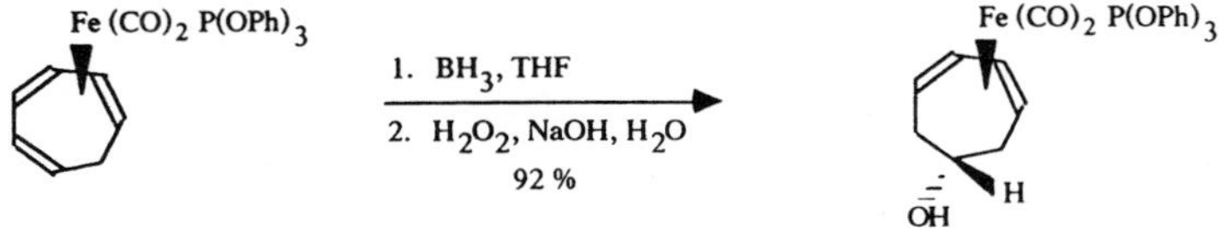 $Fe(CO)_2P(OPh)_3$ complex	1. BH_3, THF; 2. H_2O_2, NaOH, H_2O; 92 % → $Fe(CO)_2P(OPh)_3$ complex, OH, H	154

H. Nucleophilic Additions of Organometallics

TYPE OF REACTION	NUCLEOPHILIC ADDITION OF ORGANOMETALLICS ON IRON COMPLEXED DIENALS OR DIENONES 1/8	References
1.	R^1-diene-CHO ($Fe(CO)_3$) $\xrightarrow{R^2M}$ Ψ-exo 2 + Ψ-endo 3	see the tables

R^1	R^2M	Yield	2 / 3	ref.
- CO_2Me	CH_3MgI	80 %	60/40	75,162
- CH_3	CH_3MgI	57 %	53/47	116
- CO_2Me	CH_3Li	85 %	80/20	75
- CO_2Me	PhLi	94 %	66/34	75
- CO_2Me	$tBuO_2CCH_2Li$	94 %	66/34	139a
- CO_2Me	$PhSCH_2Li$	88 %	72/28	75
- CO_2Me	$CH_3(CH_2)_2C\equiv CLi$	96 %	70/30	130
- CO_2Me	$LiO_2C(CH_2)_2C\equiv CLi$	42 %	60/40	75
- CO_2Me	$BrMgO(CH_2)_2C\equiv CMgBr$	87 %	70/30	75
- CO_2Me	$ClCH_2Li$	76 %	54/46	96

R^1	R^2M	Yield	2 / 3	ref.
- CO_2Me	allyl boronate of diisopropyl tartrate (CO_2iPr, CO_2iPr)	100 %	59/41	9
- CH_3	allyl boronate of diisopropyl tartrate (CO_2iPr, CO_2iPr)	100 %	62/38	9
- CO_2Me	$CH_3Ti(OiPr)_3$	94 %	25/75	75
- CO_2Me	$HC\equiv C\,CH_2Al_{2/3}Br$	94 %	25/75	75
- CO_2Me	$CH_2=C=(Al_{2/3}Br)(CH_2)_4CH_3$	80 %	43/57	75
- CO_2Me	$=\bullet=C(C_5H_{11})(SiMe_3)$	65 %	0/100	107

TYPE OF REACTION	NUCLEOPHILIC ADDITION OF ORGANOMETALLICS ON IRON COMPLEXED DIENALS OR DIENONES 2/8	References
COMMENTS	The nucleophilic additions are high yielding reactions giving two chromatographically well-resolved Ψ-exo and Ψ-endo alcohols. The diastereoselectivity of the reaction is directly linked to the s-cis s-trans conformational equilibrium of the iron complexed dienal **1** , but also to the nature of the metal. Organolithium and organomagnesium compounds adding on the *s-cis* conformer *anti* to the organometallic moiety lead to the major Ψ-exo alcohol **2**. The observed diastereoselectivities are rather moderate. On the other hand, organotitanium and organoaluminium nucleophiles have been suggested to react in a major *endo* fashion through an intramolecular transfer from iron or a carbonyl ligand[11]. Particularly useful is the complete Ψ-endo diastereoselectivity observed with the allenylsilane nucleophile, the origin of this selectivity is unclear at this time.	

TYPE OF REACTION	NUCLEOPHILIC ADDITION OF ORGANOMETALLICS ON IRON COMPLEXED DIENALS OR DIENONES 3/8	References

2.

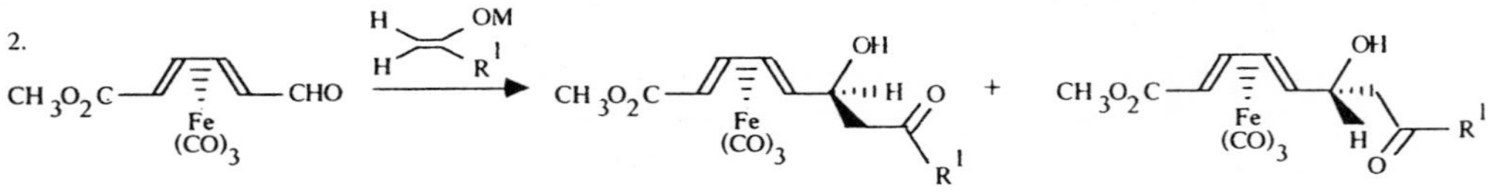

CH_3O_2C–(diene)$Fe(CO)_3$–CHO + H(H)C=C(OM)R^1 → CH_3O_2C–(diene)$Fe(CO)_3$–CH(OH)–CH_2–CO–R^1

Ψ- exo 1 + Ψ- endo 2

References: 90a

R^1	M	Experimental conditions	Overall yield	1/2
- tBu	- Li	THF, - 70°C, 10 min	92 %	87/13
- OtBu	- Li	THF, - 70°C, 10 min	57 %	74/26
- tBu	- $SiMe_2tBu$	BF_3, CH_2Cl_2, - 78°C, 1 h	69 %	92/8
- OtBu	- $SiMe_2tBu$	BF_3, CH_2Cl_2, - 78°C, 1 h	50 %	80/20

COMMENTS

The observed moderate diastereoselectivity is analogous to the one observed with organolithium nucleophiles.

TYPE OF REACTION	NUCLEOPHILIC ADDITION OF ORGANOMETALLICS ON IRON COMPLEXED DIENALS OR DIENONES 4/8	References
3. 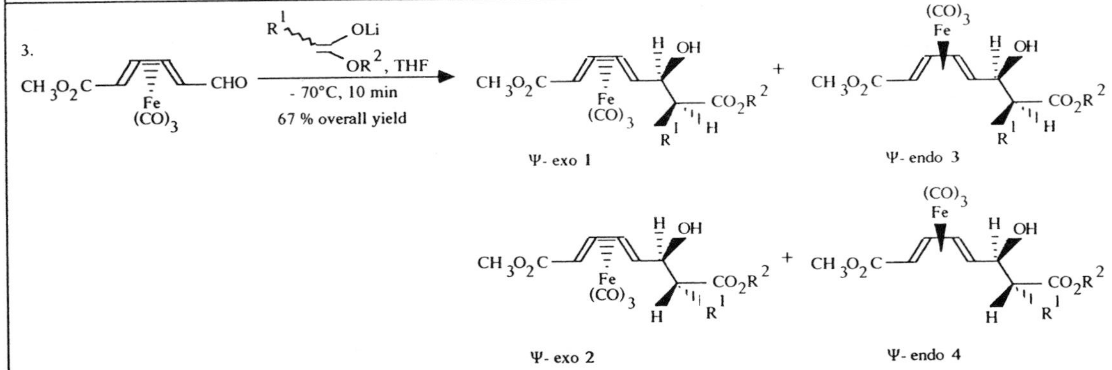		90a

R^1	R^2	Ψ- exo (1 + 2)/ Ψ- endo (3 + 4)	syn (1 + 3)/ anti (2 + 4)
$-CH_3$	$-C(CH_3)_3$	76/24	43/57
$-OCH_3$	$-CH_3$	79/21	36/64

COMMENTS

The *four* aldolisation adducts are isolated and each Ψ-exo and Ψ-endo couples can be chromatographically separated. The depicted syn/anti relationships are established, after decomplexation, using vicinal coupling constants in agreement with previous litterature data. It is interesting to note that the diastereoselectivity Ψ-exo and Ψ-endo of the aldolisation is moderate with a poor syn/anti selectivity.

TYPE OF REACTION	NUCLEOPHILIC ADDITION OF ORGANOMETALLICS ON IRON COMPLEXED DIENALS OR DIENONES 5/8	References
4. CH_3O_2C–(diene-$Fe(CO)_3$)–CHO → (OLi-cyclohexene, THF; - 70°C, 10 min; 76 % overall yield)	Ψ- exo **1** : 42 % + Ψ- exo **2** : 29 % + Ψ- endo **3** : 5 %	90a
COMMENTS	*Three* aldolisation adducts are isolated and the depicted relative configurations have been established as in the previous case. The ratio Ψ-exo (**1**+**2**)/Ψ-endo (**3**) : 93/7 is now excellent.	

TYPE OF REACTION	NUCLEOPHILIC ADDITION OF ORGANOMETALLICS ON IRON COMPLEXED DIENALS OR DIENONES 6/8	References
5.		20

CH_3O_2C–[diene–$Fe(CO)_3$]–CHO (**1**) $\xrightarrow{R^1M}$ CH_3O_2C–[diene–$Fe(CO)_3$]–CH(R^1)OH (**2**) + CH_3O_2C–[diene–$Fe(CO)_3$]–CH(OH)R^1 (**3**)

R^1M	Overall yield	2/3
CH_3MgI	77 %	1.7/1
CH_3Li	74 %	4.3/1
$CH_3Ti(OiPr)_3$	77 %	1/10

COMMENTS

The organomagnesium and the organolithium nucleophiles give a major alcohol 2 arising from the addition *anti* to the organometallic unit on the major s-cis conformer of 1 (X-ray analysis of 1 shows that it crystallizes as the s-cis conformer). The relative configuration of the alcohol 2 has been determined by chemical correlation with a known compound[77]. In the same manner as observed previously, the organotitanium nucleophile gives a major alcohol 3 with a reverse stereoselectivity. It is interesting to point out that the (E,Z)-stereochemistry of 1 is retained during the nucleophilic addition.

TYPE OF REACTION	NUCLEOPHILIC ADDITION OF ORGANOMETALLICS ON IRON COMPLEXED DIENALS OR DIENONES 7/8	References

6.

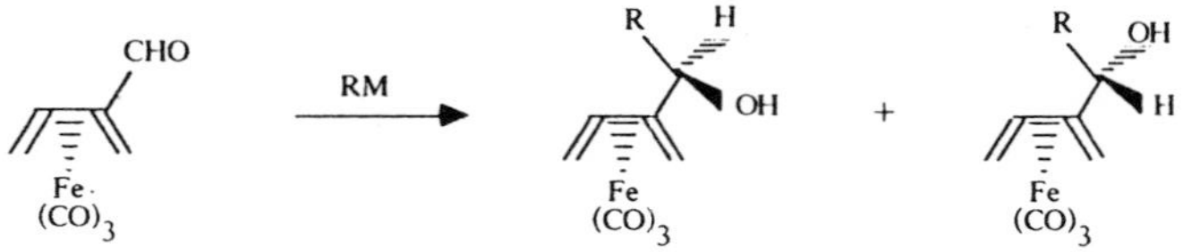

1

RM	Yield	Diastereoselectivity : ratio of the more polar to the less polar diastereoisomer
CH_3MgI	97 %	63/37
Ph Mg Br	88 %	50/50
iPr Mg Br	80 %	63/37
CH_3Li	98 %	80/20
CH_3Li (LiBr)	85 %	91/9
nBuLi	90 %	92/8
$CH_3Ti(OiPr)_3$	100 %	42/58
$(CH_3)_2CuLi$	40 %	<10/>90

References: 11, 116

COMMENTS

The nucleophilic addition is generally high-yielding giving two well chromatographically separated alcohols. It is interesting to note that the relative stereochemistry of the *more* or *less* polar diastereoisomer has not been rigourously established. The diastereoselectivity of the nucleophilic addition is moderate (organomagnesium and organotitanium reagents) showing the lack of a preferred conformer for 1. The organolithium nucleophiles are slightly more selective.

TYPE OF REACTION	NUCLEOPHILIC ADDITION OF ORGANOMETALLICS ON IRON COMPLEXED DIENALS OR DIENONES 8/8	References
7.	CH_3–diene–$Fe(CO)_3$ ketone **1** $\xrightarrow{R^2M}$ **2** + **3**	19, 26, 153

R^1	R^2M	Overall yield	2/3
$-CH_3$	Ph Mg Br	70 %	> 98/2
$-CH_3$	PhLi	95 %	> 98/2
- Ph	CH_3Li	91 %	> 98/2
- i - amyl	CH_3Li	95 %	> 98/2
$-CH_3$	C_2H_5Li	90 %	> 98/2
$-C_2H_5$	CH_3Li	100 %	> 98/2
$-CH_3$	Ph - C≡CLi	100 %	> 98/2
- C≡C - Ph	CH_3Li	100 %	75/25
- C≡C - Ph	PhLi	98 %	87/13

COMMENTS

As observed previously during the reduction of **1** by hydrides, these nucleophilic additions are *highly* stereoselective except in the case of iron complexed alkynyl ketones. The nucleophile adds *anti* to the organometallic unit on the *major s-cis* conformer of 1.

I. Aldolization Reactions

TYPE OF REACTION	ALDOLISATION REACTION USING IRON COMPLEXED ENOXYSILANES 1/2	References
1.		105

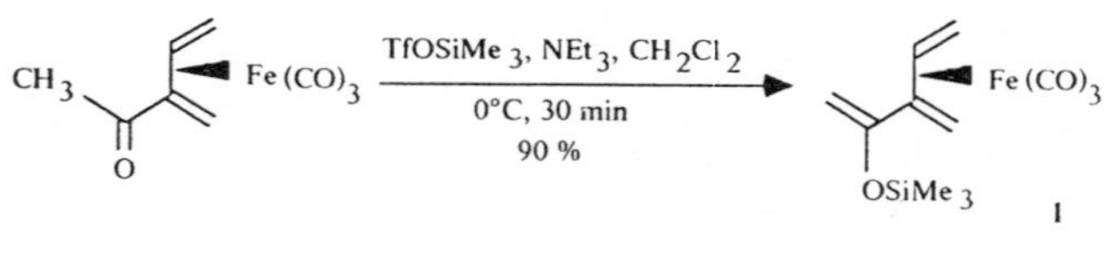

CH 3
O
Fe (CO)3
TfOSiMe 3, NEt 3, CH 2Cl 2
0°C, 30 min
90 %
Fe (CO)3
OSiMe 3
1

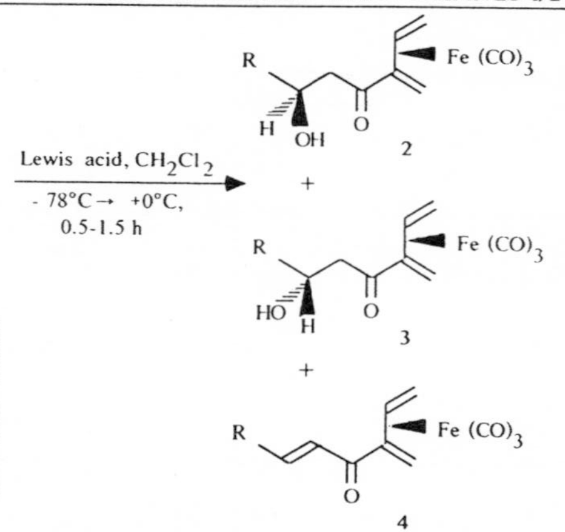

Lewis acid, CH2Cl2
- 78°C → +0°C,
0.5-1.5 h
R
Fe (CO)3
H
OH
O
2
+
R
Fe (CO)3
HO
H
O
3
+
R
Fe (CO)3
O
4

R	Lewis acid (1 equiv.)	Yield 2 + 3	ratio 2 / 3	Yield 4
- Ph	$TiCl_4$	90 %	60/40	3 %
- $CH(CH_3)_2$	$TiCl_4$	86 %	60/40	14 %
- $CH(CH_3)_2$	BF_3-Et_2O	92 %	45/55	0 %

COMMENTS

The best way to prepare the iron complexed enoxysilane in a good yield is the Mukaiyama's method. The obtained aldol products **2** and **3** are often accompanied by the deshydratation one **4** in variable amounts. The aldolisation reaction shows a *lack* of diastereoselectivity in these experimental conditions. Most likely, the lack of a thermodynamically favored s-trans or s-cis conformation of the enol **1** accounts for this disappointing result. The relative configuration of each aldolisation ketol 2 and 3 has been established by ^{1}H- and ^{13}C-high field NMR of the syn and anti acetonides obtained after their diastereospecific reduction.

TYPE OF REACTION	ALDOLISATION REACTION USING IRON COMPLEXED ENOXYSILANES 2/2	References
2. 1 + 2, $TiCl_4$, CH_2Cl_2; - 78°C, 1h; 50 % → 3 + 4	3/4 : 4.5 / 5.5	105
COMMENTS	The aldolisation reaction is not complete since *35 %* of the iron complexed 2-acetyl butadiene is isolated. The absolute stereochemistry of the aldolisation products 3 and **4** is determined using a combination of high-field ^{1}H-NMR, optical rotation analysis (signs and values) of 3 and **4** and optical rotation analysis of the complexed 2-acetyl butadienes coming from a retroadol step performed on **3** and **4**. It is interesting to note two facts : - The chirality of the new hydroxyl bearing asymmetric centre is controlled by the chiral centre of 2, - The aldolisation step allows the resolution of the iron complexed dienyl moiety. Such examples of facial diastereoselectivity have been described recently [9, 13b, 118].	
3. $TiCl_4$, CH_2Cl_2; - 78°C, 30 min; 93 %	less polar : more polar : 65 / 35	105
COMMENTS	The relative configuration of each aldol product has not been definitely established.	

J. Michael Reactions

TYPE OF REACTION	MICHAËL REACTION ON ELECTROPHILIC IRON COMPLEXED TRIENES 1/2	References
1. CH_3O_2C, $(CO)_3Fe$, CH_3O_2C, CO_2CH_3	$(CH_3)_2\overset{\oplus}{S}\text{-}\overset{\ominus}{C}HCO_2CH_3$, PhH, 20°C, 18 hrs, 94 % → CH_3O_2C, $(CO)_3Fe$, H, H, CO_2CH_3, CO_2CH_3, CH_3O_2C + CH_3O_2C, $(CO)_3Fe$, H, CO_2CH_3, H, CO_2CH_3, CH_3O_2C; 55/45	10
COMMENTS	The nucleophilic addition of the sulfur ylide is *anti* to the organometallic unit. The absolute configurations of each cyclopropane have been determined by chemical correlation with the known corresponding formyl cyclopropanes. The obtained optical purities are better than 90 %.	
2. CH_3O_2C, $Fe(CO)_3$	CH_3MgI, THF, - 40°C, 3 hrs, 93 % → CH_3O_2C, $Fe(CO)_3$, CH_3, H	36
COMMENTS	Only *one* addition adduct arising from the *anti* attack of the organomagnesium compound on the *s-trans* conformation of the complexed triene is isolated. Its absolute configuration is deduced by correlation with optical data of (-)- verbenalol.	

TYPE OF REACTION	MICHAËL REACTION ON ELECTROPHILIC IRON COMPLEXED TRIENES 2/2	References
3. CH_3O_2C … Fe $(CO)_3$	CH_3NO_2, CH_3CN / DBU, 20°C, 24 hrs → CH_3O_2C … Fe $(CO)_3$, NO_2, H	77
COMMENTS	*One* diastereoisomer is obtained of which the relative configuration is attributed based on previous data [36,77].	
4. CH_3 … Fe $(CO)_3$ … NO_2	CH_3NO_2, cat. piperidine / PhH, 20°C, no yield reported → CH_3 … Fe $(CO)_3$ … NO_2, NO_2	31
5. Ph … Fe $(CO)_3$ … O … Fe	1. H_2N-NH_2, C_2H_5OH reflux, 2 hours; 2. $(CH_3CO)_2O$ reflux, 15 min; 79 % → Ph … Fe $(CO)_3$, $COCH_3$, N — N, Fe + Ph … Fe $(CO)_3$, $COCH_3$, N — N, Fe; diastereoisomeric ratio : 1/78	163
COMMENTS	The relative structures of each isomeric pyrazolines has not been established.	
6. O … Fe $(CO)_3$ **1**	$NaCH(CO_2Et)_2$ / - 78°C to 20°C / 74 % → O, EtO_2C, CO_2Et, Fe $(CO)_3$	157
COMMENTS	*One* diastereoisomer is obtained. (1,4)-additions of various organocopper reagents failed. This reaction is a further illustration of the enone character of **1**.	

K. Hydroxyl Assisted Epoxidation Reactions

TYPE OF REACTION	HYDROXYL ASSISTED EPOXIDATION REACTION OF IRON COMPLEXED TRIENES 1/2	References
1. CH_3O_2C … $Fe(CO)_3$ … OH, H; Ψ- exo	VO $(acac)_2$, $PhCH_3$; tBuOOH, 20°C, 43 h; 85 % → CH_3O_2C … $Fe(CO)_3$ … OH, H, O, H **1** + CH_3O_2C … $Fe(CO)_3$ … OH, H, O, H **2**	130
COMMENTS	The epoxidation conditions are compatible with the organometallic unit. *Two* epoxide-alcohols **1** and **2** (1/2 : 3/2) are obtained. Their relative stereochemistries are deduced from an X-ray analysis of **2**. The *threo* major diastereoisomer **1** is in agreement with Sharpless's Z olefins epoxidation.	
2. CH_3O_2C … $Fe(CO)_3$ … H, OH; Ψ- endo	VO $(acac)_2$, $PhCH_3$; tBuOOH, 20°C, 30 h; 78 % → CH_3O_2C … $Fe(CO)_3$ … HO, H, H, O, H **1** + CH_3O_2C … $Fe(CO)_3$ … HO, H, H, O, H **2**	130
COMMENTS	Two epoxide-alcohols **1** and **2** (1/2 : 3/2) are obtained. Their relative stereochemistries are based on previous results and on the analysis of their high-field ^{1}H-NMR spectra.	
3. CH_3O_2C … $Fe(CO)_3$ … OH, H	VO $(acac)_2$, $PhCH_3$; tBuOOH, 20°C, 6 h; 77 % → CH_3O_2C … $Fe(CO)_3$ … OH, H, O, CH_3 **1** + CH_3O_2C … $Fe(CO)_3$ … OH, H, O, CH_3 **2**	129
COMMENTS	The reaction is highly stereoselective (1/2 : 97/3) and furnishes the expected major *erythro* epoxide-alcohol.	

TYPE OF REACTION	HYDROXYL ASSISTED EPOXIDATION REACTION OF IRON COMPLEXED TRIENES 2/2	References
4. CH_3O_2C … Fe $(CO)_3$ … H … OH	VO (acac)$_2$, PhCH$_3$ / tBuOOH, 20°C, 6 h / 72 % → CH_3O_2C … Fe $(CO)_3$ … H … OH … CH_3 … O **1** + CH_3O_2C … Fe $(CO)_3$ … O … CH_3 … O **2**	129
COMMENTS	The erythro epoxide-alcohol **1** is isolated (72 %) accompanied by the epoxide-ketone **2** (13 %). The absolute stereochemistry of **1** is deduced from previous data.	
5. HO … Me … H … Fe $(CO)_3$ … CO_2Me **1**	VO (acac)$_2$, PhCH$_3$ / tBuOOH, 20°C, 6 h / 71 % → HO … O … H … Me … Fe $(CO)_3$ … CO_2Me **2**	
COMMENTS	The hydroxyl assisted epoxidation of the iron complexed Ψ-endo alkenol **1** is highly stereoselective since only *one* epoxide is isolated. The relative configuration of **2** is deduced from an X-ray cristallographic analysis. It is interesting to note that the isomeric Ψ-exo alkenol gives only degradation products. The (Z,E)-geometry of **1** is retained during the epoxidation step.	129

L. Formation of Epoxides

TYPE OF REACTION	FORMATION OF EPOXIDES LINKED TO IRON COMPLEXED DIENES 1/6	References
1.	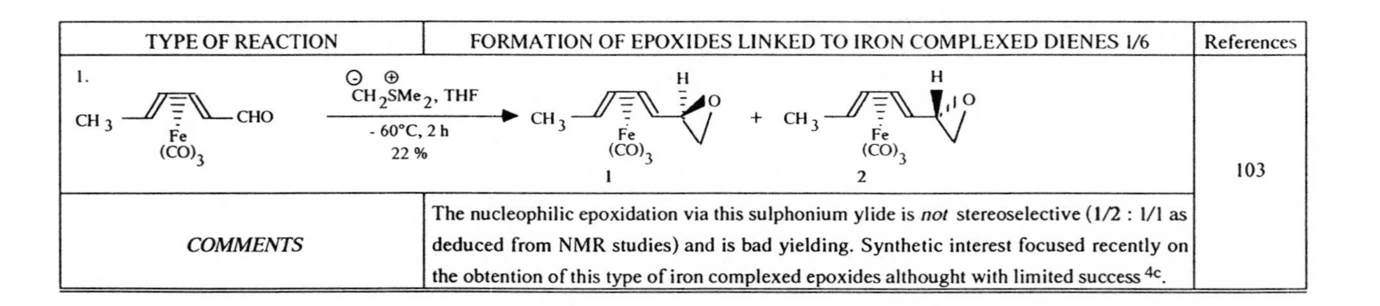	103
COMMENTS	The nucleophilic epoxidation via this sulphonium ylide is *not* stereoselective (1/2 : 1/1 as deduced from NMR studies) and is bad yielding. Synthetic interest focused recently on the obtention of this type of iron complexed epoxides althought with limited success [4c].	

TYPE OF REACTION	FORMATION OF EPOXIDES LINKED TO IRON COMPLEXED DIENES 2/6	References

2.

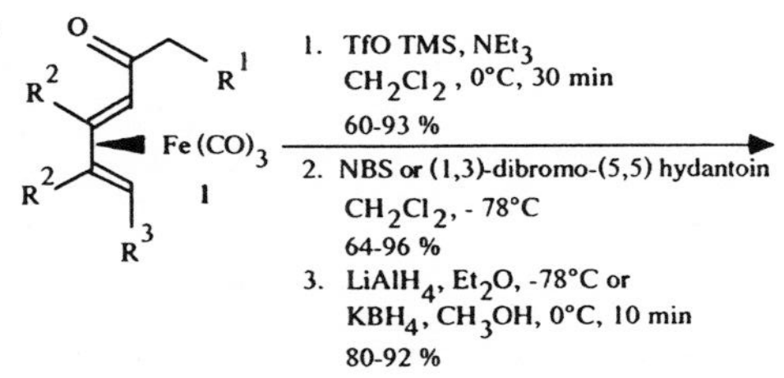

1. TfO TMS, NEt_3
   CH_2Cl_2 , 0°C, 30 min
   60-93 %
2. NBS or (1,3)-dibromo-(5,5) hydantoin
   CH_2Cl_2, - 78°C
   64-96 %
3. $LiAlH_4$, Et_2O, -78°C or
   KBH_4, CH_3OH, 0°C, 10 min
   80-92 %

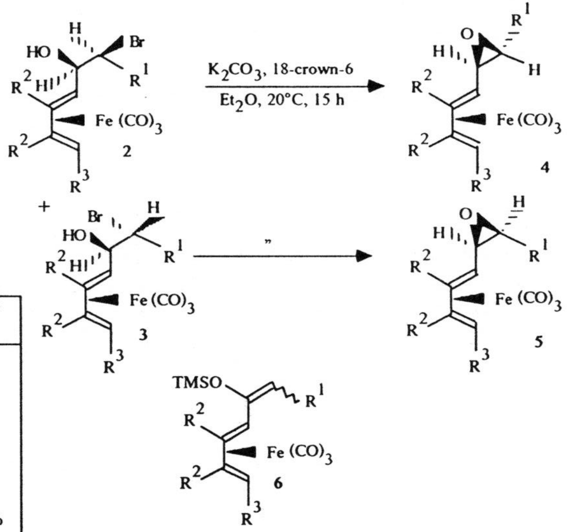

K_2CO_3, 18-crown-6
Et_2O, 20°C, 15 h

R^1	R^2	R^3	Yield of 4	Yield of 5
- H	- H	- H	90 %	90 %
- Me	- Me	- H	87 %	–
- Me	- Me	- H	–	84 %
- H	- H	- Me	67 %	67 %
- H	- H	- CO_2Me	95 %	95 %
- $(CH_2)_3CO_2Me$	- H	- CO_2Me	80-90 %	–
- $(CH_2)_3CO_2Me$	- H	- CO_2Me	–	80-90 %

References: 104, 105

COMMENTS

The electrophilic bromination of the intermediate enoxy silanes 6 leads to chromatographically separable diastereoisomeric iron complexed halogeno ketones which are the precursors of the halogenohydrines 2 and 3 in a ratio of about 1/1. The reduction is stereospecific and furnishes only the Ψ-endo alcohols 2 or 3 [26]. A high-yielding cyclisation of the two types of halogenohydrins 2 and 3 is highly dependent on the use of the crown-ether 18-crown-6.

TYPE OF REACTION	FORMATION OF EPOXIDES LINKED TO IRON COMPLEXED DIENES 3/6	References
3.	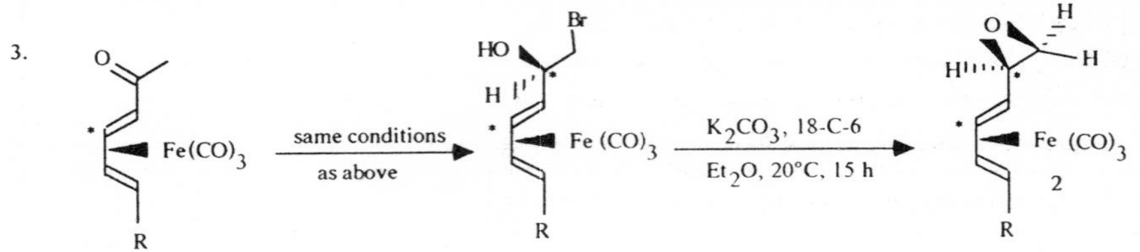 Fe(CO)$_3$ complex → (same conditions as above) → HO, H, Br, Fe (CO)$_3$ complex → (K_2CO_3, 18-C-6; Et_2O, 20°C, 15 h) → epoxide, Fe (CO)$_3$ complex **2**	105
COMMENTS	No racemisation of the optically active complexes is observed through the sequence since the optical purity of these two iron complexed epoxides measured by ^{1}H-NMR in the presence of Eu (hfc)$_3$ exceeds 96 %.	
4.	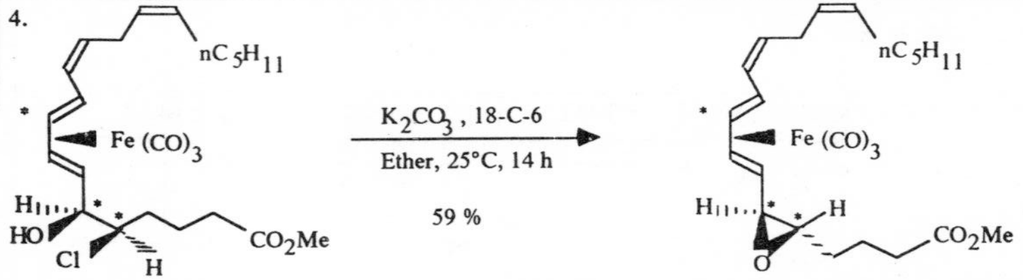 Chlorohydrin (HO, Cl, H, nC_5H_{11}, CO_2Me), Fe (CO)$_3$ complex → (K_2CO_3, 18-C-6; Ether, 25°C, 14 h; 59 %) → epoxide (nC_5H_{11}, CO_2Me), Fe (CO)$_3$ complex	105
COMMENTS	The epoxidation is not complete in these conditions since 36 % of the starting chlorhydrine is recovered.	

R	Yield of cyclisation
- CH_3	67 %
- CO_2Me	95 %

TYPE OF REACTION	FORMATION OF EPOXIDES LINKED TO IRON COMPLEXED DIENES 4/6	References
5. H_3C 2 O 1 4 Fe (CO)$_3$	1. TfOTMS, NEt_3 CH_2Cl_2, 0°C, 30 min 2. (1,3)-dibromo-(5,5)-hydantoin CH_2Cl_2, 78°C 3. DIBAH, Ether, -78°C 84 % → Br H OH Fe (CO)$_3$ 1 + Br HO H Fe (CO)$_3$ 2 1/2 : 95/5 KOH, 18-C-6 / Ether, 20°C, 8 h 93 % → O H Fe (CO)$_3$ 3 + O H Fe (CO)$_3$ 4 3/4 : 95/5	105
COMMENTS	The same strategy is efficient toward the synthesis of epoxides linked to iron complexed dienes at the position 2 of the complex.	

TYPE OF REACTION	FORMATION OF EPOXIDES LINKED TO IRON COMPLEXED DIENES 5/6	References
6. $Fe(CO)_3$ O 1. TfOTMS, NEt_3 CH_2Cl_2, 0°C 2. (1,3)-dibromo-(5,5)-hydantoin CH_2Cl_2, -78°C 77 %	CH_3 H Br O $Fe(CO)_3$ 1 + H CH_3 Br O $Fe(CO)_3$ 2 1/2 : 25/75 1 → $LiAlH_4$, Et_2O, - 78°C, 5 min, 95 % → CH_3 H Br H OH $Fe(CO)_3$ 3 + CH_3 H Br HO H $Fe(CO)_3$ 4 (3/4 : 55/45) 2 → " 95 % → H CH_3 Br H OH $Fe(CO)_3$ 5 3 → KOH, 18-C-6, Ether, 20°C, 15 h, > 98 % → CH_3 H O H $Fe(CO)_3$; 4 → same conditions > 98 % → CH_3 H O H $Fe(CO)_3$; 5 → same conditions > 98 % → H CH_3 O H $Fe(CO)_3$	105
COMMENTS	The relative configurations of the bromodienones 1, 2, bromohydrines 3, 4 and 5 are deduced from the geometry of the oxiranes assuming that the cyclisation is *trans* diaxial. It is interesting to note that the reduction of 2 gives only one diastereoisomeric alcohol 5.	

TYPE OF REACTION	FORMATION OF EPOXIDES LINKED TO IRON COMPLEXED DIENES 6/6	References
7. $Fe(CO)_3$ — 1. TfOTMS, NEt_3, CH_2Cl_2, 0°C, 30 min; 2. NCS, CH_2Cl_2, - 78°C, 65 % → CH_3, Cl, H, O, $Fe(CO)_3$ **1** + H, Cl, CH_3, O, $Fe(CO)_3$ **2**; 1/2 : 52/48; **2** — CH_3Li, THF, - 78 °C, 30 min, 83 % → CH_3, H, O, CH_3, $Fe(CO)_3$		105
COMMENTS	The nucleophilic addition of methyllithium on the chloroketone 2 is highly diastereoselective since only one oxirane is obtained. It is useful to note that this modification leads to trisubstituted iron complexed epoxides.	

M. Osmylation Reactions

TYPE OF REACTION	OSMYLATION OF IRON COMPLEXED TRIENES 1/2	References
1. 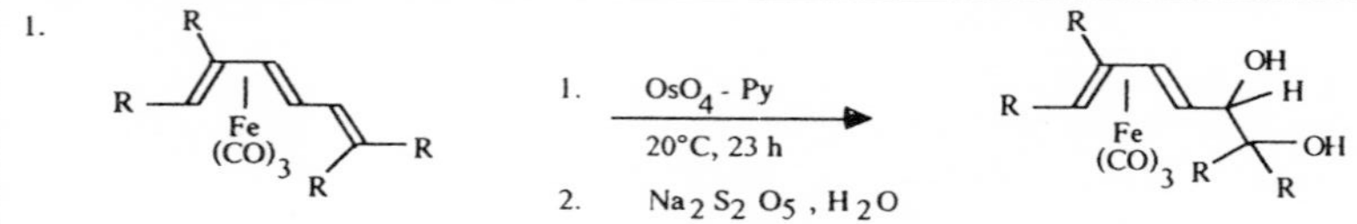	1. OsO_4 - Py, 20°C, 23 h 2. $Na_2S_2O_5$, H_2O, 20°C, 23 h R = -H : 10 % R = $-CH_3$: 69 %	160
COMMENTS	These results were given *without* indications about the diastereoselectivity of the osmylation reaction.	
2. 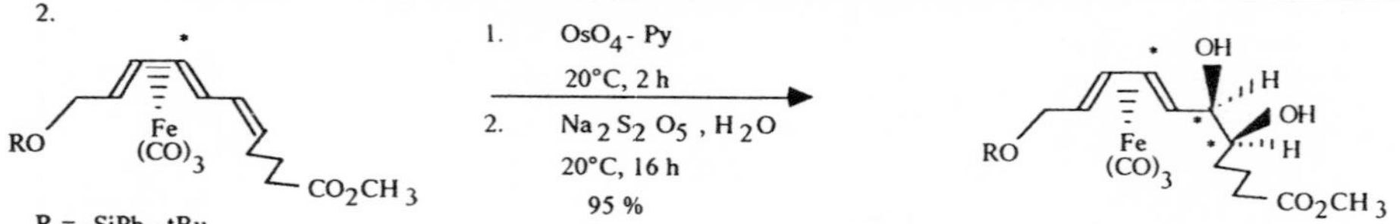 R = $-SiPh_2tBu$	1. OsO_4 - Py, 20°C, 2 h 2. $Na_2S_2O_5$, H_2O, 20°C, 16 h 95 %	112
COMMENTS	Only, *one* complexed diol is isolated and characterized. The absolute configuration of its cyclic carbonate has been unambiguously established by X-Ray cristallography [115].	
3. 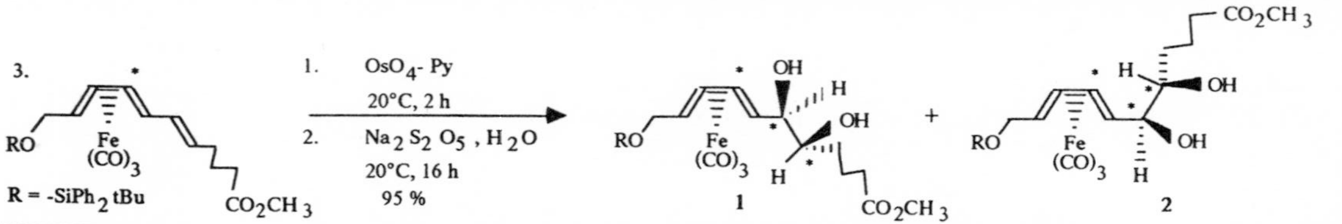 R = $-SiPh_2tBu$	1. OsO_4 - Py, 20°C, 2 h 2. $Na_2S_2O_5$, H_2O, 20°C, 16 h 95 % 1 + 2	112
COMMENTS	*Two* complexed diols 1 and 2 (1/2 : 90/10) are obtained. 1 arising from the anti nucleophilic attack by OsO_4 of the *s-trans* form of the alkene is the major one. Analoguously, the absolute configurations of their cyclic carbonates have been unambiguously established by X-Ray cristallography [115].	112

TYPE OF REACTION	OSMYLATION OF IRON COMPLEXED TRIENES 2/2	References
4. RO, Fe $(CO)_3$ — idem, 93 % → RO, Fe $(CO)_3$, OH, H, OH, H		113
COMMENTS	*One* complexed diol is obtained. The absolute configuration has been ascribed as drawn on the basis of previous literature data. It is particularly noteful to remark that the osmylation step is completely chemioselective for the double bond even in the presence of an excess OsO_4.	
5. $(CO)_3$ Fe, OTBS — α. 1. OsO_4, Py., THF 2. $NaHSO_3$, H_2O 99 % ; β. cat. OsO_4, tBuOOH, Et_4NOAc, CH_3COCH_3 20°C, 6 h →	$(CO)_3$ Fe, HO, OH, OTBS **1** + $(CO)_3$ Fe, O, OH, OTBS **2**	161
COMMENTS	In contrast of the stoechiometric osmylation, the catalytic version is also compatible with the organometallic unit with a good yield of the diol 1 (80 % yield) but always contaminated by the ketol 2 (20 % yield). It is worth to note the regioselectivity of this over oxidation, offering thus, interesting synthetic potentialities.	

N. Cyclopropanation Reactions

TYPE OF REACTION	CYCLOPROPANATION REACTION OF IRON COMPLEXED TRIENES 1/5	References
1. CH_3O_2C … Fe $(CO)_3$ … CH_3, CH_3	$CHCl_3$, $Et_3BnN^{\oplus}$, $Cl^{\ominus}$ / 50 % NaOH, 20°C, 45 min / 63 % → CH_3O_2C … Fe $(CO)_3$ … H, Cl, Cl, CH_3, CH_3	10b
COMMENTS	The carbene addition *anti* to the organometallic unit on the depicted s-trans conformation is stereospecific since *one* dichlorocyclopropane is isolated.	
2. CH_3O_2C … $(CO)_3$ Fe … CH_3, CH_3	$N_2CHCO_2CH_3$, $PhCH_3$ / Cu, 90°C / 66 % → CH_3O_2C … $(CO)_3$ Fe … H, H, CO_2CH_3, CH_3, CH_3 + CH_3O_2C … $(CO)_3$ Fe … H, CO_2CH_3, H, CH_3, CH_3 (1/1)	10
COMMENTS	These two chromatographically separable cyclopropanes when analyzed by ^1H- and ^{13}C-NMR shows the presence of only *one* diastereoisomer at the new C-C bond which is formed in the vicinity of the complex. The optical purity of these compounds is measured after chemical derivatization and is found to be greater than 90 %.	

TYPE OF REACTION	CYCLOPROPANATION REACTION OF IRON COMPLEXED TRIENES 2/5	References
3. Me, Fe $(CO)_3$, CO_2Me, 1 → $(CH_3)_2C^{\ominus}-N^{\oplus}{\equiv}N$, Ether, - 20°C, 1 h	3 : 74 % + 2 : 25 %; PhEt, 120°C, 30 min, 98 % → 5 : 78 % + 4 : 18 %	11,116, 133
COMMENTS	During the cycloaddition, only *one* Δ^1-pyrazoline **3** is characterized together with some Δ^2-pyrazoline **2**. This step is highly stereoselective and correlates well with the previous results in terms of conformational equilibrium of **1**. On the other hand, the thermal decomposition of the Δ^1-pyrazoline **3**, most likely through diradical intermediates is *not* stereospecific since the *trans* cyclopropane **4** is isolated together with the expected *cis* one **5**. High-field NMR experiments confirm that **4** and **5** are unique diastereoisomers.	

TYPE OF REACTION	CYCLOPROPANATION REACTION OF IRON COMPLEXED TRIENES 3/5	References
4. Me, Fe $(CO)_3$, CO_2Me, **1** → CH_3, CH_3, $N \equiv N$, Ether, - 20°C, 1 h	Me, Me, Me, N, H, N, Fe $(CO)_3$, CO_2Me, **2** : 26 % + Me, H, N = N, Fe $(CO)_3$, Me, Me, CO_2Me, H, **3** : 58 % + Me, N, N, H, Me, Me, Fe $(CO)_3$, CO_2Me, H, **4** : 8 % ; **3** → PhEt, 120°C, 30 min, 98 % → Me, H, Me, Me, Fe $(CO)_3$, CO_2Me, H, **5** ; **4** → " → Me, Me, Me, H, CO_2Me, Fe $(CO)_3$, H, **6**	11, 116, 133
COMMENTS	The cycloaddition of diazopropane on the s-trans enoate **1** is stereoselective (3/4 : 88/12). The thermolysis of the major adduct **3** gives only *one* chiral cyclopropane 5, while the thermolysis of the **3** and **4** mixture gives a mixture of 5 and **6**.	

TYPE OF REACTION	CYCLOPROPANATION REACTION OF IRON COMPLEXED TRIENES 4/5	References
5.	CH_3, CH_3, H, N, N, CO_2CH_3, Fe, $(CO)_3$ **2** : 54 % + CH_3, CH_3, N, N, H, CO_2CH_3, H, Fe, $(CO)_3$ **3** + CH_3 CH_3, H, N, N, H_3CO_2C, H, Fe, $(CO)_3$ **4**; **1** (CO_2Me, Fe, $(CO)_3$) → CH_3, CH_3, ⊕, ⊖, $N{\equiv}N$, Ether, 0°C; **3** + **4** : 26 % ; **3**/**4** : 4/1; **3** + **4** → PhEt, 120°C, 3 h., 70 % → **5** (Me, Me, H, CO_2Me, H, Fe, $(CO)_3$) + **6** (CO_2Me, Me, Me, H, H, Fe, $(CO)_3$); **5**/**6** : 4/1	11, 116
COMMENTS	The cycloaddition on this type of complex is moderately stereoselective. It is interesting to note that the free electrophilic double bond must not be sterically very encumbered since the Michaël adduct 2 is the major one. The thermolysis of the mixture of Δ^1 - pyrazolines **3**+**4** (**3**/**4** : 4/1) gives a mixture of the two cyclopropanes **5**+**6** in the same ratio **5**/**6** : 4/1. In this case, it is more likely that this step is highly stereoselective.	

TYPE OF REACTION	CYCLOPROPANATION REACTION OF IRON COMPLEXED TRIENES 5/5	References
6. Fe(CO)$_3$ — CH_2I_2, Ether; Zn / Cu; 18 h, 35°C; 23 % → Fe(CO)$_3$ (H, H)	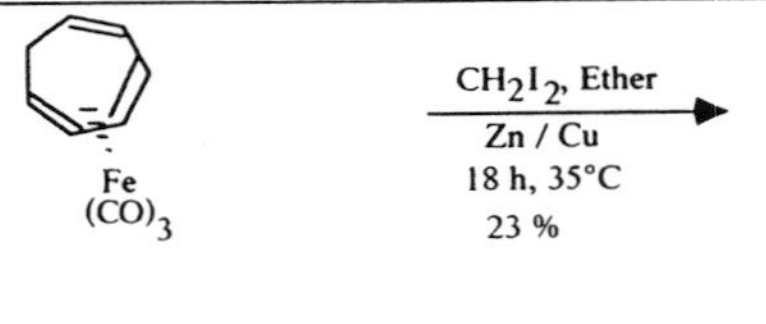	65
COMMENTS	The methylene addition takes place on the less sterically hindered face on the ring.	
7. Fe(CO)$_3$ **1** — CH_3CHN_2, Ether; 0°C, 12 h; 74 % → [N=N, H, CH_3, O, Fe(CO)$_3$ **2** + N=N, CH_3, H, O, Fe(CO)$_3$ **3**; **2**/**3** : 2.7/1] — Ph CH_3; 86°C, 1h; 100 % → CH_3, H, O, Fe(CO)$_3$ **4** + H, CH_3, O, Fe(CO)$_3$ **5**; 4/5 : 93/7	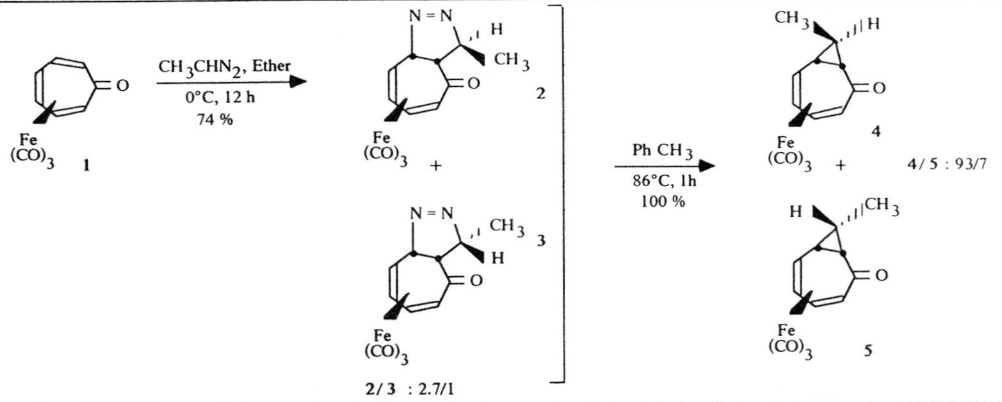	164
COMMENTS	The cycloaddition takes place on the less sterically hindered π face of the enone *anti* to the organometallic unit. The thermolysis step is *not* stereospecific when conducted on each Δ^1 - pyrazoline **2** or **3** since variable mixtures of **4** and **5** are obtained in each case.	

TYPE OF REACTION	[2 + 2] CYCLOADDITION OF AN IRON COMPLEXED IMINE WITH A KETENE 1/1	References
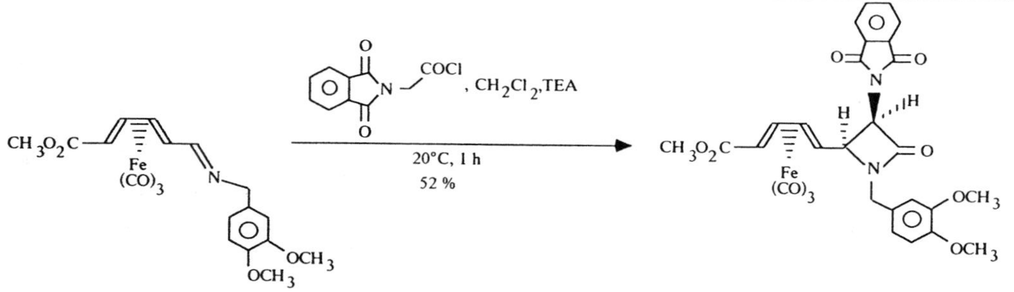 CH_3O_2C … Fe $(CO)_3$ … N … OCH_3, OCH_3	COCl, CH_2Cl_2, TEA; 20°C, 1 h; 52 % → CH_3O_2C … Fe $(CO)_3$ … H, H, O, N, O, O, N, OCH_3, OCH_3	10b
COMMENTS	The cycloaddition is highly stereoselective. One β-lactam cycloadduct is obtained with a *cis* stereochemistry. Based on previous results, the phtalimido ketene is thought to approach the imine in its *s-trans* conformation and *anti* to the organometallic unit.	

P. [3+2] Cycloadditions

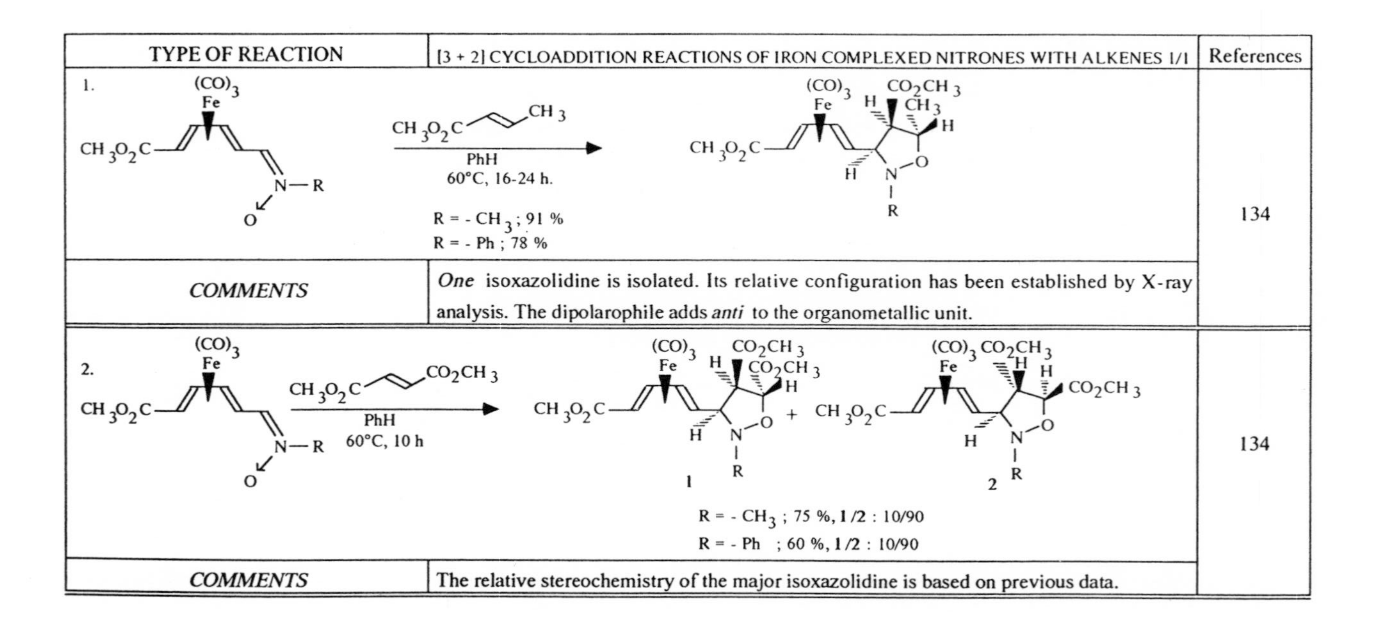

TYPE OF REACTION	[3 + 2] CYCLOADDITION REACTIONS OF IRON COMPLEXED NITRONES WITH ALKENES 1/1	References
1.	R = - CH_3 ; 91 % R = - Ph ; 78 %	134
COMMENTS	*One* isoxazolidine is isolated. Its relative configuration has been established by X-ray analysis. The dipolarophile adds *anti* to the organometallic unit.	
2.	R = - CH_3 ; 75 %, 1/2 : 10/90 R = - Ph ; 60 %, 1/2 : 10/90	134
COMMENTS	The relative stereochemistry of the major isoxazolidine is based on previous data.	

TYPE OF REACTION	[3 + 2] CYCLOADDITION OF AN IRON COMPLEXED TRIENE WITH AN AZOMETHINE YLIDE	References
CH_3O_2C–(triene)–Fe$(CO)_3$, CO_2CH_3, CO_2CH_3	(H, D), Ph, H, H, CO_2Et, Ph (aziridine); PhH, reflux, 14 h; 78 % → CH_3O_2C–(diene)–Fe$(CO)_3$ pyrrolidine: H, CO_2Et, N–Ph, H, Ph, CO_2CH_3, CO_2CH_3	10b
COMMENTS	*Two* diastereoisomeric pyrrolidines are formed in a ratio 1/2. The regioselectivity of the cycloaddition has been established par ^{1}H-NMR spectroscopy of the deuterated cycloadducts using the deuterium labelled aziridine. The relative stereochemistry of the cycloadducts has not been established.	

TYPE OF REACTION	CYCLOADDITION REACTION OF NITRILE OXIDES WITH IRON COMPLEXED TRIENES 1/1	References
1. R–triene–Fe$(CO)_3$ **1**	α. $R^1CH_2NO_2$, PhNCO; Cat. Et_3N, PhH; 20°C, 22 h. β. Cl(R[1])C=N - OH, Ether; Et_3N; 20°C, 2 h. → R–Fe$(CO)_3$ O–N R[1] **2** + R–Fe$(CO)_3$ O–N R[1] **3**	35
COMMENTS	The relative stereochemistry of each major cycloadduct is based on a single X-Ray analysis of **2c**. As expected, the nitrile oxides attack *anti* to the organometallic moiety on the complexed triene in its *s-trans* conformation. The two sets of experimental conditions α and β gave the same results in terms of yields and stereoselectivity.	
CH_3–triene–Fe$(CO)_3$ **1**	BnO–C$_6$H$_3$(CH$_3$O)–CH$_2$CH$_2$CH$_2$–NO_2, PhNCO; Cat. Et_3N, PhH; 20°C, 48 h; 75 % in **2** (**2**/**3** : 89/11). → CH_3–Fe$(CO)_3$ O–N –C$_6$H$_3$(OCH$_3$)(OBn) **2** + CH_3–Fe$(CO)_3$ O–N –C$_6$H$_3$(OCH$_3$)(OBn) **3**	135
COMMENTS	The absolute stereochemistry of the major cycloadduct **2** is based on previous results and on the sign of the optical rotation of the final product (+)-(S)-[6]-gingerol.	

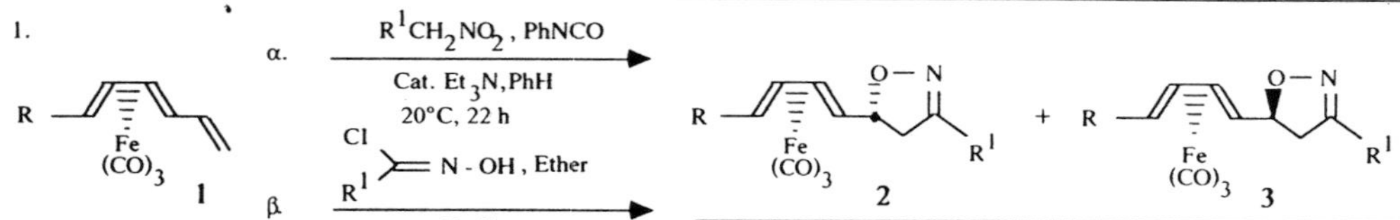

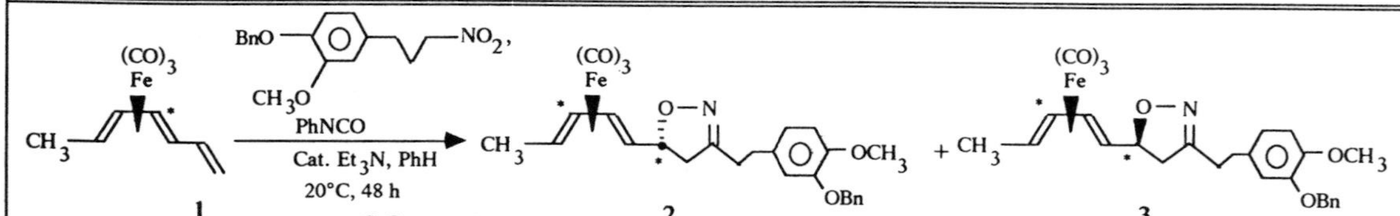

Entry	R	R[1]	2/3 ratio	Yield of 2
a	- CH_3	- CH_3	88/12	72 %
b	- CH_2OSiPh_2tBu	- CH_3	90/10	70 %
c	- CO_2CH_3	- CH_3	88/12	75 %
d	- CH_3	- Et	86/14	65 %
e	- CH_3	- tBu	90/10	83 %
f	- CH_3	- Ph	91/9	71 %

TYPE OF REACTION	DIELS-ALDER REACTION OF IRON COMPLEXED POLYENES 1/1	References
1. 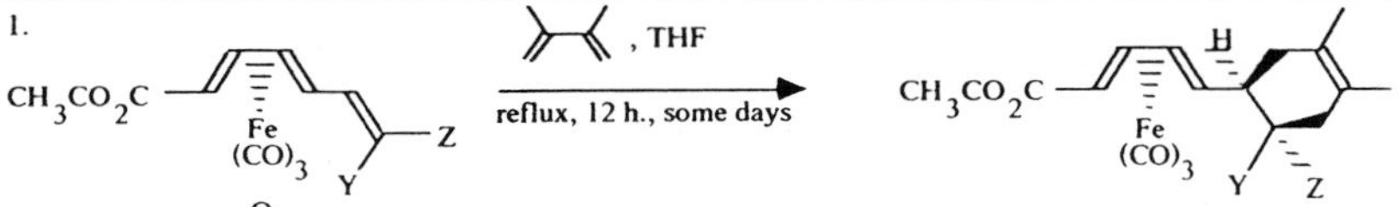 CH_3CO_2C — Fe $(CO)_3$ — Y, Z **1a** : Y = Z 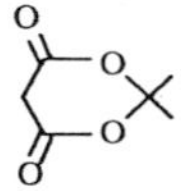 = (2,2-dimethyl-1,3-dioxane-4,6-dione) **1b** : Y = Z = - CO_2CH_3 **1c** : Y = Z = - CN **1d** : Y = - CN, Z = - CO_2CH_3 **1e** : Y = - H, Z = - CO_2CH_3	(2,3-dimethylbutadiene), THF; reflux, 12 h., some days → CH_3CO_2C — Fe $(CO)_3$ — H, Y, Z **2a** : 70 % **2b** : 8-10 % **2c** : 8-10 % **2d** : 8-10 % **2e** : 0 %	37
COMMENTS	*One* cycloadduct is prepared in each case. The relative configuration of **2a** established by its X-ray analysis confirm an addition of the diene *anti* to the organometallic moiety.	
2. 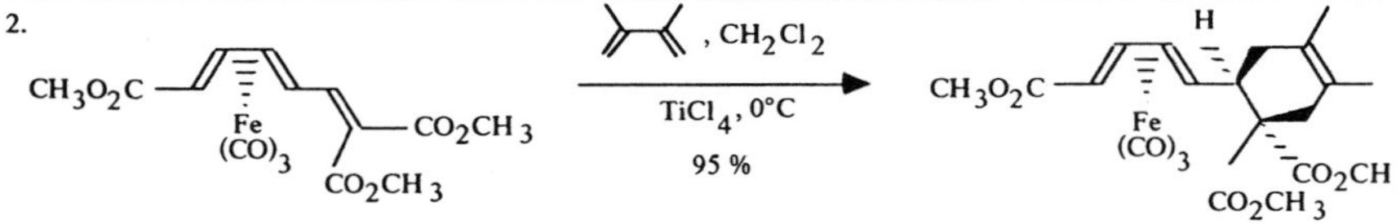 CH_3O_2C — Fe $(CO)_3$ — CO_2CH_3, CO_2CH_3	(2,3-dimethylbutadiene), CH_2Cl_2; $TiCl_4$, 0°C; 95 % → CH_3O_2C — Fe $(CO)_3$ — H, CO_2CH_3, CO_2CH_3	37
COMMENTS	Only *one* isomer is observed. The nature and the quantity of the Lewis acid (one equivalent) appear to be critical for the success of the cycloaddition.	

TYPE OF REACTION	DIELS-ALDER REACTION OF IRON COMPLEXED TRIENES 1/1	References
3.	CH_2Cl_2, high pressure (8-12 kbar), 20°C	157

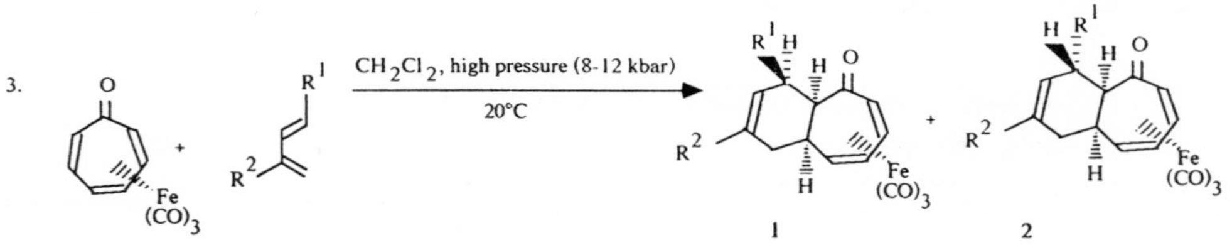

R^1	R^2	1 / 2	1 + 2
- OTMS	- H	3/1	96 %
- (3,5)-dinitro benzoate	- H	4/3	74 %
- H	- OTMS	1/0	47 %
- OAc	- H	1/1	75 %

COMMENTS

Usual Diels-Alder conditions failed contrary to the high-pressure variant. As expected for a normal enone, only *one* regioisomer is formed and the cycloaddition proceeds via an *endo* major transition state. The relative stereochemical assignements are based on a combination of X-ray crystallographic analysis and chemical correlation.

TYPE OF REACTION	DIEL-ALDER REACTION ON A LINEAR IRON COMPLEXED TETRAENE 1/1	References
CH_3O_2C — Fe $(CO)_3$	N - CH_3, THF; reflux, 4 h; 70 % → CH_3O_2C — Fe $(CO)_3$, H, CH_3, N, O, O	37
COMMENTS	A *single* adduct has been caracterized (X-Ray analysis). The dienophile reacts from the face *anti* to the organometallic unit with an expected *endo* transition state. No bond shift of the organometallic unit along the tetraene is observed before the cycloaddition.	

R. Hetero Diels-Alder Reaction

TYPE OF REACTION	HETERO DIELS-ALDER REACTION OF AN IRON COMPLEXED DIENAL 1/1	References
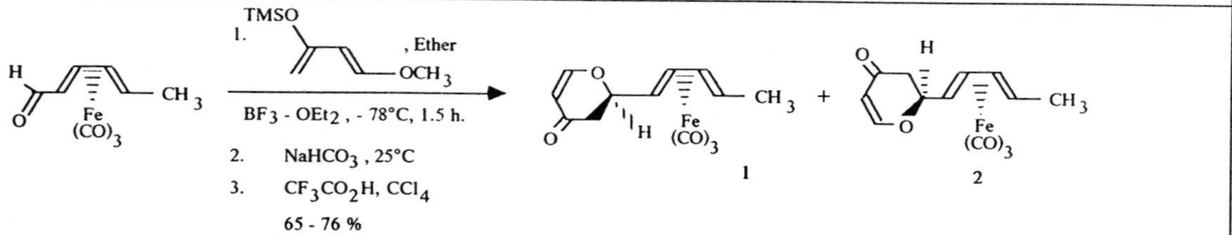 H, O, Fe, $(CO)_3$, CH_3 — 1. TMSO, OCH_3, Ether; $BF_3 - OEt_2$, - 78°C, 1.5 h. 2. $NaHCO_3$, 25°C 3. CF_3CO_2H, CCl_4 65 - 76 %	O, O, H, Fe, $(CO)_3$, CH_3 **1** + O, H, O, Fe, $(CO)_3$, CH_3 **2**	101
COMMENTS	*Two* separable iron complexed dihydropyrones 1 and 2 (1/2 : 3/1) have been isolated. The major one arises from the attack of the Danishevsky's diene on the carbonyl in the *s-cis* conformation, *anti* to the organometallic moiety. The relative stereochemistry of 2 has been established by an X-Ray analysis.	

TYPE OF REACTION	ALKYLATION OF IRON COMPLEXED METYL ESTER ENOLATE	References
		165

CO_2CH_3 (Fe(CO)₃ complex) → 1. LDA, THF, - 70°C, 10 min; 2. R^1X → 1 + 2

R^1	X	R^2	R^3	1 / 2	yield 1 + 2
$-CH_3$	-I	-H	$-CH_3$	93 / 7	97 %
$-CH_2Ph$	-Br	-H	$-CH_3$	91 / 9	51 %
$-CH_2CH=CH_2$	-Br	-H	$-CH_3$	87 / 13	48 %

COMMENTS

This *first* successfull case of alkylation of an anion adjacent to the organometallic group appears to be kinetically controlled. The X-Ray analysis of **2** (R^1 = $-CH_3$, $R^2 = R^3$ = -H) combined with chromatographic and spectroscopic data of related compounds established the depicted relative configuration of the *major* alkylated complex **1**. These results are consistent with an electrophilic attack on the *major s-trans* enolate *anti* to the bulky $Fe(CO)_3$ group.

ACKNOWLEDGMENTS

We thank Mrs. A.M. Moustier, Drs. A. Monpert and C. Guillou for useful comments. Professor W.A. Donaldson is gratefully acknowledged for his suggestions and fruitful exchange of information. We particularly thank Mrs. N. Voisin for her patience during the typewriting of this manuscript.

ADDENDUM

Since submission of this chapter, many new interesting developments in organoiron chemistry have been published (until beginning of 1995). They are briefly analyzed in this addendum, using the same organization as used in this chapter.

Section I: General Data

New acyclic organoiron complexes have been obtained in optically pure form using different strategies:

1. Separation of diastereoisomeric mixtures using chiral amides[166] in the case of sorbic acid derivatives or via chiral oxazolidines of imines for new complexes containing electron donor substituents.[167]
2. Biochemical resolution is not only compatible with such complexes but is also particularly efficient, both in reduction by bakers yeast and during reaction with lipases.[168,169]
3. Asymmetric complexation of dienes has also been reported[170] with, for the first time, an excellent diastereoselectivity in the case of a chiral dienamide.[171]
4. Reactions of symmetrical (pentadienyl) $Fe(CO)_3^+$ cations with a chiral optically pure phosphine give diastereoisomeric complexed phosphonium salts which could be separated.[172] This is a new example of a second-order asymmetric transformation in organometallic chemistry.[173]
5. New quantitative results have been obtained in terms of structure and stereodynamics on (diene) $Fe(CO)_2L$ complexes.[174] Particularly important, from a synthetic view point, are the strong effects of the substituents on diene and of the iron ligands on the racemization processes.[175]

Section II: η^5-Pentadienyl Cations

Studies dealing with the reactivity of these cations are actively pursued as follows:

1. The regioselectivity of the addition of carbo- or heteronucleophiles on such isolated cations has been discussed.[176]
2. The use of the *in situ* method appears of much synthetic interest since it is highly stereoselective[177,178] and occurs also with overall retention of configuration.[178] Interesting modifications of reactivity have also been observed

during the addition of alcohols on such cations in the presence of molecular sieves.[179]

3. Cross-conjugated pentadienyl cations complexed to $Fe(CO)_3$ are new starting points for diquinanes and eight-membered carbocycles.[180] Interestingly, on the trimethylenemethane intermediates, ozonolysis is nondestructive.
4. The intramolecular trapping, by sulfur nucleophiles, of these acyclic pentadienyl-complexed cations gives a new stereoselective synthesis of tetrahydrothiopyrans;[181] furthermore, the oxidation of these derivatives by oxone is compatible with the organometallic moiety and occurs also with a good diastereoselectivity giving the corresponding sulfoxides.

Sections III and IV: Synthetic Applications

A new, shorter, synthesis of optically active 5-HETE has been reported.[182] An erythro carbonate vicinal to an $Fe(CO)_3$ unit was already used as a key intermediate in the preparation of (5*R*, 6*S*)-diHETE. However, an unusual epimerization process was observed with this molecule, pointing out a new role for the $Fe(CO)_3$ group as a relay of chemical information.[183] A very interesting new approach to (9*Z*)-retinoic acid has been reported: in this synthesis the $Fe(CO)_3$ moiety plays a key role to control the required geometry of the double bond.[184] The highly diastereoselective Michael addition to alkylidene malonate, already used in the synthesis of (–)-verbenalol was developed in an elegant synthesis of the *as*-indacene unit of ikaguramycin.[185] Organoiron methodologies for the stereoselective synthesis of fragments of macrolactin A have been reported.[186] Interestingly, new applications of these complexes have also appeared recently in the field of liquid crystals[187] and amphiphilic compounds.[188]

Section V: Reactions Compatible with the Organometallic Unit

New developments in the Friedel–Crafts acylation include a one-carbon functionalization[189] and the synthesis of α-ketoesters.[190] Acyl halides adjacent to (η^4-diene) $Fe(CO)_3$ units react with allylsilanes to give the corresponding ketones.[191] They are also starting materials for the synthesis of new cyclopentadienes and fulvenes.[192] The 1,2 nucleophilic addition of functionalized zinc–copper reagents to a complexed aldehyde gives the expected secondary alcohol,[193] while addition of lithium alkyls on (*E,Z*)-complexed dienones produces the tertiary alcohols[194] in a highly stereoselective manner. This organometallic addition on imines is also stereoselective, especially in the case of organocerium reagents.[195] The 1,4-addition on alkylidene malonates substituted by the η^4-(diene) $Fe(CO)_3$ group is highly stereoselective, whatever the nucleophilic reagent used.[196] The hetero Diels–Alder reaction of Danishefsky's diene with a complexed benzylimine is the key step in the total synthesis of a biologically active piperidine alkaloid.[197] A closely related 4-piperidone has been obtained by an intramolecular Mannich type cyclization.[198] The classical methods for the preparation of allenes are also compatible with these complexes; they have been used for the synthesis of η^4-dienyl $Fe(CO)_3$ complexed γ-dienyl allenes.[199] A novel highly stereoselective method for the synthesis of chiral fluorides has been discovered recently; it uses, as key intermediates, these organoiron complexes.[200]

Interesting new cyclocarbonylation reactions have been reported.[201] An intramolecular version of this reaction appears especially useful from a synthetic point of view.[202] It is important to note that hydride[203] and carbon nucleophiles[204] may add directly on diene-$Fe(CO)_3$ complexes.

REFERENCES AND NOTES

1. (a) Reihlen, H.; Gruhl, A.; Von Hessling, G.; Pfrengle, O. *Liebigs Ann. Chem.* **1930**, *482*, 161; (b) Hallam, B.F.; Pauson, P.L. *J. Chem. Soc.* **1958**, 642.
2. (a) Koerner Von Gustorf, E.A.; Grevels, F.W.; Fischer, I. *The Organic Chemistry of Iron*; Academic Press: New York, 1978, Vol. 1; (b) Koerner Von Gustorf, E.A.; Grevels, F.W.; Fischer, I. *The Organic Chemistry of Iron*; Academic Press: New York, 1981, Vol. 2.
3. (a) Davies, S.G. *Organotransition Metal Chemistry: Applications to Organic Synthesis*; Pergamon Press: Oxford, 1982; (b) Pearson, A.J. *Metallo-Organic Chemistry*; John Wiley & Sons: New York, 1985.
4. (a) Franck-Neumann, M. *Organometallics in Organic Synthesis*; de Meijere, A.; tom Dieck, H., Eds.; Springer-Verlag: Berlin, 1987, 247; (b) Pearson, A.J. *Acc. Chem. Res.* **1980**, *13*, 463; Pearson, A.J. *Trans. Met. Chem.* **1981**, *6*, 67; Pearson, A.J. *Pure Appl. Chem.* **1983**, *55*, 1767; (c) Grée, R. *Synthesis* **1989**, 341; (d) Knölker, H.J. *Synlett.* **1992**, 371.
5. Whitlock, H.W. Jr.; Markezich, R.L. *J. Am. Chem. Soc.* **1971**, *93*, 5290; Markezich, R.L. Ph.D. Thesis, University of Wisconsin, 1971; *Diss. Abst. Int. B., 32*, 4, 2075B.
6. The complexes obtained from cyclic derivatives, such as cyclohexadienes or tropone have been already reviewed: see the refs. 3 and 4b.
7. (a) Musco, A.; Palumbo, R.; Paiaro, G. *Inorg. Chim. Acta* **1971**, *5*, 157; (b) de Montarby, L.; Tourbah, H.; Grée, R. *Bull. Soc. Chim. Fr.* **1989**, 419.
8. Franck-Neumann, M.; Martina, D.; Heitz, M.P. *Tetrahedron Lett.* **1982**, 3493.
9. (a) Roush, W.R.; Chan Park, J. *Tetrahedron Lett.* **1990**, *31*, 4707; (b) Howell, J.A.S.; Palin, M.G.; El Hafa, H.; Top, S.; Jaouen, G. *Tetrahedron: Asymmetry* in print.
10. (a) Monpert, A.; Martelli, J.; Grée, R.; Carrié, R. *Tetrahedron Lett.* **1981**, *22*, 1961; (b) Monpert, A. Docteur Ingenieur Thesis, University of Rennes, 1983.
11. Franck-Neumann, M.; Martina, D.; Heitz, M.P. *J. Organomet. Chem.* **1986**, *301*, 61.
12. (a) Gigou-Barbedette, A. Docteur Ingenieur Thesis, University of Rennes, 1990; (b) Mangeney, P.; Alexakis, A.; Normant, J.F. *Tetrahedron Lett.* **1988**, *29*, 2677.
13. (a) Pinsard, P. Docteur Ingenieur Thesis, University of Rennes, 1989; (b) Pinsard, P.; Lellouche, J.P.; Beaucourt J.P.; Grée, R. *Tetrahedron Lett.* **1990**, *31*, 1137.
14. Alcock, N.W.; Crout, D.H.G.; Henderson, C.M.; Thomas, S.E. *J. Chem. Soc., Chem. Commun.* **1988**, 746.
15. Franck-Neumann, M.; Briswalter, C.; Chemla, P.; Martina, D. *Synlett.* **1990**, 637.
16. Kappes, D.; Gerlach, H.; Zbinden, P.; Dobler, M.; König, W.A.; Krebber, R.; Wenz, G. *Angew. Chem. Int. Ed.* **1989**, *28*, 1657.
17. Solladié-Cavallo, A.; Suffert, A. *Magn. Reson. Chem.* **1985**, *23*, 739.
18. Xu, M.; Tran, C.D. *J. Chromatography* **1991**, *543*, 233.
19. Chemla, P. Ph.D. Thesis, University of Strasbourg, 1988.
20. Morey, J.; Grée, D.; Mosset, P.; Toupet, L.; Grée, R. *Tetrahedron Lett.* **1987**, *28*, 2959.
21. Donaldson, W.A.; Tao, C. *Synlett.* **1991**, 895. The total synthesis of (+)-5-HETE methyl ester has been recently performed using a partially modified route: Donaldson W.A., personal communication.

22. Djedaini, F.; Grée, D.; Martelli, J.; Grée, R.; Leroy, L.; Bolard, J.; Toupet, L. *Tetrahedron Lett.* **1989**, *30*, 3781.
23. Maglio, G.; Musco, A.; Palumbo, R.; Sirigu, A. *Chem. Comm.* **1971**, 100.
24. Kappes, D.; Gerlach, H.; Zbinden, P.; Dobler, M. *Helv. Chim. Acta* **1990**, *73*, 2136.
25. Howard, P.W.; Stephenson, G.R.; Taylor, S.C. *J. Chem. Soc., Chem. Commun.* **1990**, 1182; Stephenson, G.R.; Howard, P.W.; Taylor, S.C. *J. Chem. Soc., Chem. Commun.* **1991**, 127.
26. Clinton, N.A.; Lillya, C.P. *J. Am. Chem. Soc.* **1970**, *92*, 3058.
27. Schio, L. Ph.D. Thesis, University of Rennes, 1990.
28. Graf, R.E.; Lillya, C.P. *J. Organomet. Chem.* **1976**, *122*, 377.
29. Greaves, E.O.; Knox, G.R.; Pauson, P.L.; Thomas, S.; Sim, G.A.; Woodhouse, D.I. *J. Chem. Soc. Chem. Commun.* **1974**, 257.
30. Franck-Neumann, M.; Sedrati, M.; Mokhi, M. *J. Organomet. Chem.* **1987**, *326*, 389.
31. (a) Hafner, A.; Von Philipsborn, N.; Salzer, A. *Angew. Chem. Int. Ed.* **1985**, *97*, 136; (b) Salzer, A.; Schmalle, H.; Stauber, R.; Streiff, S. *J. Organomet. Chem.* **1991**, *408*, 403.
32. Pinsard; P.; Lellouche, J.P.; Beaucourt, J.P.; Toupet, L.; Schio, L.; Grée, R. *J. Organomet. Chem.* **1989**, *371*, 219.
33. Whitlock, H.W.; Reich, C.; Woessner, W.D. *Angew. Chem. Int. Ed.* **1971**, *93*, 2483; Whitlock, H.W.; Chua, H.Y.N. *J. Am. Chem. Soc.* **1965**, *87*, 3606.
34. Martina, D.; Brion, F. *Tetrahedron Lett.* **1982**, *23*, 865.
35. Le Gall, T.; Lellouche, J.P.; Toupet, L.; Beaucourt, J.P. *Tetrahedron Lett.* **1989**, *30*, 6517.
36. Laabassi, M.; Grée, R. *Tetrahedron Lett.* **1988**, *29*, 611.
37. Benvegnu, T.; Martelli, J.; Grée, R.; Toupet, L. *Tetrahedron Lett.* **1990**, *31*, 3145.
38. Colson, P.J.; Franck-Neumann, M.; Sedrati, M. *Tetrahedron Lett.* **1989**, *30*, 2393.
39. Franck-Neumann, M.; Sedrati, M.; Mokhi, M. *Tetrahedron Lett.* **1986**, *27*, 3861.
40. Franck-Neumann, M.; Sedrati, M.; Mokhi, M. *J. Organomet. Chem.* **1987**, *326*, 389.
41. Franck-Neumann, M.; Sedrati, M.; Mokhi, M. *New J. Chem.* **1990**, *14*, 471.
42. Nunn, K.; Mosset, P.; Grée, R.; Saalfrank, R.W.; Peters, K.; Von Schnering, H.G. *Angew. Chem. Int. Ed. Engl.* **1992**, *31*, 224.
43. Laabassi, M.; Grée, R. *Bull. Soc. Chim. Fr.* **1992**, *129*, 151.
44. Hill, L.; Richards, C.J.; Saberi, S.P.; Thomas, S.E. *Pure Appl. Chem.* **1992**, *64*, 371; Richards C.J.; Thomas, S.E. *Tetrahedron Asymmetry* **1992**, *3*, 143.
45. (a) Fischer, E.O.; Fischer, R.D. *Angew. Chem.* **1960**, *72*, 919; (b) Palotai, I.M.; Stephenson, G.R.; Ross, W.J.; Tupper, D.E. *J. Organomet. Chem.* **1989**, *364*, C11; (c) Pearson, A.J. *J. Chem. Soc. Chem. Commun.* **1977**, 339.
46. Mahler, J.E.; Pettit, R. *J. Am. Chem. Soc.* **1963**, *85*, 3955.
47. Dauben, H.J.; Gadecki, F.A.; Harmon, K.M.; Pearson, D.L. *J. Am. Chem. Soc.* **1957**, *79*, 4557.
48. Pearson, A.J.; O' Brien, M.K. *J. Org. Chem.* **1989**, *54*, 4663; Pearson, A.J.; Ong, C.W. *J. Am. Chem. Soc.* **1981**, *103*, 6686.
49. Meng, W.D.; Stephenson, G.R. *J. Organomet. Chem.* **1989**, *371*, 355.
50. Birch, A.J.; Haas, M.A. *J. Chem. Soc.* C, **1971**, 2465; Birch, A.J.; Chauncy, B.; Kelly, L.F.; Thompson, D.J. *J. Organomet. Chem.* **1985**, *286*, 37; Birch, A.J.; Haas, M.A. *Tetrahedron Lett.* **1968**, 3705.
51. Owen, D.A.; Stephenson, G.R.; Finch, H.; Swanson, S. *Tetrahedron Lett.* **1989**, *30*, 2607.
52. Stephenson, G.R.; Owen, D.A. *Tetrahedron Lett.* **1991**, *32*, 1291.
53. Donaldson, W.A.; Ramaswamy, M. *Synth. React. Inorg. Met. Org. Chem.* **1987**, *17*, 49.
54. Bayoud, R.S.; Biehl, E.R.; Reeves, P.C. *J. Organomet. Chem.* **1979**, *174*, 297.
55. Sorensen, T.S.; Jablonski, C.R. *J. Organomet. Chem.* **1970**, *25*, C62.
56. Lillya, C.P.; Sahatjian, R.A. *J. Organomet. Chem.* **1970**, *25*, C67.
57. (a) Clinton, N.A.; Lillya, C.P. *Chem. Commun.* **1968**, 579; (b) Clinton, N.A.; Lillya, C.P. *J. Am. Chem. Soc.* **1970**, *92*, 3065.
58. Bayoud, R.S.; Biehl, E.R.; Reeves, P.C. *J. Organomet. Chem.* **1978**, *150*, 75.

59. Foreman, M.I. *J. Organomet. Chem.* **1972**, *39*, 161.
60. Donaldson, W.A. *J. Organomet. Chem.* **1990**, *395*, 187.
61. Mahler, J.E.; Gibson, D.H.; Pettit, R. *J. Am. Chem. Soc.* **1963**, *85*, 3959.
62. (a) Bonner, T.G.; Holder, K.A.; Powell, P. *J. Organomet. Chem.* **1974**, *77*, C37; (b) Bonner, T.G.; Holder, K.A.; Powell, P.; Styles, E. *J. Organomet. Chem.* **1977**, *131*, 105. For two recent examples in the η^5-cyclohexadienyl iron complexes series, see: (c) Knölker, H.J.; Bauermeister, M. *J. Chem. Soc. Chem. Commun.* **1989**, 1468; (d) ibid. *J. Chem. Soc., Chem. Commun.* **1990**, 664; (e) Knölker, H.J.; Boese, R.; Hartmann, K. *Angew. Chem. Int. Ed. Engl.* **1989**, *28*, 1678.
63. (a) Niemer, B.; Breimair, J.; Wagner, B.; Polborn, K.; Beck, W. *Chem. Ber.* **1991**, *124*, 2227; (b) Lehmann, R.E.; Kochi, J.K. *Organometallics* **1991**, *10*, 190.
64. Hafner, A.; Bieri, J.H.; Prewo, R.; Von Philipsborn, W.; Salzer, A. *Angew. Chem.* **1983**, *95*, 736; Salzer, A.; Hafner, A. *Helv. Chim. Acta.* **1983**, *66*, 1774; Hafner, A.; Von Philipsborn, W.; Salzer, A. *Helv. Chim. Acta* **1986**, *69*, 1757.
65. Hafner, A.; Bieri, J.H.; Prewo, R.; Von Philipsborn, W.; Salzer, A. *Angew. Chem. Int. Ed. Engl.* **1983**, *22*, 713.
66. Pinsard, P.; Lellouche, J.P.; Beaucourt, J.P.; Grée, R. *J. Organomet. Chem.* **1988**, *354*, 193.
67. (a) Birch, A.J.; Pearson, A.J. *J. Chem. Soc. Perkin Trans. I* **1976**, 954. Cadmium organometallics have been also added to η^5-cyclohexadienyl iron complexed cations: Birch, A.J.; Pearson , A.J. *Tetrahedron Lett.* **1975**, 2379.
68. Whitesides, T.H.; Neilan, J.P. *J. Am. Chem. Soc.* **1976**, *98*, 63.
69. Mc Ardle, P.; Sherlock, H. *J. Chem. Soc. Chem. Commun.* **1976**, 537.
70. Maglio, G.; Palumbo, R. *J. Organomet. Chem.* **1974**, *76*, 367; Maglio, G.; Musco, A.; Palumbo, R. *J. Organomet. Chem.* **1971**, *32*, 127.
71. (a) Yeh, M.C.P.; Sun, M.L.; Lin, S.K. *Tetrahedron Lett.* **1991**, *32*, 113. The nucleophilic additions of organometallics on iron complexed cations are rather rare, even in the cyclohexadienyl series: see for example (b) Pearson, A.J. *Aust. J. Chem.* **1976**, *29*, 1101 (lithium dimethylcuprate); (c) Pearson, A.J. *Aust. J. Chem.* **1977**, *30*, 345 (dialkylcuprates); (d) Owen, D.A.; Stephenson, G.R.; Finch, H.; Swanson, S. *Tetrahedron Lett.* **1990**, *31*, 3401 (phenyllithium, lithium dimethylcuprate, sodiomalonate); (e) Birch, A.J.; Croes, P.E.; Lewis, J.; White, D.A.; Wild, S.B. *J. Chem. Soc. (A)* **1968**, 332 (methyllithium, sodiomalonate); (f) Bandara, B.M.R.; Birch, A.J.; Khor, T.C. *Tetrahedron Lett.* **1980**, *21*, 3625 (alkyllithiums); (g) Howard, P.W.; Stephenson, G.R.; Taylor, S.J. *J. Organomet. Chem.* **1989**, *370*, 97 (sodiomalonate).
72. Donaldson, W.A.; Ramaswamy, M. *Tetrahedron Lett.* **1989**, *30*, 1343.
73. Pearson, A.J.; Ray, T. *Tetrahedron* **1985**, *41*, 5765; Pearson, A.J.; Ray, T.; Richards, I.C.; Clardy, J.; Silveira, L. *Tetrahedron Lett.* **1983**, *24*, 5827; Donaldson, W.A.; Ramaswamy, M. *Tetrahedron Lett.* **1988**, *29*, 1343.
74. Mc Daniel, K.F.; Kracker II, L.R.; Thamburaj, P.K. *Tetrahedron Lett.* **1990**, *31*, 2373.
75. Mosset, P. Doctor Ingenieur Thesis, University of Rennes, 1984.
76. Grée, R.; Laabassi, M.; Mosset, P.; Carrié, R. *Tetrahedron Lett.* **1985**, *26*, 2317.
77. Laabassi, M. Ph.D. Thesis, University of Rennes, 1991.
78. Stephenson, G.R.; Voyle, M.; Williams, S. *Tetrahedron Lett.* **1991**, *32*, 5265.
79. Theoretical considerations dealing with this question are discussed in the following reference: Davies, S.G.; Green, M.L.H.; Mingos, D.M.P. *Tetrahedron* **1978**, *34*, 3047.
80. Nucleophilic additions at internal carbon atoms of η^5-pentadienyl complexed cations have been reported for several cyclic complexes: Edwards, R.; Howell, J.A.S.; Johnson, B.F.G.; Lewis, J. *J. Chem. Soc.* **1974**, 2105; Aumann, R. *J. Organomet. Chem.* **1973**, *47*, C29; Pearson, A.J.; Kole, S.L.; Ray, T. *J. Am. Chem. Soc.* **1984**, *106*, 6060; Pearson, A.J.; Holden M.S. *J. Organomet. Chem.* **1990**, *383*, 307; Pearson, A.J.; Burells, M.P. *J. Chem. Soc., Chem. Commun.* **1989**, 1332; Pearson, A.J. *Synlett.* **1990**, 10; Cotton, F.A.; Deeming, A.J.; Josry, P.L.; Ullah, S.S.; Domingos, A.J.P.; Johnson, B.F.G.; Lewis, J. *J. Am. Chem. Soc.* **1971**, *93*, 4624; Domingos, A.J.P.; Johnson, B.F.G.; Lewis, J. *J. Chem. Soc., Dalton Trans.* **1975**, 2288 (M = Ru); Sosinsky, B.A.; Knox, S.A.R.;

Stone, F.G.A. *J. Chem. Soc., Dalton Trans.* **1975**, 1633 (M = Ru); Evans, J.; Johnson, B.F.G.; Lewis, J. *J. Chem. Soc., Dalton Trans.* **1972**, 2668 (M = Co, Ir, Rh); Edwards, R.; Howell, J.A.S.; Johnson, B.F.G.; Lewis, J. *J. Chem. Soc., Dalton Trans.* **1974**, 2105; Deeming, A.J.; Ullah, S.S.; Domingos, A.J.P.; Johnson, B.F.G.; Lewis, J. *J. Chem. Soc., Dalton Trans.* **1974**, 2093 (M = Fe, Os, Ru).

81. This type of nucleophilic additions is rare in the case of acyclic complexes: (a) Whitesides, T.H.; Neilan, J.P. *J. Am. Chem. Soc.* **1976**, *98*, 63; (b) Bleeke, J.R.; Rauscher, D.J. *J. Am. Chem. Soc.* **1989**, *111*, 8972 (M = Ru); (c) Bleeke, J.R.; Hays, M.K. *Organometallics* **1987**, *6*, 1367; (d) Bleeke, J.R.; Wittenbrink, R.J.; Clayton, T.W. Jr.; Chiang, M.Y. *J. Am. Chem. Soc.* **1990**, *112*, 6539; (e) Roell Jr., B.C.; Mc Daniel, K.F. *J. Am. Chem. Soc.* **1990**, *112*, 9004 (M = Mn).
82. Donaldson, W.A.; Ramaswany, M. *Tetrahedron Lett.* **1989**, *30*, 1339.
83. Riley, P.E.; Davis, R.E. *Acta. Cryst.* **1976**, *B32*, 381.
84. O'Brien, M.K.; Pearson, A.J.; Pinkerton, A.A.; Schmidt, W.; Willman, K. *J. Am. Chem. Soc.* **1989**, *111*, 1499.
85. Uemura, M.; Minami, T.; Yamashita, Y.; Hiyoshi, K.; Hayashi, Y. *Tetrahedron Lett.* **1987**, *28*, 641; silyl enol ethers in the presence of BF_3-Et_2O react in a similarly regiospecific manner with η^2-tetracarbonyliron complexes of γ-acetoxy-α,β-unsaturated esters: see ref. 86.
86. Green, J.R.; Carroll, M.K. *Tetrahedron Lett.* **1991**, *32*, 1141.
87. To the best of our knowledge, two examples of this type of intramolecular trapping of iron complexes cations were known previously; see the references 4 (preparation of an iron complexed lactone) and 88 (synthesis of a *cis*-hydrindane derivative). An interesting case of intramolecular trapping of an η^3-π-allyl molybdenum cation has also been reported: Pearson, A.J. *Synlett.* **1990**, 10.
88. Pearson, A.J.; Chandler, M. *J. Chem. Soc. Perkin Trans. 1* **1982**, 2641.
89. As part of polyether antibiotics and marine natural products, oxygenated heterocycles renewed the interest in novel and versatile methodologies for their preparation, especially in the optically active form: Boivin, T.L.B. *Tetrahedron* **1987**, *43*, 3309; Moore, R.E. *Marine Natural Products*; Scheuer, P.J., Ed.; Academic Press: New York, 1978, Vol. 1, 43; ibid., Scheuer, P.J., Ed.; Academic Press: New York, 1983, Vol. 5, 201; Lin, Y.Y.; Risk, M.; Rays, S.M.; Van Engen, D.; Clardy, J.; Golik, J.; James, J.C.; Nakanishi, K. *J. Am. Chem. Soc.* **1981**, *103*, 6773; Nicolaou, K.C.; Prasad, C.V.C.; Ogilvie, W.W. *J. Am. Chem. Soc.* **1990**, *112*, 4988; Shimizu, Y.; Chou, H.N.; Bando, H.; Van Duyne, G.; Clardy, J.C. *J. Am. Chem. Soc.* **1986**, *108*, 514; Pawlak, J.; Tempesta, M.S.; Golik, J.; Zagorski, M.G.; Lee, M.S.; Nakanishi, K.; Iwashita, T.; Gross, M.L.; Tomer, K.B. *J. Am. Chem. Soc.* **1987**, *109*, 1144.
90. (a) Teniou, A. Ph.D. Thesis, University of Rennes, 1991; (b) Teniou, A.; Toupet, L.; Grée, R. *Synlett.* **1991**, 195.
91. Grée, D.; Grée, R.; Lowinger, T.B.; Martelli, J.; Negri, J.T.; Paquette, L.A. *J. Am. Chem. Soc.* **1992**, *114*, 8841.
92. Cordonnier, M.A. Diplôme d'Etudes Approfondies, University of Rennes, 1991.
93. Hachem, A.; Teniou, A.; Grée, R. *Bull. Soc. Chim. Belg.* **1991**, *100*, 625.
94. Paquette, L.A.; Negri, J.T. *J. Am. Chem. Soc.* **1991**, *113*, 5072; *ibid.* **1992**, *114*, 8835.
95. Grée, D.; Martelli, J.; Grée R., unpublished results.
96. Lellouche, J.P.; Bulot, E.; Beaucourt, J.P.; Martelli, J.; Grée, R. *J. Organomet. Chem.* **1988**, *342*, C21.
97. See for instance: Nelson, N.A.; Kelly, R.C.; Johnson, R.A. *Chem. Eng. News* **1982**, 30; Moore, P.K. *Prostanoids: Pharmacological Physiological and Clinical Relevance,* Cambridge University Press: London, 1985; Johnson, M.; Carey, F.; Mc Millan, R.M. In *Assays in Biochemistry*; Campbell, P.N.; Marshall, R.D., Eds., Academic Press: London, 1983, Vol. 19, p. 40; Watkins, D.W.; Myron, B.; Peterson, J.; Fletcher, R. *Prostaglandins in Clinical Practice*; Raven Press: New York, 1989; Mitra, A. *The Synthesis of Prostaglandins*; John Wiley and Sons: New York, 1977, and cited references.

98. *Leukotrienes and Lipoxygenases: Chemical, Biological and Clinical Aspects*; Rokach, J., Ed.; Elsevier: New York, 1989; *SRS-A and Leukotrienes*, Sheard, P.; Piper, P.J., Ed.; John Wiley and Sons: Chichester, 1981; *The Leukotrienes, Chemistry and Biology*; Chakrin, L.W.; Bailey, D.M., Ed.; Academic Press: Orlando, 1984; Samuelsson, B. *Angew. Chem. Int. Ed. Engl.* **1982**, *21*, 902; ibid., *Science* **1983**, *220*, 568.
99. Fitzpatrick, F.A.; Murphy, R.C. *Pharmacol. Rev.* **1989**, *40*, 229; Capdevila, J.H.; Falck, J.R.; Estabrook, R.W.; Faseb, J. **1992**, *6*, 731.
100. Spector, A.A.; Gordon, J.A.; Moore, S.A. *Prog. Lip. Res.* **1988**, *27*, 271.
101. Donaldson, W.A.; Tao, C.; Bennett, D.W.; Grubisha, D.S. *J. Org. Chem.* **1991**, *56*, 4563.
102. Gigou, A.; Lellouche, J.P.; Grée, R., unpublished results.
103. Stephenson, G.R.; Hastings, M. *J. Organomet. Chem.* **1989**, *375*, C27.
104. Franck-Neumann, M.; Abdali, A.; Colson, P.J.; Sedrati, M. *Synlett.* **1991**, 331.
105. Franck-Neumann, M.; Colson, P.J. *Synlett.* **1991**, 891.
106. Franck-Neumann, M.; Colson, P.J.; Geoffroy, P.; Taba, K.M. *Tetrahedron Lett.* **1992**, *33*, 1903.
107. Nunn, K.; Mosset, P.; Grée, R.; Saalfrank, R.W. *Angew. Chem. Int. Ed. Engl.* **1988**, *27*, 1188.
108. Danheiser, R.L.; Carini, D.J.; Kwasigroch, C.A. *J. Org. Chem.* **1986**, *51*, 3870.
109. Guindon, Y.; Delorme, D.; Lau, C.K.; Zamboni, R. *J. Org. Chem.* **1988**, *53*, 267; Leblanc, Y.; Fitzsimmons, B.J.; Zamboni, R.; Rokach, J. *J. Org. Chem.* **1988**, *53*, 265.
110. Yadagiri, P.; Lumin, S.; Falck, J.R.; Karara, A.; Capdevila, J. *Tetrahedron Lett.* **1989**, *30*, 429.
111. Nunn, K.; Mosset, P.; Grée, R.; Saalfrank, R.W. *J. Org. Chem.* **1992**, *57*, 3359.
112. Gigou, A.; Lellouche, J.P.; Beaucourt, J.P.; Toupet, L.; Grée, R. *Angew. Chem. Int. Ed. Engl.* **1989**, *28*, 755; Lellouche, J.P.; Gigou-Barbedette, A.; Grée, R. *Bull. Soc. Chim. Fr.* **1992**, *129*, 605.
113. Gigou, A.; Beaucourt, J.P.; Lellouche, J.P.; Grée, R. *Tetrahedron Lett.* **1991**, *32*, 635.
114. For earlier examples see: Johnson, B.F.G.; Lewis, J.; Parker, D.G.; Postle, S.R. *J. Chem. Soc. Dalton Trans.* **1977**, 794, and ref. 3b.
115. Toupet, L.; Grée, R.; Gigou-Barbedette, A.; Lellouche, J.P.; Beaucourt, J.P. *Acta Cryst.* **1991**, *C47*, 1173.
116. Heitz, M.P. Ph.D. Thesis, University of Strasbourg, 1983.
117. Adams, J.; Fitzsimmons, B.J.; Girard, Y.; Leblanc, Y.; Evans, J.F.; Rokach, J. *J. Am. Chem. Soc.* **1985**, *107*, 464.
118. Pinsard, P.; Lellouche, J.P.; Beaucourt, J.P.; Grée, R. *Tetrahedron Lett.* **1990**, *31*, 1141.
119. Jaouen, G.; Vessieres, A.; Top, S.; Ismail, A.A.; Butler, I.S. *C.R. Acad. Sci.* Paris, Ser. II **1984**, *298*, 683; Jaouen, G.; Vessieres, A.; Top, S. *Pure Appl. Chem.* **1985**, *57*, 1865; Jaouen, G.; Vessieres, A.; Top, S. *J. Am. Chem. Soc.* **1985**, *107*, 4778; Tondue, S.; Top, S.; Vessieres, A.; Jaouen, G. *J. Chem. Soc., Chem. Commun.* **1985**, 326; Savignac, M.; Jaouen, G.; Rodger, C.A.; Perrier, R.E.; Sayer, B.J.; Mc Glinchey, M.J. *J. Org. Chem.* **1986**, *51*, 2328; Jaouen, G.; Vessieres, A.; Top, S.; Savignac, M.; Ismail, A.A. *Organometallics* **1987**, *6*, 1985.
120. Teniou, A.; Grée, R. *Bull. Soc. Chim. Belg.* **1991**, *100*, 411.
121. Namiki, M.; Igarashi, Y.; Sakamoto, K.; Nakamura, T.; Koga, Y. *Biochem. Biophys. Res. Commun.* **1986**, *138*, 540; Sumitomo Chemicals Patent, Eur. Pat. Appl. 1986, n°*183*, 177.
122. Schio, L.; Grée, R., unpublished results.
123. Grée, R.; Kessabi, J.; Mosset, P.; Martelli, J.; Carrié, R. *Tetrahedron Lett.* **1984**, *25*, 3697.
124. Knox, G.R.; Thom, I.G. *J. Chem. Soc., Chem. Commun.* **1981**, 373.
125. Streinz, L.; Vrkoc, J.; Konecny, K.; Ramanuk, M. *Chem. Abs.* **1991**, *114*: 122717g; ibid., Eur. Pat. Appl. 1989, n°*378*, 228.
126. Semmelhack, M.F.; Fewkes, E.J. *Tetrahedron Lett.* **1987**, *28*, 1497.
127. Boehm, M.F.; Prestwich, G.D. *J. Org. Chem.* **1987**, *52*, 1349; ibid. *J. Labelled Comp. Radiopharm.* **1988**, *XXV*, 653.
128. Brion, J.D.; Le Baut, G.; Ducrey, P.; Plessard-Robert, S.; Cudennec, C.; Seurre, G. Eur. Pat. Appl. 1989, n°337, 885.
129. Laabassi, M.; Toupet, L.; Grée, R. *Bull. Soc. Chim. Fr.* **1992**, *129*, 47, and cited references.

130. Lellouche, J.P.; Breton, P.; Beaucourt, J.P.; Toupet, L.; Grée, R. *Tetrahedron Lett.* **1988**, *29*, 2449.
131. Taylor, G.A. *J. Chem. Soc. Perkin Trans. 1* **1979**, 1716.
132. Monpert, A.; Martelli, J.; Grée, R.; Carrie, R. *Nouv. J. Chim.* **1983**, *7*, 345.
133. Franck-Neumann, M.; Martina, D.; Heitz, M.P. *Tetrahedron Lett.* **1982**, *23*, 3493.
134. Monpert, A.; Martelli, J.; Grée, R. *J. Organomet. Chem.* **1981**, *210*, C45.
135. Le Gall, T.; Lellouche, J.P.; Beaucourt, J.P. *Tetrahedron Lett.* **1989**, *30*, 6521.
136. Martelli, J.; Grée, D.; Kessabi, J.; Grée, R.; Toupet, L. *Tetrahedron Lett.* **1989**, *45*, 4213.
137. Murai, F.; Tagawa, M.; Damtoft, S.; Jensen, S.R.; Nielsen, B.J. *Chem. Pharm. Bull.* **1984**, *32*, 2809.
138. Laabassi, M.; Grée, R. *Synlett.* **1990**, 265.
139. (a) Boulaajaj, S.; Le Gall, T.; Vaultier, M.; Grée, R.; Toupet, L.; Carrié, R. *Tetrahedron Lett.* **1987**, *28*, 1761; (b) Boulaajaj, S. Doctorate Thesis, University of Rennes, 1985.
140. Knouzi, N.; Vaultier, M.; Carrié, R. *Bull. Soc. Chim. Fr.* **1985**, 815.
141. Hudlicky, T.; Seoane, G.; Price, J.D.; Gadamasetti, K.G. *Synlett.* **1990**, 433.
142. Still, W.C.; Galynker, I. *Tetrahedron* **1981**, *37*, 3981.
143. Corey, E.J.; Weigel, L.O.; Chamberlin, A.R.; Cho, H.; Hua, D.H. *J. Am. Chem. Soc.* **1980**, *102*, 6615; Corey, E.J.; Kim, S.; Yoo, S.E.; Nicolaou, K.C.; Melvin Jr., L.S.; Brunelle, J.J.; Falck, J.R.; Trybulski, E.J.; Lett, R.; Sheldrake, P.W. *J. Am. Chem. Soc.* **1978**, *100*, 4620; see also Graham, R.J.; Weiler, R. *Tetrahedron Lett.* **1991**, *32*, 1027, and cited references.
144. Still, W.C.; Novack, V.J. *J. Am. Chem. Soc.* **1984**, *106*, 1148.
145. Saunders, M.; Kouk, K.N.; Wu, Y.D.; Still, W.C.; Lipton, M.; Chang, G.; Guida, W.C. *J. Am. Chem. Soc.* **1990**, *112*, 1419.
146. Omura, S. *Macrolides Antibiotics Chemistry, Biology and Practice*; Academic Press: New York, 1984.
147. Boden, E.P.; Keck, G.E. *J. Org. Chem.* **1985**, *50*, 2394; Kurihara, T.; Nakajima, Y.; Mitsunobu, O. *Tetrahedron Lett.* **1976**, *28*, 2455.
148. Franck-Neumann, M.; Chemla, P.; Martina, D. *Synlett.* **1990**, 641.
149. Greaves, E.O.; Knox, G.R.; Pauson, P.L. *J. Chem. Soc., Chem. Commun.* **1969**, 1124; see also Hardy, A.D.U.; Sim, G.A. *J. Chem. Soc., Dalton Trans.* **1972**, 2305 where the X-ray analysis of the following π-allyl complex has been described.

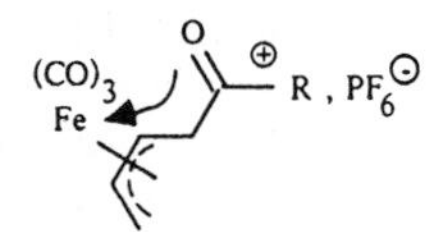

150. Frederiksen, J.S.; Graf, R.E.; Gresham, D.G.; Lillya, C.P. *J. Am. Chem. Soc.* **1979**, *101*, 3863.
151. Graf, R.E.; Lillya, C.P. *J. Organomet. Chem.* **1979**, *166*, 53.
152. Adams, C.M.; Cerioni, G.; Hafner, A.; Kalchhauser, H.; Von Philipsborn, W.; Prewo, R.; Schwenk, A. *Helv. Chim. Acta.* **1988**, *71*, 1116.
153. Kuhn, D.E.; Lillya, C.P. *J. Am. Chem. Soc.* **1972**, *94*, 1682.
154. Pearson, A.J.; Chang. K. *J. Chem. Soc., Chem. Commun.* **1991**, 394.
155. Ashby, E.C.; Prather, J. *J. Am. Chem. Soc.* **1986**, *88*, 729.
156. Stüber, S.; Ugi, I. *Synthesis* **1974**, 437.
157. Rigby, J.H.; Ogbu, C.O. *Tetrahedron Lett.* **1990**, *31*, 3385.
158. Barton, D.H.R.; Gunatilaka, A.A.L.; Nakasishi, T.; Patin, H.; Widdowson, D.A.; Worth, B.R. *J. Chem. Soc. Perkin Trans. 1* **1976**, 821; Evans, G.; Johnson, B.F.G.; Lewis, J. *J. Organomet. Chem.* **1975**, *102*, 507.
159. Banthorpe, D.V.; Fitton, H.; Lewis, J. *J. Chem. Soc. Perkin Trans. 1* **1973**, 2051.
160. Johnson, B.F.G.; Lewis, J.; Parker, D.G.; Postle, S.R. *J. Chem. Soc., Dalton Trans.* **1977**, 794.
161. Pearson, A.J.; Srinivasan, K. *J. Chem. Soc., Chem. Commun.* **1991**, 393.

162. Grée, R.; Laabassi, M.; Mosset, P.; Carrié, R. *Tetrahedron Lett.* **1984**, *25*, 3693.
163. Postnov, V.N.; Andrianov, V.G.; Struchkov, Y.T.; Baran, A.M.; Sazonova, V.A. *J. Organomet. Chem.* **1984**, *262*, 201.
164. Martina, D. Ph.D. Thesis, University of Strasbourg, 1977; Franck-Neumann, M.; Martina, D. *Tetrahedron Lett.* **1975**, *22 + 23*, 1759.
165. Donaldson, W.A.; Graig, R.; Spanton, S. *Tetrahedron Lett.* **1992**, *33*, 3967.
166. Nakanishi, S.; Yamamoto, H.; Otsuji, Y.; Nakazumi, H. *Tetrahedron Asymmetry* **1993**, *4*, 1969.
167. Howell, J.A.S.; Bell, A.G.; O'Leary, P.J.; Mc Ardle, P.; Cunningham, D.; Stephenson, G.R.; Hastings M. *Organometallics* **1994**, *13* (5), 1806.
168. (a) Howell, J.A.S.; Palin, M.G.; Jaouen, G.; El Hafa, H.; Top, S. *Tetrahedron Asymmetry* **1992**, *3*, 1355; (b) Howell, J.A.S.; Palin, M.G.; Jaouen, G.; Top, S.; El Hafa, H.; Cense, J.-M. *Tetrahedron Asymmetry* **1993**, *4*, 1241.
169. (a) Uemura, M.; Nishimura, H.; Yamada, S.; Nakamura, K.; Hayashi, Y. *Tetrahedron Lett.* **1993**, *34*, 6581; (b) Uemura, M.; Nishimura, H.; Yamada, S.; Hayashi, Y.; Nakamura, K.; Ishihara, K.; Ohno, A. *Tetrahedron Asymmetry* **1994**, *5*, 1673.
170. Schmalz, H.-G.; Heßler, E.; Bats, J.W.; Dürner, G. *Tetrahedron Lett.* **1994**, *35*, 4543.
171. Pearson, A.J.; Chang, K.; Mc Conville, D.B.; Youngs, W.J. *Organometallics* **1994**, *13*, 4.
172. Donaldson, W.A.; Shang, L.; Rogers, R.D. *Organometallics* **1994**, *13*, 6.
173. (a) For older references in this area see: Howell, J.A.S.; Thomas, M.J. *J. Chem. Soc., Dalton Trans.* **1983**, 1401; (b) Kane-Maguire, L.A.P.; Honig, E.D.; Sweigart, D.A. *Chem. Rev.* **1984**, *84*, 525.
174. (a) Howell, J.A.S.; Bell, A.G.; Cunningham, D.; Mc Ardle, P.; Albright, T.A.; Goldschmidt, Z.; Gottlieb, H.E.; Hezroni-Langerman, D. *Organometallics* **1993**, *12*, 2541; (b) Claire, K.S.; Howarth, O.W.; Mc Camley, A. *J. Chem. Soc., Dalton Trans.* **1994**, *18*, 2615; *Chem. Abstr.* 122: 10228g.
175. Howell, J.A.S.; Squibb, A.D.; Bell, A.G.; Mc Ardle, P.; Cunningham, D.; Goldschmidt, Z.; Gottlieb, H.E.; Hezroni-Langerman, D. *Organometallics* **1994**, *13*, 4336.
176. (a) Donaldson, W.A.; Bell, P.T.; Jin, M.-J. *J. Org. Chem.* **1992**, *441*, 449; (b) Donaldson, W.A.; Jin, M.-J. *Bull. Soc. Chim. Belg.* **1993**, *102*, 297; (c) Donaldson, W.A.; Jin, M.-J.; Bell, P.T. *Organometallics* **1993**, *12*, 1174; (d) Donaldson, W.A.; Jin, M.-J. *Tetrahedron* **1993**, *49*, 8787; (e) Donaldson, W.A.; Shang, L. *Tetrahedron Letters* **1995**, *36*, 1575.
177. Heßler, E.; Schmalz, H.-G.; Dürner, G. *Tetrahedron Lett.* **1994**, *35*, 4547.
178. Roush, W.R.; Wada, C.K. *Tetrahedron Lett.* **1994**, *35*, 7347.
179. Quirosa-Guillou, C.; Lellouche, J.-P. *J. Org. Chem.* **1994**, *59*, 4693.
180. Franck-Neumann, M.; Kastler, A. *Synlett* **1995**, 61.
181. Hachem, A.; Toupet, L.; Grée, R. *Tetrahedron Lett.* **1995**, *36*, 1849.
182. Tao, C.; Donaldson, W.A. *J. Org. Chem.* **1993**, *58*, 2134.
183. Lellouche, J.-P.; Gigou-Barbedette, A.; Grée, R. *J. Organomet. Chem.* **1993**, *461*, 167.
184. Wada, A.; Hiraishi, S.; Ito, M. *Chem. Pharm. Bull.* **1994**, *42*, 757.
185. Roush, W.R.; Wada, C.K. *J. Am. Chem. Soc.* **1994**, *116*, 2151.
186. (a) Benvegnu, T.; Schio, L.; Le Floc'h, Y.; Grée, R. *Synlett* **1994**, 505; (b) Donaldson, W.A.; Bell, P.T.; Wang, Z.; Bennett, D.W. *Tetrahedron Lett.* **1994**, *35*, 5829.
187. Ziminski, L.; Malthête, J. *J. Chem. Soc., Chem. Commun.* **1990**, 1496.
188. Nakanishi, S.; Kataoka, M.; Otsuji, Y. *Chem. Express* **1992**, *7*, 921; *Chem. Abstr.* **1993**, 118: 255077c.
189. Franck-Neumann, M.; Bissinger, P.; Geoffroy, P. *Tetrahedron Lett.* **1993**, *34*, 4643.
190. Franck-Neumann, M.; Geoffroy, P. *Tetrahedron Lett.* **1994**, *35*, 7027.
191. Nakanishi, S.; Kumeta, K.; Otsuji, Y. *Tetrahedron Lett.* **1994**, *35*, 3727.
192. Nakanishi, S.; Kumeta, K.; Terada, K. *Synthesis* **1995**, 33.
193. Wang, J.L.; Ueng, C.H.; Yeh, M.C.P. *J. Chin. Chem. Soc.* **1994**, *41*, 129; *Chem. Abstr.* **1994**, 121: 35797j.

194. Takemoto, Y.; Takeuchi, J.; Iwata, C. *Tetrahedron Lett.* **1993**, *34*, 6067.
195. Takemoto, Y.; Takeuchi, J.; Iwata, C. *Tetrahedron Lett.* **1993**, *34*, 6069.
196. Wada, C.K.; Roush, W.A. *Tetrahedron Lett.* **1994**, *35*, 7351.
197. Takemoto, Y.; Ueda, S.; Takeuchi, J.; Nakamoto, T.; Iwata, C. *Tetrahedron Lett.* **1994**, *35*, 8821.
198. Ripoche, I.; Gelas, J.; Grée, D.; Grée, R.; Troin, Y. *Tetrahedron Lett.* **1995**, in press.
199. Nunn, K.; Mosset, P.; Grée, R.; Saalfrank, R.W. *Tetrahedron* **1993**, *49*, 9775.
200. Grée, D.; Kermarrec, C.; Martelli, J.; Grée, R.; Toupet, L.; J.P. Lellouche, to be published.
201. (a) Franck-Neumann, M.; Michelotti, E.L.; Simler, R.; Vernier, J.-M. *Tetrahedron Lett.* **1992**, *33*, 7361; (b) Franck-Neumann, M.; Vernier J.-M. *Tetrahedron Lett.* **1992**, *33*, 7365.
202. (a) Yeh, M-C.P.; Sheu, B.-A.; Fu, H.-W.; Tau, S.-I.; Chuang, L.-W. *J. Am. Chem. Soc.* **1993**, *115*, 5941; (b) Wang, J.-L.; Ueng, C-H.; Cheng, S-J.; Yeh, M-C.P. *Organometallics* **1994**, *13*, 4453.
203. Chang, S.; White, P.S.; Brookhart, M. *Organometallics* **1993**, *12*, 3636.
204. (a) Chou, S-S.P.; Hsu, C-H.; Yeh, M-C. *Tetrahedron Lett.* **1992**, *33*, 643; (b) Hwu, C.C.; Yeh, M.C.P. *Huaxue* **1993**, *51*, 417; *Chem. Abstr.* **1994,** 121: 133238x.

NOVEL CARBONYLATION REACTIONS CATALYZED BY TRANSITION METAL COMPLEXES

Masanobu Hidai and Youichi Ishii

Advances in Metal-Organic Chemistry
Volume 4, pages 275–309.

ISBN: 1-55938-709-2

I. INTRODUCTION

Carbonylation reactions catalyzed by transition metal complexes have long been studied and recognized as one of the most versatile reactions in organic syntheses.[1,2] Furthermore, they are also of great importance in industrial processes. In spite of their long history, much effort is still being paid to develop both new carbonylation reactions and catalyst systems. In this chapter we would like to review our recent work on the development of novel types of carbonylation reactions. The first part describes the cyclocarbonylation of allylic compounds, and the second part summarizes our recent results on carbonylation reactions by using homogeneous multimetallic catalysts.

II. NOVEL CYCLOCARBONYLATION OF ALLYLIC COMPOUNDS

A. Synthesis of Naphthyl Acetates

Much attention has recently been focused on carbonylation of aromatic C–H bonds. In the later 1980s, rhodium- or iridium-catalyzed photocarbonylation of aromatic hydrocarbons to the corresponding aldehydes was developed by Tanaka[3] and Eisenberg,[4] and palladium catalyzed oxidative carbonylation to acids by Fujiwara[5] (Scheme 1).

The former reaction proceeds by oxidative addition of the C–H bond to the metal center of the catalyst, while the latter reaction by electrophilic attack of a Pd(II) species on the C–H bond. In both reactions the first step of the catalytic cycle includes the aromatic C–H bond fission by an active metal species to form a metal–carbon bond. The following CO insertion into the aryl–metal bond and reductive elimination of the aroyl ligand thus formed yield the product (Scheme 2, A). On the other hand, another possible process for the carbonylation of aromatic C–H bonds includes the reaction of an aromatic system with a preformed acyl complex (Scheme 2, B). In the transition metal-catalyzed carbonylation reactions, acyl complexes play important roles as intermediates in many cases. They react with alcohols, amines, olefins, and dihydrogen to yield organic products such as

$$C_6H_6 + CO \xrightarrow[h\nu]{RhCl(CO)(PMe_3)_2} C_6H_5CHO$$

$$C_6H_6 + CO + {}^tBuOOH \xrightarrow{Pd(OAc)_2} C_6H_5COOH$$

Scheme 1.

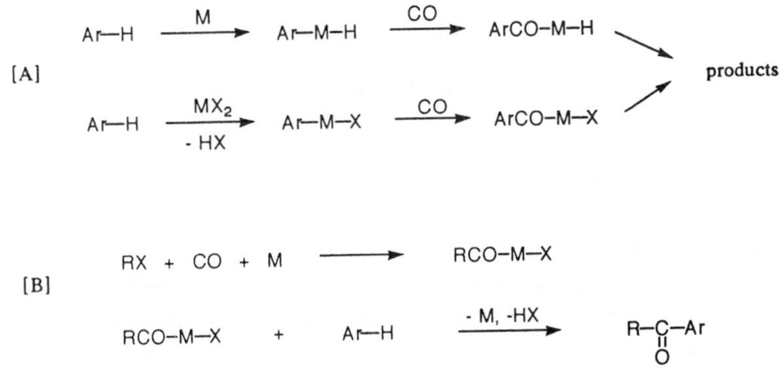

[A]

Ar—H —M→ Ar—M—H —CO→ ArCO—M—H ↘ products

Ar—H —MX_2, - HX→ Ar—M—X —CO→ ArCO—M—X ↗

[B]

RX + CO + M ——→ RCO—M—X

RCO—M—X + Ar—H —- M, -HX→ R—C(=O)—Ar

Scheme 2.

esters, amides, ketones, and aldehydes. However, until recently aromatic systems have failed to be noticed as reagents to attack the acyl intermediates.

With the intention of developing a novel carbonylation reaction of aromatic C–H bonds, we have embarked on an investigation of catalytic carbonylation via the reaction of aromatic C–H bonds with acyl intermediates. We focused our attention on intramolecular carbonylation of substrates with two functional groups in the same molecule: a functional group which can react with a metal species and CO to form an acyl complex, and an aromatic group which undergoes intramolecular reaction with the acyl complex. In such a reaction, one carbon elongation by the carbonylation and the following cyclization (cyclocarbonylation) leads to a new fused aromatic system. In spite of the potential versatility of the cyclocarbonylation of aromatic systems, only a few such examples have been reported (Scheme 3).[6–9]

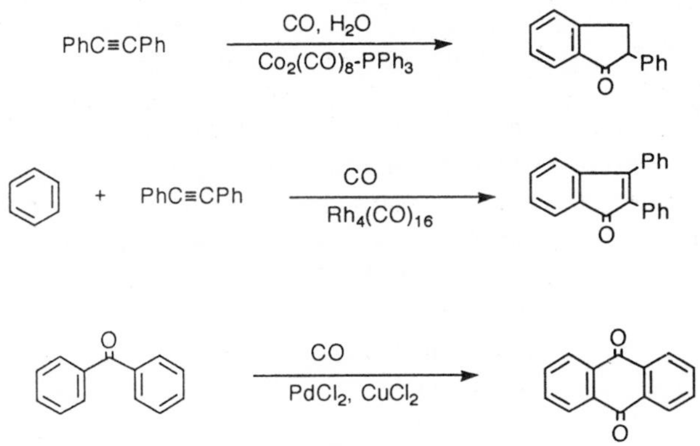

PhC≡CPh —CO, H_2O / $Co_2(CO)_8$-PPh_3→ [2-phenylindan-1-one]

[benzene] + PhC≡CPh —CO / $Rh_4(CO)_{16}$→ [2,3-diphenylinden-1-one]

[benzophenone] —CO / $PdCl_2$, $CuCl_2$→ [anthraquinone]

Scheme 3.

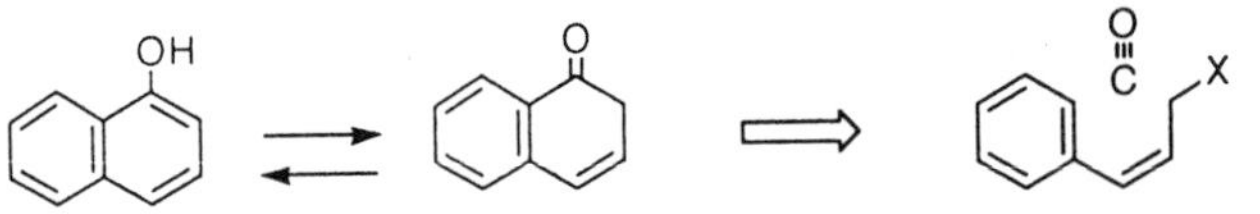

OH O O C X

Scheme 4.

The products of these reactions have been limited to a narrow range of compounds such as indanones and quinones. In addition, very drastic reaction conditions are required in most cases. These limitations have prevented the cyclocarbonylations of aromatic systems from systematic application in organic syntheses.

Based on the background described above, we selected 1-naphthol as the first synthetic target molecule because it can be regarded as a hiding member of cyclic aromatic ketone (Scheme 4). Simple retrosynthetic analysis leads to an idea that readily available cinnamyl compounds are suitable substrates.

Although attempted carbonylation of cinnamyl acetate (**1a**) in the presence of NEt_3 and a catalytic amount of $PdCl_2(PPh_3)_2$ gave no isolable and characterizable product, we found that addition of Ac_2O is very effective to make the reaction clean. 1-Naphthol, the expected product was isolated as the corresponding acetate. Thus, **1a** and cinnamyl bromide (**1b**) were smoothly cyclocarbonylated to form 1-naphthyl acetate (**2a**) in 74% and 41% yield, respectively, when heated at 160 °C under 60 atm of CO in the presence of Ac_2O, NEt_3, and a catalytic amount of $PdCl_2(PPh_3)_2$ (Eq. 1).[10]

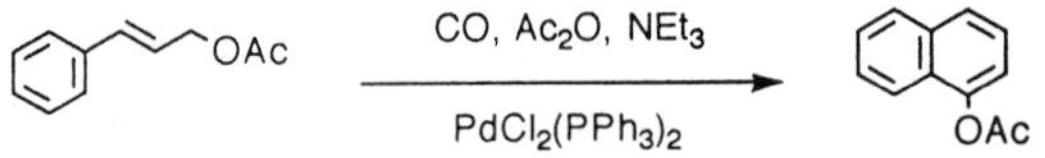

OAc — CO, Ac_2O, NEt_3 / $PdCl_2(PPh_3)_2$ → OAc (1)

1a **2a** 74%

Synthesis of 1-naphthyl acetate. Cinnamyl acetate (**1a**, 10 mmol), Ac_2O (20 mmol), NEt_3 (20 mmol), $PdCl_2(PPh_3)_2$ (0.07 mmol), and benzene (8 mL) were charged in a stainless steel autoclave under a nitrogen atmosphere. The reactor was closed, pressurized to 60 atm with CO, heated quickly up to 160 °C, and kept at this temperature for 1 h with magnetic stirring. The reactor was then cooled to room temperature and the gas purged. GC analysis of the reaction mixture indicated that 1-naphthyl acetate was formed in 74% yield based on the **1a** charged (92% conversion). The reaction mixture was washed with 1 N aqueous HCl, aqueous $NaHCO_3$, and water, and the resulting solution was evaporated to give a brown oil. The oil was purified by silica gel column chromatography followed by bulb-to-bulb distillation to give 1-naphthyl acetate in 46% yield.

Addition of NEt_3 is essential to obtain **2a** in good yields. The base is necessary to quench the acetic acid (and HBr) formed, and to promote the acetylation of

Table 1. Effect of Catalyst on the Cyclocarbonylation of Cinnamyl Acetate (**1a**)[a]

Catalyst	*Conv. (%)*	*Yield (%)*[b]
$PdCl_2(PPh_3)_2$	92	74
$Pd(CO)(PPh_3)_3$	88	88
$Pd(PPh_3)_4$	73	70
$PtCl_2(PPh_3)_2$	66	44
$NiBr_2(PPh_3)_2$	14	14
$Ru_3(CO)_{12}$	9	9
$Co_2(CO)_8$	7	5
$RhCl(CO)(PPh_3)_2$	2	2

Notes: [a]Reaction conditions: cinnamyl acetate 10 mmol, catalyst 0.07 mmol, Ac_2O 20 mmol, NEt_3 20 mmol, benzene 8 mL, CO 60 atm, 160 °C, 1 h.
[b]Based on the cinnamyl acetate charged.

1-naphthol. The reaction proceeded smoothly at 160 °C to give **2a** in good yield, but lower reaction temperatures resulted in a drastic decrease in the yield of **2a** and formation of a complex mixture of unidentified high-boiling compounds.

A variety of palladium monophosphine complexes were found to be effective catalysts for cyclocarbonylation of **1a** (Table 1). Zero-valent palladium–phosphine complexes such as $Pd(PPh_3)_4$ and $Pd(CO)(PPh_3)_3$ were the most effective catalysts. $PdCl_2(PPh_3)_2$ was also conveniently used under usual conditions (0.7–5 mol %), but analogous complexes such as $PdCl_2(PMePh_2)_2$ and $PdCl_2(PMe_2Ph)_2$ showed somewhat higher catalytic activity under low catalyst concentration conditions. In contrast, none of $Pd(OAc)_2$, $PdCl_2(dppe)$ (dppe = $Ph_2PCH_2CH_2PPh_2$), and $PdCl_2(AsPh_3)_2$ showed catalytic activity. Not only palladium complexes but also $PtCl_2(PPh_3)_2$ was an effective catalyst. Compound **2a** was also formed by the catalysis of $NiBr_2(PPh_3)_2$, $Co_2(CO)_8$, $Ru_3(CO)_{12}$, and $RhCl(CO)(PPh_3)_2$, but the yields were quite low. Based on these results, we concluded that smooth cyclocarbonylation of **1a** to give **2a** requires a reaction temperature of about 160 °C and the presence of Ac_2O, NEt_3, and a catalytic amount of palladium or platinum monophosphine complexes.

R OAc **1f-h** —(CO, Ac_2O, NEt_3 / $PdCl_2(PPh_3)_2$)→ R OAc **2f-h** + R OAc **2f-h'** (2)

Table 2. Synthesis of Naphthyl Acetates[a]

Substrate	*Product*	*Conv. (%)*	*Yield(%)*[b]
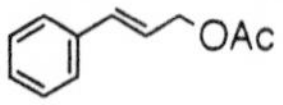 1a (OAc)	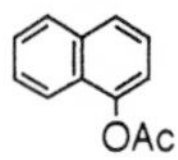 2a (OAc)	92	74
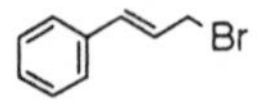 1b (Br)	2a	100	41
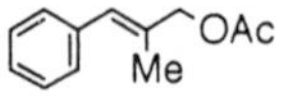 1c (Me, OAc)	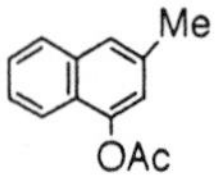 2c (Me, OAc)	69	59
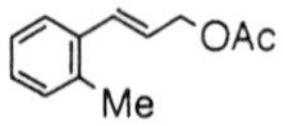 1d (Me, OAc)	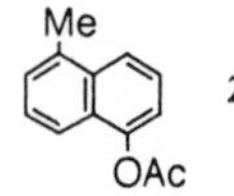 2d (Me, OAc)	81	58
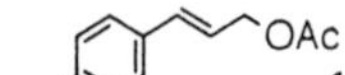 1e (Me, OAc)	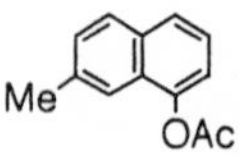 2e (Me, OAc)	90	76
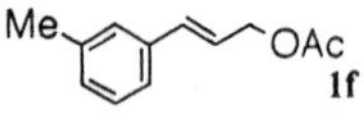 1f (Me, OAc)	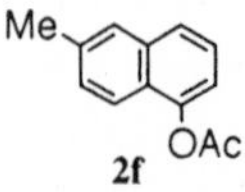 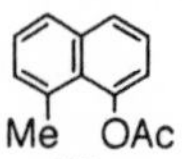 2f (Me, OAc); 2f' (Me, OAc)	83	69 (58 : 42)
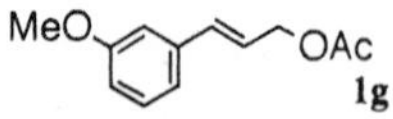 1g (MeO, OAc)	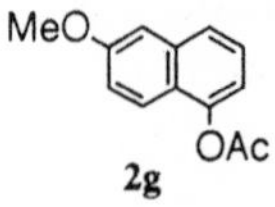 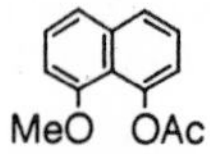 2g (MeO, OAc); 2g' (MeO, OAc)	77	77 (78 : 22)
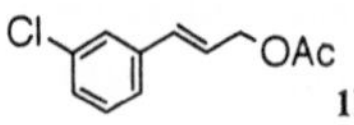 1h (Cl, OAc)	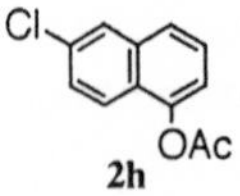 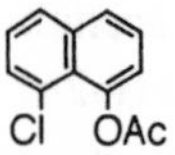 2h (Cl, OAc); 2h' (Cl, OAc)	91	88 (74 : 26)

Notes: [a]Reaction conditions: substrate 10 mmol, $PdCl_2(PPh_3)_2$ 0.07 mmol, Ac_2O 20 mmol, NEt_3 20 mmol, benzene 8 mL; CO 60 atm, 160 °C, 1 h.

The applicability of this cyclocarbonylation to the preparation of a series of substituted 1-naphthyl acetates was examined as shown in Table 2. Among the substrates examined, α- or γ-methylcinnamyl acetate failed to yield the corresponding naphthyl acetate due to the palladium-catalyzed fast elimination of acetic acid to give phenylbutadienes and subsequent polymerization. Substituents on the β-position of the allylic chain and on the phenyl group did not interfere the reaction. *meta*-substituted (**1f–1h**) substrates were cyclocarbonylated to yield a mixture of

Table 3. Synthesis of Phenanthryl Acetates[a]

Substrate	*Product*	*Isolated Yield (%)*
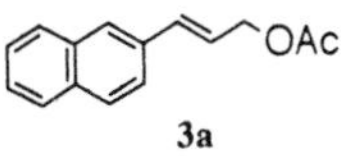 **3a**	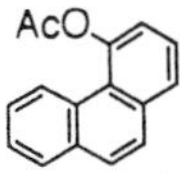 **4a**	73
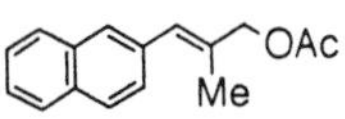 **3b**	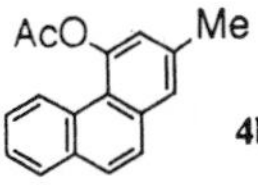 **4b**	76
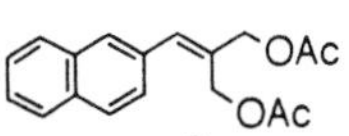 **3c**	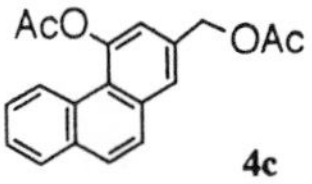 **4c**	64
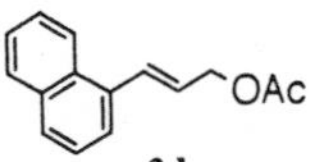 **3d**	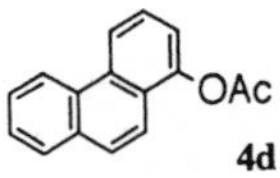 **4d**	50
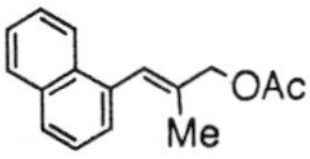 **3e**	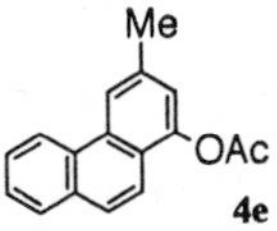 **4e**	70

Note: [a]Reaction conditions: substrate 10 mmol, $PdCl_2(PPh_3)_2$ 0.5 mmol, Ac_2O 20 mmol, NEt_3 20 mmol, benzene 8 mL, CO 70 atm. 170 °C, 1.5 h.

6- and 8-substituted 1-naphthyl acetates (Eq. 2). In each case, the more sterically favorable 6-substituted isomer (**2f–2h**) was the major product. The ratio of the products varied with the starting compounds from 58:42 to 78:22.

The use of naphthylallyl acetate instead of cinnamyl acetate as a substrate has provided a novel route to construct a tricyclic system (Table 3).[11] 3-(2-Naphthyl)allyl acetate (**3a**) was smoothly cyclocarbonylated under similar reaction conditions to those for **1a**. Interestingly, the more sterically unfavorable α-position of the naphthalene ring was selectively attacked to yield 4-phenanthryl acetate (**4a**) in 80% yield and no 1-anthryl acetate (**5a**), the cyclization product at the β-position, was detected by GC at all (Scheme 5).

The cyclization at the β-position giving the anthracene structure seems strongly disfavored. No palladium monophosphine complexes other than $PdCl_2(PPh_3)_2$ nor $PtCl_2(PPh_3)_2$ as a catalyst changed the selectivity. Furthermore, no anthracene derivative was observed even in an attempted reaction of 3-(1-methyl-2-naphthyl)allyl acetate (**3f**), which cannot cyclize to yield the phenanthrene skeleton.

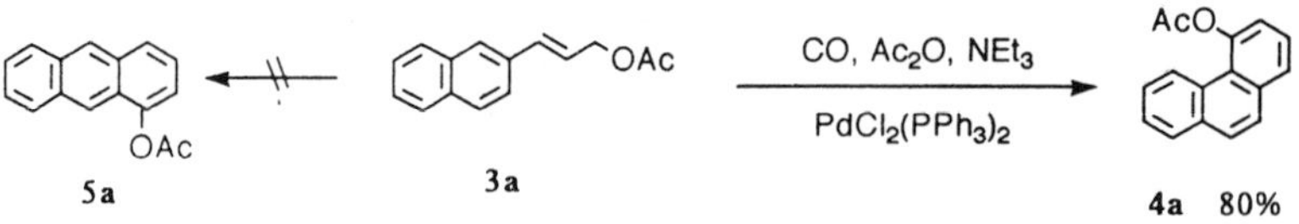

5a ←/— 3a $\xrightarrow[PdCl_2(PPh_3)_2]{CO,\ Ac_2O,\ NEt_3}$ 4a 80%

Scheme 5.

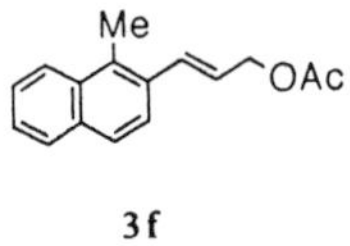

3f

Chart 1.

On the other hand, 3-(1-naphthyl)allyl acetates smoothly cyclized at the β-position to give the corresponding 1-phenanthryl acetates. Therefore the present cyclocarbonylation can be used as a selective synthetic method for phenanthryl acetates.

B. Synthesis of Fused-Ring Heteroaromatic Compounds

Fused heteroaromatics, such as benzofurans, benzothiophenes, indoles, and carbazoles, are among the most attractive group of compounds as targets in organic synthesis because of their wide occurrence in natural products and biologically active compounds. Especially desirable is the development of a selective synthetic route to fused heteroaromatics having functional groups at specific positions. It is well known that benzofuran, benzothiophene, and indole readily undergo substitution reactions at the five-membered heterocyclic part, but regioselective functionalization of the six-membered carbocyclic part is much more difficult. In this respect, cyclocarbonylation of heteroaromatic compounds seems to be promising, because such a reaction is expected to afford various fused heteroaromatic compounds having an acetoxy group at specific positions of the six-membered carbocyclic part. From these points of view, the cyclocarbonylation of 3-(heteroaryl)allyl acetates was extensively studied,[12] although no examples of the catalytic cyclocarbonylation of aromatic heterocycles have appeared in the literature.

As expected, 3-furylallyl and 3-thienylallyl acetates were cyclocarbonylated by $PdCl_2(PPh_3)_2$ catalyst in the presence of Ac_2O and NEt_3 to give acetoxybenzofurans and acetoxybenzothiophenes, respectively, in high yields (Eq. 3). The results are summarized in Table 4.

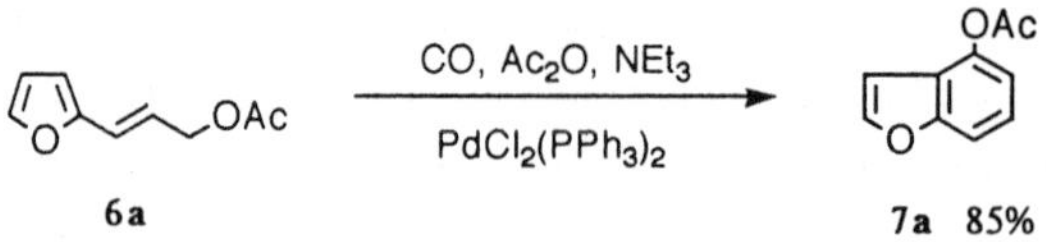

$$\mathbf{6a} \xrightarrow[PdCl_2(PPh_3)_2]{CO,\ Ac_2O,\ NEt_3} \mathbf{7a}\ 85\% \quad (3)$$

Table 4. Synthesis of Fused-Ring Heteroaromatic Compounds[a]

Substrate	*Product*	*Isolated Yield (%)*
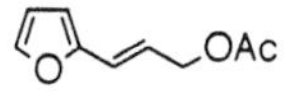 6a	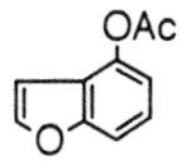 7a	85
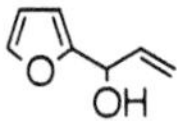 6b	7a	56
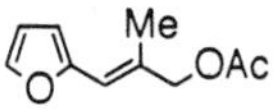 6c	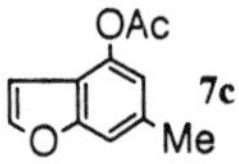 7c	78
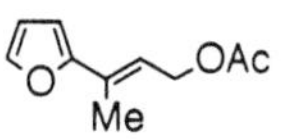 6d	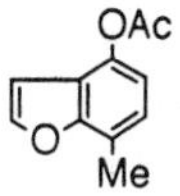 7d	56
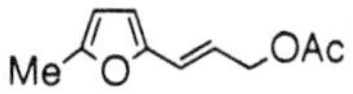 6e	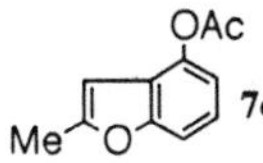 7e	76
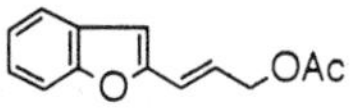 6f	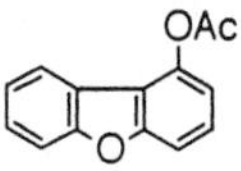 7f	70
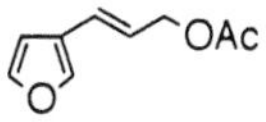 6g	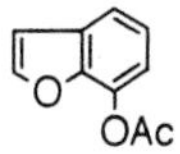 7g	89
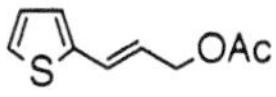 8a	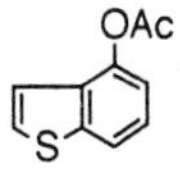 9a	79
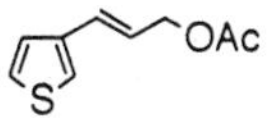 8b	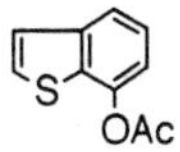 9b	86

Notes: [a]Reaction conditions: substrate 10 mmol, $PdCl_2(PPh_3)_2$ 0.5 mmol, Ac_2O 20 mmol, NEt_3 20 mmol, benzene 10 mL, CO 70 atm, 170 °C, 1.5 h.

1-(2-Furyl)-2-propen-1-ol (**6b**) also gave 4-acetoxybenzofuran (**7a**) in a moderate yield. Here, acetylation apparently preceded carbonylation. The dibenzofuran skeleton was also constructed from 3-(2-benzofuranyl)allyl acetate (**6f**). It should be pointed out that 3-(3-furyl)allyl acetate (**6g**) and 3-(3-thienyl)allyl acetate (**8b**) selectively cyclized at the 2-position of the heterocyclic nucleus to give 7-acetoxybenzofuran (**7g**) and 7-acetoxybenzothiophene (**9b**), respectively, as the only products. No 3-acetoxyisobenzofuran (**10**), 3-acetoxyisobenzothiophene, or related compounds formed by cyclization at the 4-position of the heterocyclic nucleus were detected (Scheme 6).

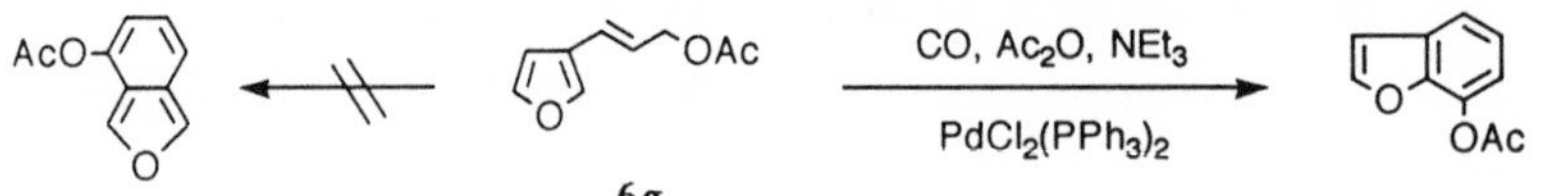

AcO- 10 ← (not) 6g OAc; CO, Ac_2O, NEt_3 / $PdCl_2(PPh_3)_2$ → 7g 89% OAc

Scheme 6.

In order to compare the reactivities of furan and benzene rings, effects of reaction temperature on the cyclocarbonylation was examined by using **1a** and **6a**. As shown in Table 5, both substrates gave the cyclocarbonylation products in high yields at 170 °C. Reactions at lower temperatures resulted in the decrease of yields because of side reactions affording unidentified high-boiling by-products. At 100 °C, **1a** gave **2a** only in 2% yield, while **6a** gave **7a** in 48% yield. This result suggests that a furan ring is more reactive in the cyclocarbonylation than a benzene ring. In contrast, no isolable cyclocarbonylation product was obtained in the reactions of 3-(2-pyridyl)- and 3-(3-pyridyl)allyl acetates. Therefore, the order of the reactivity in the cyclocarbonylation is considered to be furan > benzene > pyridine ring.

The synthetic utility of the cyclocarbonylation was demonstrated in a facile synthesis of cannabifuran (**14**), a naturally occurring dibenzofuran derived from *Cannabis sativa* L.[13] Benzofuran **12** was easily obtained by DIBAH reduction and subsequent acetylation of the condensation product of ethyl heptanoate and aldehyde **11**, the latter of which was in turn prepared from isothymol by a known procedure.[14] Cyclocarbonylation of **12** proceeded smoothly to give acetoxydibenzofuran (**13**) in 74% yield. Hydrolysis of **13** by KOH afforded **14** in 94% yield

Table 5. Reactivity of Aromatic Systems[a]

Substrate	*Reaction Temperature (°C)*	*Conv. (%)*	*Yield (%)*[b]
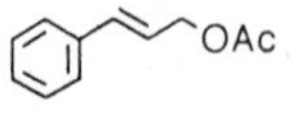 Ph-CH=CH-CH₂OAc	170	100	81
	100	99	2
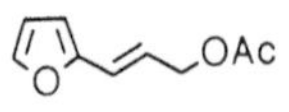 2-furyl-CH=CH-CH₂OAc	170	100	(85)
	130	100	98 (83)
	100	97	48 (41)
	70	33	5
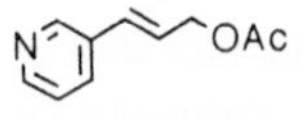 3-pyridyl-CH=CH-CH₂OAc	170	100	0

Notes: [a]Reaction conditions: substrate 10 mmol, $PdCl_2(PPh_3)_2$ 0.5 mmol, Ac_2O 20 mmol, NEt_3 20 mmol, benzene 10 mL, CO 70 atm, 1.5 h.

[b]GC yield based on the starting substrate. Isolated yield in parentheses.

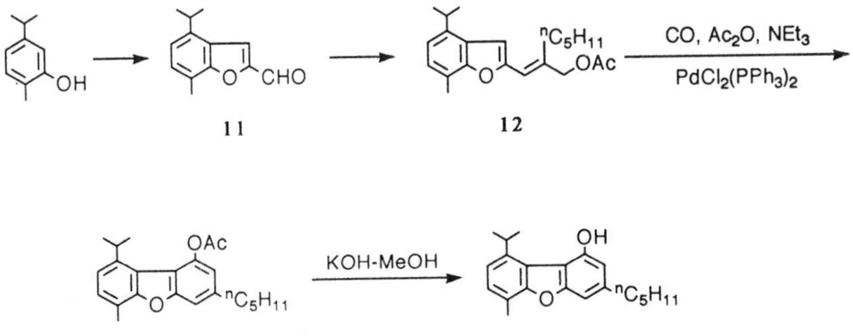

Scheme 7.

(Scheme 7). It is worth mentioning that this synthesis is advantageous in that **13** is the only regioisomer which can be formed by the reaction. Obviously this point is important in synthesis of multisubstituted compounds such as **14** in terms of avoiding complicated purification procedures.

Synthesis of Cannabifuran. A mixture of allylic acetate (**12**, 1.027 g, 3 mmol), $PdCl_2(PPh_3)_2$ (0.15 mmol), Ac_2O (6 mmol), NEt_3 (6 mmol), and benzene (3 mL) in a stainless steel autoclave was pressurized with CO (70 kg/cm^2) and was heated at 170 °C for 1.5 h with stirring. The autoclave was cooled and CO was discharged. The mixture was diluted with ether, washed successively with dilute aqueous HCl, saturated aqueous $NaHCO_3$, and water, and dried over $MgSO_4$. Solvent was evaporated to give a brown oil, which was purified by column chromatography on silica gel (hexane/ether, 30:1) to give 1-acetoxy-9-isopropyl-6-methyl-3-pentyldibenzofuran (**13**, 0.784 g, 74%) as a pale yellow oil. Acetate **13** (336 mg), KOH (440 mg), and MeOH (10 mL) were stirred for 1 h at room temperature. The mixture was acidified with 6 N aqueous HCl (5 mL) and extracted with ether, and the extract was dried over $MgSO_4$. Evaporation of solvent gave crude 9-isopropyl-6-methyl-3-pentyldibenzofuran-1-ol (cannabifuran **14**, 277 mg, 94%) as a pale yellow solid. Recrystallization from hexane gave colorless prisms: mp 80–81 °C (lit.[14] mp 78–79 °C).

Allylic acetates substituted with a pyrrole or indole ring underwent cyclocarbonylation at 130 °C. Thus, 3-(2-pyrrolyl)allyl acetate (**15**) and 3-(3-indolyl)allyl acetate (**18**) gave 4-acetoxyindole (**16**) and 1-acetoxycarbazole (**19**), respectively (Scheme 8). The yields were somewhat lower than those obtained from furan or thiophene derivatives, probably because of the relative instability of five-membered nitrogen heterocycles. However, this reaction is interesting as a novel

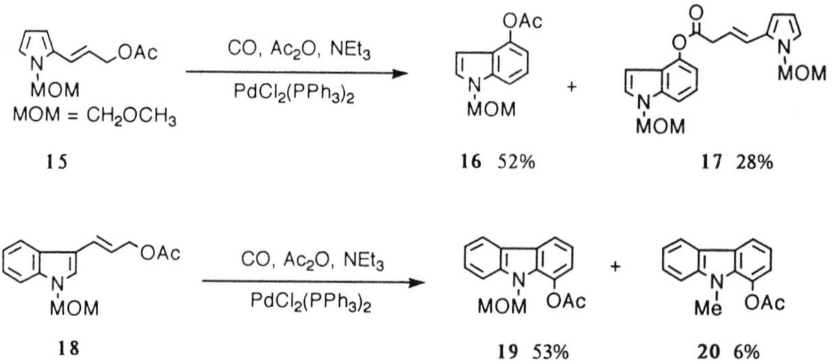

OAc, CO, Ac_2O, NEt_3, $PdCl_2(PPh_3)_2$, MOM, MOM = CH_2OCH_3, 15, 16 52%, 17 28%, 18, 19 53%, 20 6%

Scheme 8.

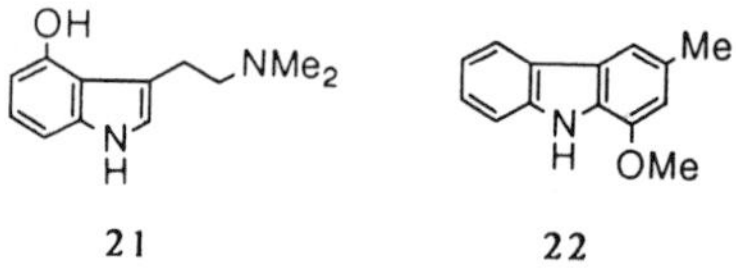

OH, NMe_2, N H, Me, OMe, 21, 22

Chart 2.

synthetic route to oxyindoles and oxycarbazoles because these structures are widely found in alkaloids such as psilocin (**21**) and murrayafolin-A (**22**).

As shown in Scheme 8, substrate **15** gave not only the expected product **16** but also a dimeric product **17**. Compound **17** was possibly formed from the reaction of 4-hydroxy-1-methoxymethylindole and an intermediary acylpalladium complex (*vide infra*). The formation of **17** was suppressed either by performing the reaction at a higher temperature or by using a larger excess amount of Ac_2O and NEt_3. In the cyclocarbonylation of **18**, in addition to the expected product **19**, a small amount of the *N*-methylcarbazole **20** was obtained. The mechanism of formation of this by-product is not clear.

C. Mechanism of Cyclocarbonylation

In order to obtain information about the catalytic intermediates and elucidate the mechanism of the cyclocarbonylation, reaction of $Pd(CO)(PPh_3)_3$ with *trans*- or *cis*-cinnamyl bromide was investigated in detail (Scheme 9).[10b,15] The reaction of $Pd(CO)(PPh_3)_3$ with *trans*-cinnamyl bromide (**1b**) in toluene at room temperature under 20 atm of CO gave an acyl complex, *trans*-[{(*E*)-PhCH=CHCH$_2$CO)}-PdBr(PPh$_3$)$_2$] (**23a**), in 74% yield. By similar treatment of $Pt(CO)(PPh_3)_3$ under 1 atm of CO, *trans*-[{(*E*)-PhCH=CHCH$_2$CO)}PtBr(PPh$_3$)$_2$] (**23b**) was obtained in

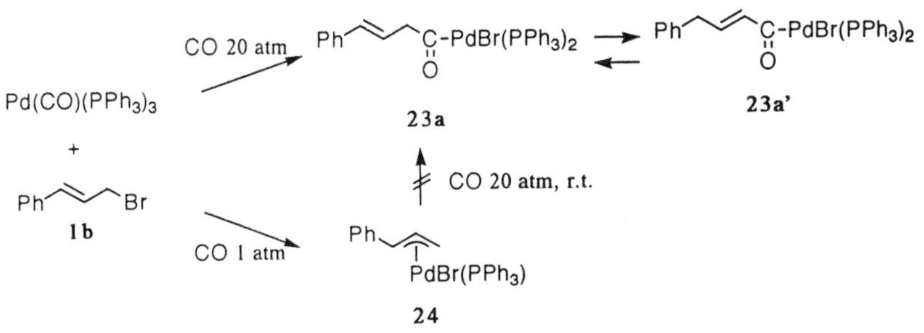

Pd(CO)(PPh3)3 + Ph⌒⌒Br (1b) —CO 20 atm→ Ph⌒⌒C(=O)-PdBr(PPh3)2 (23a) ⇌ Ph⌒⌒C(=O)-PdBr(PPh3)2 (23a')

1b —CO 1 atm→ Ph-π-allyl-PdBr(PPh3) (24) —CO 20 atm, r.t. ⇸ 23a

Scheme 9.

78% yield. ^{1}H NMR spectrum of **23a** in C_6D_6 showed two sets of signals attributable to two *trans*-CH=CHCH$_2$ groups, indicating that **23a** exists in a solution as a 1:1 mixture with *trans*-[{(*E*)-PhCH$_2$CH=CHCO)}PdBr(PPh$_3$)$_2$] (**23a′**) at room temperature. Complex **23a′** seems to be formed by C=C double-bond migration in **23a**, but analogous migration was not observed for **23b** by ^{1}H NMR under similar conditions. On the other hand, when Pd(CO)(PPh$_3$)$_3$ was allowed to react with **1b** under 1 atm of CO, a π-allyl complex **24** was formed in a high yield, which was not converted to **23a** at room temperature under 20 atm of CO in the presence of 1 equiv of PPh$_3$.[16] These results indicate that, in the reaction of Pd(CO)(PPh$_3$)$_3$ with **1b** at room temperature under high CO pressure, insertion of CO into the palladium–carbon bond of the initially formed σ-allyl complex proceeds smoothly before the σ–π equilibrium is attained.

It should be also mentioned that when Pd(CO)(PPh$_3$)$_3$ was treated with *cis*-cinnamyl bromide under 20 atm of CO, **23a** was isolated in 70% yield instead of the expected acyl complex *trans*-[{(*Z*)-PhCH=CHCH$_2$CO)}PdBr(PPh$_3$)$_2$] (**25a**). In this case, complex **23a** is supposed to be formed by C=C double-bond migration in **25a** (Scheme 10) because the initially formed σ-allyl complex should be smoothly carbonylated under high CO pressure before σ–π equilibrium is attained as mentioned above. The equilibrium among acyl complexes **23a**, **23a′**, **25a**, and **25a′** lies exclusively to more sterically favorable **23a** and **23a′** at room temperature.

Next, reactivities of **23a** and **23b** under catalytic cyclocarbonylation conditions were investigated. Treatment of **23a** and **23b** with excess Ac$_2$O and NEt$_3$ under 60 atm of CO at 160 °C resulted in the formation of **2a** in 54% and 40% yield, respectively (Eq. 4). Complex **23a** was also an active catalytic precursor in the cyclocarbonylation of **1b**. The reaction of **23a** and **23b** to give **2a** is supposed to

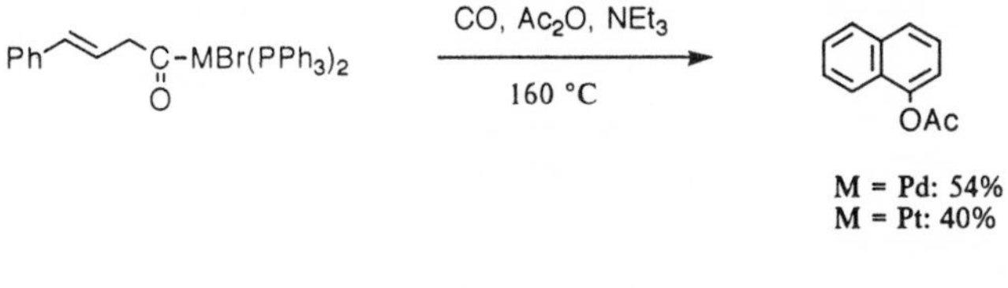

Ph⌒⌒C(=O)-MBr(PPh3)2 —CO, Ac2O, NEt3 / 160 °C→ 1-naphthyl-OAc (4)

M = Pd: 54%
M = Pt: 40%

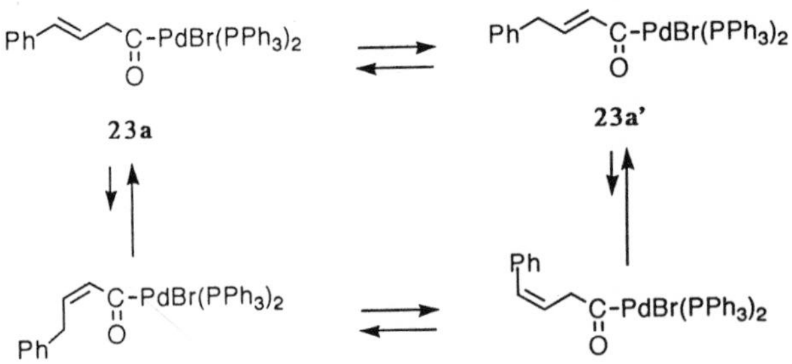

Scheme 10.

proceed through (*Z*)-acyl complexes such as **25a** and **25a′** by C=C double-bond migration.

Based on the results of stoichiometric model cyclocarbonylations of cinnamyl bromide, it is reasonable to assume that, also in the catalytic cyclocarbonylation of 3-arylallyl acetates, unsaturated acylpalladium complexes such as [(*Z*)-$ArCH{=}CHCH_2CO]Pd(OAc)(PPh_3)_n$ or [(*Z*)-$ArCH_2CH{=}CHCO]Pd(OAc)(PPh_3)_n$ (n = 1 or 2), which could be formed by successive oxidative addition of allylic acetate, CO insertion, and C=C double bond *E-Z* isomerization, are key intermediates. The subsequent intramolecular cyclization of the intermediary acylpalladium complexes would produce bicyclic ketones, which tautomerize to 1-naphthol, hydroxybenzofuran and so on. Acetylation by Ac_2O yields aromatic acetates as the stable final products (Scheme 11).

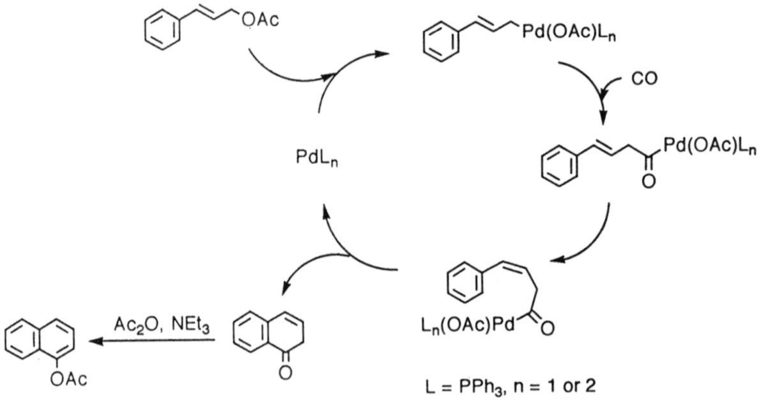

Scheme 11.

Participation of unsaturated acylpalladium complexes and phenolic compounds such as 1-naphthol in the catalytic cycle was further substantiated by the formation of **17** in the carbonylation of **15** (see Scheme 8). Reaction between an unsaturated acylpalladium complex and 4-hydroxy-1-methoxymethylindole would explain the formation of **17** (Eq. 5).

OH N MOM + N MOM O Pd(OAc)(PPh$_3$)$_n$ → O O N MOM N MOM **17** (5)

The detailed mechanism of the ring closure of the acyl complexes is of great interest, but still remains ambiguous. However, several facts helpful in elaborating the mechanism were obtained. As described above, cyclization of 3-(2-naphthyl)allyl acetate (**3a**) occurred selectively at the α-position of the aromatic nucleus, in spite of greater steric hindrance, to give 4-phenanthryl acetate. 3-(2-Furyl)- and 3-(2-thienyl)allyl acetates (**6g** and **8b**) cyclcized selectively at the 2-position. Furthermore, it was revealed that the order of reactivity of aromatic rings in the cyclocarbonylation is furan > benzene > pyridine ring. These findings strongly suggest that the ring closure step is an electrophilic reaction.

Palladation of aromatic C–H bonds by Pd(II) species proceeds via an electrophilic substitution reaction. However, palladation of the aromatic *ortho*-C–H bond in acylpalladium intermediates (Scheme 12, A) seems not involved in the present cyclocarbonylation because palladation of naphthalene is known to occur selectively at the β-position, which is inconsistent with the cyclocarbonylation of **3a** (see Scheme 5).[5,17] Other possible mechanisms include direct electrophilic attack of the acyl group on the aromatic ring (Scheme 12, B), or alternatively, electrocyclization of an alkenylketene intermediate derived from the acylpalladium complex (Scheme 12, C).

Cyclization of coordinated alkenylketene has been proposed for the benzannulation of chromium–arylcarbene complexes (Scheme 13).[18,19] Quite interestingly, Dötz reported that the benzannulation of some diaryl–carbene chromium complexes having different aryl groups with tolan gave products formed by the cyclization at the more electron-deficient aryl groups (Scheme 14).[20] In order to compare the present catalytic cyclocarbonylation with benzannulation of arylcarbene complexes, the selectivity in the cyclocarbonylation of 3,3-diarylallyl acetates was investigated.[21]

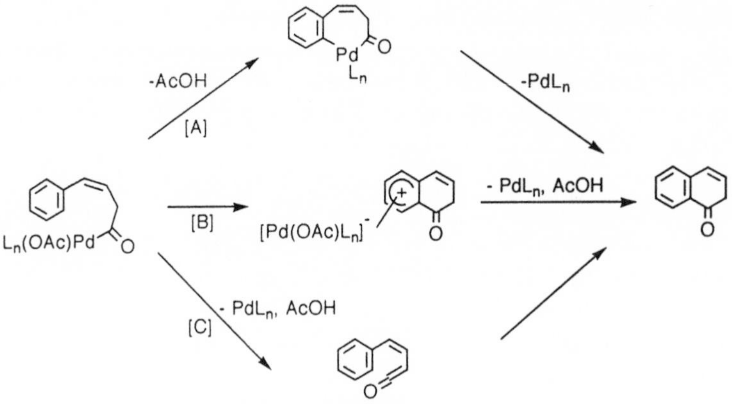

-AcOH
[A]
-PdL_n
[B]
$[Pd(OAc)L_n]^-$
- PdL_n, AcOH
$L_n(OAc)Pd$
[C]
- PdL_n, AcOH

Scheme 12.

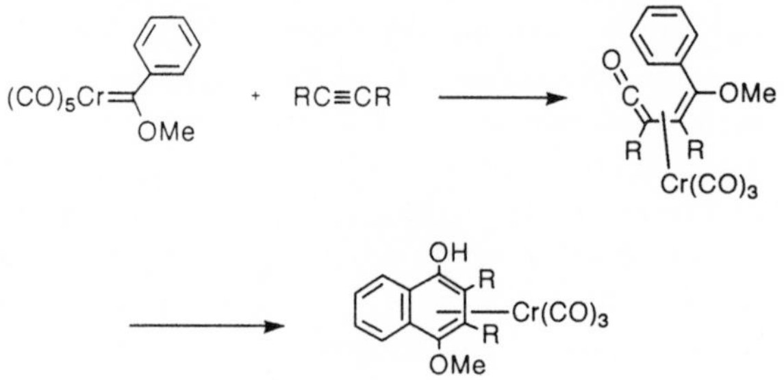

$(CO)_5Cr$ OMe + RC≡CR → OMe R R $Cr(CO)_3$

→ OH R $Cr(CO)_3$ R OMe

Scheme 13.

For example, in the case of 3-(2-furyl)-3-phenylallyl acetate (**26**, E:Z = 91:9), cyclocarbonylation occurred both at the furan and benzene ring to form 7-phenylbenzofuran and 4-(2-furyl)naphthalene skeleton, respectively, where the benzofuran derivative **27** was the major product (Eq. 6). It is clear that this selectivity did not arise from the E-Z isomer ratio of the substrate, because if there was no E-Z

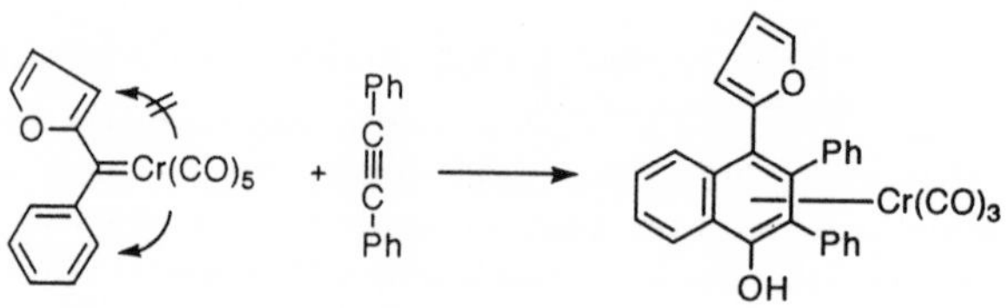

$Cr(CO)_5$ + Ph–C≡C–Ph → Ph $Cr(CO)_3$ Ph OH

Scheme 14.

$$\textbf{26}\ E:Z = 91:9 \xrightarrow[\text{PdCl}_2(\text{PPh}_3)_2]{\text{CO, Ac}_2\text{O, NEt}_3} \textbf{27}\ 55\% + \textbf{28}\ 23\% \quad (6)$$

isomerization under reaction conditions, the major isomer of the substrate (*E*-**26**) would have afforded **28**, the minor product in the actual reaction.

Similar selectivity was also found in some other reactions of 3,3-diarylallyl acetates. The selectivities obtained here are opposite to those reported for benzannulation. Considering the difference in the selectivity between both reactions, the mechanism including electrophilic attack of the acyl group (Scheme 12, B) seems to be more reasonable, although the participation of the alkenylketene intermediate (Scheme 12, C) cannot be completely ruled out.

D. Synthesis of Phenyl Esters

As described in the preceding sections, the palladium catalyzed cyclocarbonylation is considered to proceed via an acyl–palladium complex derived from an allylic compound, CO, and a palladium(0) complex. It is expected that suitable choice of substrates would make it possible for such an acyl complex to react with an organic moiety other than an aromatic ring. Among such reactions, the carbonylation of an allyl compound having another C=C double bond in the same molecule seems attractive because the double bond would insert into the acyl–palladium bond to give phenol derivatives.

Although a variety of catalytic reactions have been developed by using the acyl–palladation of olefins as a key step,[22] their applications for the synthesis of six-membered aromatic systems have not been reported.[23] As predicted, in the presence of NEt_3, Ac_2O, and a catalytic amount of $PdCl_2(PPh_3)_2$, 5-phenyl-2,4-pentadienyl acetate (**29a**) was cyclocarbonylated to give 2-acetoxybiphenyl (**30a**) (Eq. 7).[24]

$$\textbf{29a} \xrightarrow[\text{PdCl}_2(\text{PPh}_3)_2]{\text{CO, Ac}_2\text{O, NEt}_3} \textbf{30a}\ 74\% \quad (7)$$

Synthesis of 2-acetoxybiphenyl. A mixture of 5-phenyl-2,4-pentadienyl acetate (**29a**, 3 mmol), $PdCl_2(PPh_3)_2$ (0.09 mmol), Ac_2O (6 mmol), NEt_3 (6.6 mmol), and benzene (10 mL) in a stainless steel autoclave was pressurized with CO (50 atm) and heated at 140 °C for 3 h with stirring. The autoclave was cooled and CO was discharged. GC analysis of the reaction mixture revealed that 2-acetoxybiphenyl was formed in 74% yield. The reaction mixture was diluted with ether, washed with water, and dried over $MgSO_4$.

> Solvent was evaporated and the crude product was purified by silica gel column chromatography (hexane/ether, 6:1) and bulb-to-bulb distillation to give pure 2-acetoxybiphenyl (**30a**) as colorless crystals in 69% yield.

No other identifiable product including cyclopentenone derivatives was detected by GC analysis. Reaction temperatures of 120–140 °C were suitable for the reaction. Ac_2O and NEt_3 were both essential to obtain the carbonylation product in a high yield. Thus, a reaction in the absence of Ac_2O gave 2-biphenylol in 16% and **30a** in 11% yield (conv. 100%), while a reaction in the absence of NEt_3 gave **30a** in only 9% yield (based on the starting **29a**, conv. 52%). Palladium and platinum phosphine complexes such as $PdCl_2(PPh_3)_2$ and $PtCl_2(PPh_3)_2$ proved to be effective catalysts, and some ruthenium complexes such as $RuCl_2(PPh_3)_3$ showed low catalytic activities. Other group 8 metal compounds such as $Pd(OAc)_2$, $Co_2(CO)_8$, $NiBr_2(PPh_3)_2$, and $RhCl(PPh_3)_3$ were inactive.

Synthetic applicability of this reaction is demonstrated in Table 6. It should be noted that the acetate **29a** was converted to **30a** in much higher yield than the corresponding chloride (33%) or ethyl carbonate (24%). 5-Aryl-2,4-pentadienyl acetates were good substrates for this reaction, but substituents at the 2- or 4-position of the substrates somewhat lowered the yields of the products. Aliphatic substrates also gave the corresponding phenyl esters but in a little lower yields. In the reaction of *trans,trans,trans*-2,4,6-undecatrienyl acetate (**29j**), the six-membered ring formation exclusively occurred to give *o*-(1-hexenyl)phenyl acetate (**30j**), but the product was a mixture of the *cis* and *trans* isomers. It is noteworthy that the present carbonylation is applicable to the synthesis of 2,3- and 3,5-disubstituted phenyl acetates, which are difficult to be prepared by conventional electrophilic substitution reactions of phenol. This exemplifies the effectiveness of the present cyclocarbonylation as a synthetic method for substituted phenols.

A limitation of this novel carbonylation is shown in Eq. 8. When (*E*)-3,5-di(*p*-tolyl)-2,4-pentadienyl acetate (**31**) was carbonylated under similar reaction conditions, cyclization toward the tolyl group at the 3-position competed with the phenyl

CO, Ac_2O, NEt_3

$PdCl_2(PPh_3)_2$

OAc

31

OAc

+

OAc

32 17%

33 38%

(8)

Table 6. Synthesis of Phenyl Acetates[a]

Substrate	*Product*	*Isolated Yield (%)*
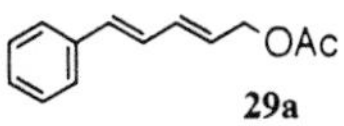 **29a**	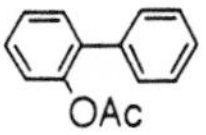 **30a**	69 (74)[b]
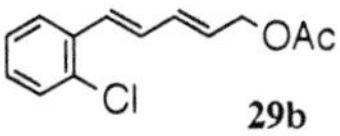 **29b**	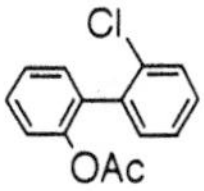 **30b**	84
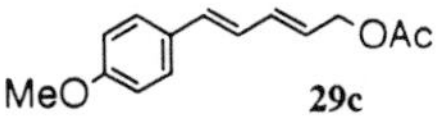 **29c**	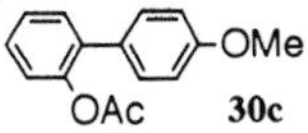 **30c**	73
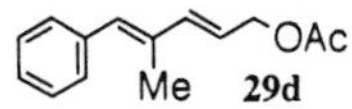 **29d**	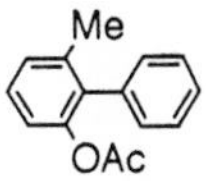 **30d**	57
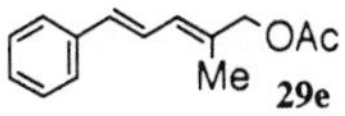 **29e**	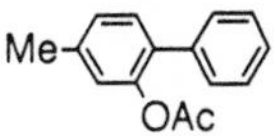 **30e**	46
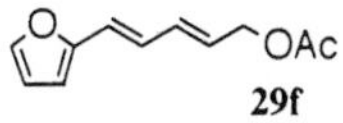 **29f**	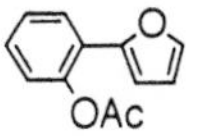 **30f**	57
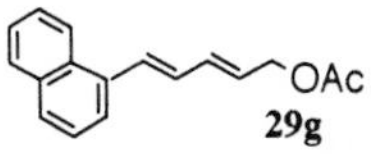 **29g**	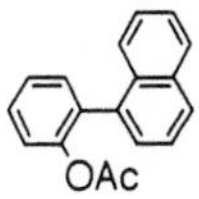 **30g**	79
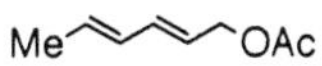 **29h**	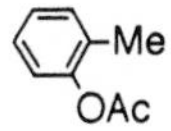 **30h**	51[c]
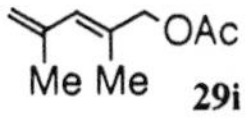 **29i**	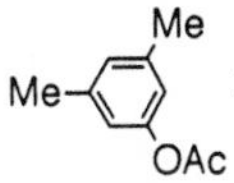 **30i**	40
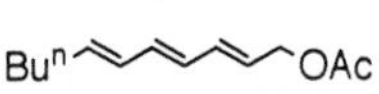 **29j**	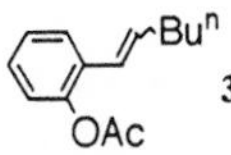 **30j**	52[d]

Notes: [a]Reaction conditions: substrate 3 mmol, $PdCl_2(PPh_3)_2$ 0.09 mmol, Ac_2O 6 mmol, NEt_3 6.6 mmol, benzene 5 mL, CO 50 atm, 140 °C, 3 h.

[b]GC yield in parentheses.

[c]Benzene 2 mL was used as solvent.

[d]*E*/*Z* = 79/21.

acetate formation to give naphthyl acetate (**32**, 17%) concurrent with the expected 2,4-di(*p*-tolyl)phenyl acetate (**33**, 38%) in spite of the *E* configuration of the substrate.

Negishi reported that palladium-catalyzed cyclocarbonylation of *cis*-2,4-pentadienyl chlorides (**34**) in the presence of MeOH and NEt_3 yields cyclopentenone derivatives **35**, and that the *cis* configuration of the substrates is required for the cyclization (Eq. 9).[22a]

CO, MeOH, NEt_3 / $PdCl_2(PPh_3)_2$ — **34** → **35** (9)

Although the catalytic systems of Eqs. 7 and 9 are closely related to each other, the cyclocarbonylation described here is in sharp contrast to Negishi's reaction in that only the six-membered products, but not the five-membered ones, are selectively obtained and that substrates of the *trans* configuration smoothly undergo the cyclization. The latter point is especially advantageous from a synthetic point of view. Thus, carbonylation of **29a** under Negishi's conditions resulted in the formation of methyl (3*E*,5*E*)-6-phenyl-3,5-hexadienoate (**36**, 60%) and methyl (2*E*,4*E*)-6-phenyl-2,4-hexadienoate (**37**, 13%) (Eq. 10).

CO, MeOH, NEt_3 / $PdCl_2(PPh_3)_2$ — **29a** → **36** 60% + **37** 13% (10)

The present reaction is considered to proceed via a hexadienoylpalladium complex, which is generated by oxidative addition of a pentadienyl acetate to a Pd(0) species followed by CO insertion (Scheme 15). This is supported by the observation that $(PhCH{=}CHCH{=}CHCH_2CO)PdCl(PPh_3)_2$ and its platinum analogue produced **30a** in 41% and 51% yield, respectively, on heating under the catalytic carbonylation conditions. In the absence of nucleophiles such as MeOH, the hexadienoylpalladium complex would undergo *E-Z* isomerization of the internal C=C double bond and intramolecular insertion of the terminal C=C double bond into the Pd–C

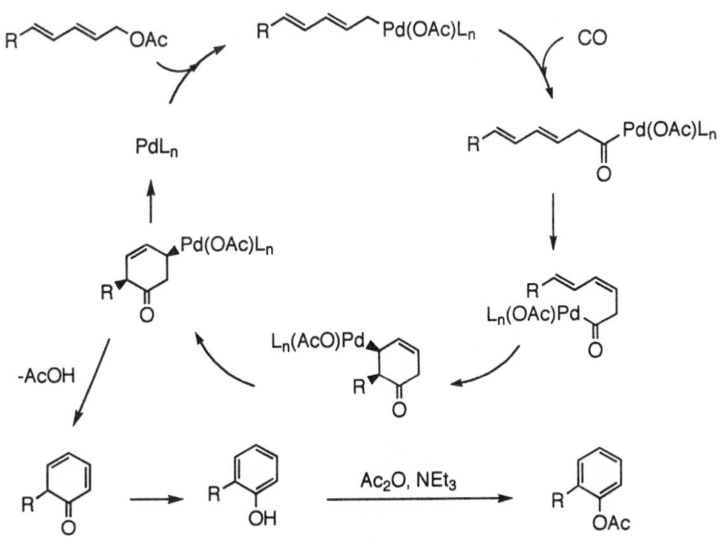

Scheme 15.

bond. Subsequent β-elimination gives a cyclohexadienone, which tautomerizes to afford the corresponding phenol and is finally acetylated by Ac_2O.

It is very interesting why the hexadienoyl palladium species selectively cyclizes to form a six-membered ring but not a five-membered one under the present reaction conditions. With the intention of obtaining information on this question, we examined carbonylation of bromomethylstyrenes (**38**), which can undergo cyclocarbonylation without *E-Z* isomerization of the C=C double bond (Scheme 16). Thus, when **38a** was carbonylated in the presence of Ac_2O, the expected six-membered product, 3-phenyl-2-naphthyl acetate (**39a**), was obtained in 20% yield concurrent with **40a** (35%). No five-membered product was observed. In contrast, carbonylation of **38a** in the absence of Ac_2O gave a five-membered compound, 1-benzyl-2-indanone (**41a**), as the only isolable product (33%).

Furthermore, carbonylation of **38b** in the absence of Ac_2O gave two types of five-membered products, indanone (**41b**, ca. 20%) and lactone (**42b**, 41%), while carbonylation in the presence of Ac_2O resulted in the formation of six-membered product **39b** (19%) in addition to **40b** (10%) and **42b** (20%).

The reactions in Scheme 16 are considered to proceed via a common acylpalladium complex **43** as a key intermediate (Scheme 17). The results shown above suggest that acetate anion delivered from Ac_2O plays an important role in controlling the products. That is, in the presence of acetate anion, the formation of the six-membered products from the intermediary complex **43** is promoted (path A), while in its absence acylpalladation of **43** to give five-membered products predomi-

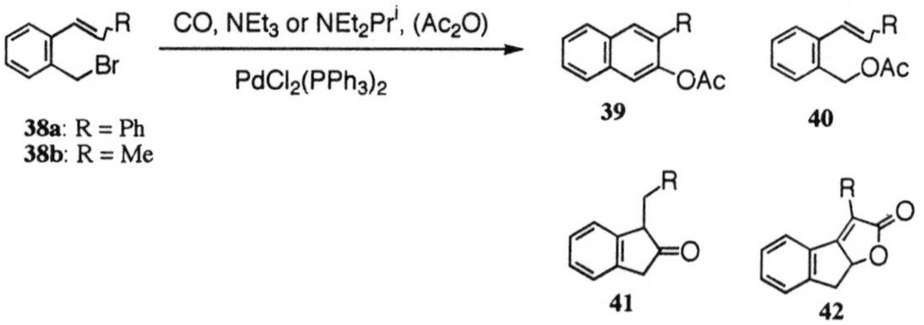

CO, NEt_3 or NEt_2Pr^i, (Ac_2O)

$PdCl_2(PPh_3)_2$

38a: R = Ph
38b: R = Me

39 **40** **41** **42**

	Ac_2O	Yield (%)			
		39	**40**	**41**	**42**
38a	added	20	35	-	-
"	-	-	-	33	-
38b	added	19	10	-	20
"	-	-	-	ca. 20	41

Scheme 16.

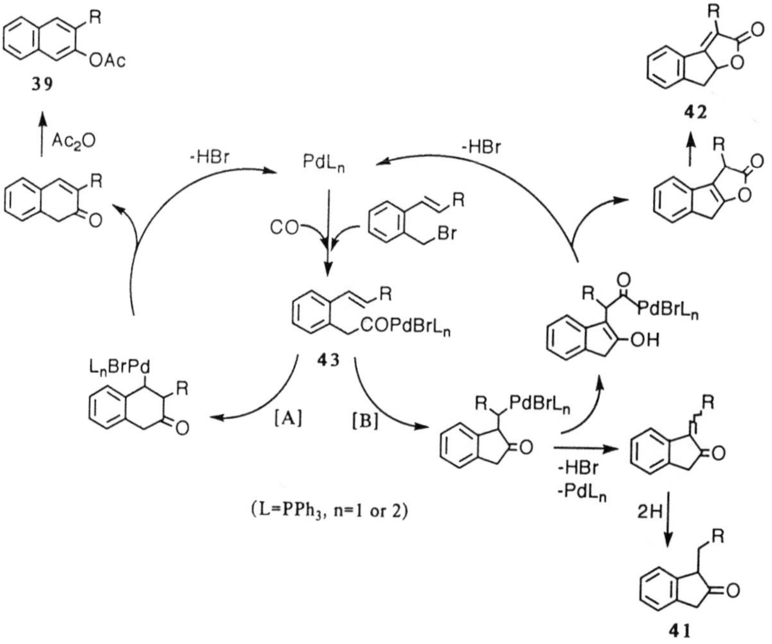

(L=PPh_3, n=1 or 2)

Scheme 17.

nantly occurs (path B). We believe that similar effect of acetate anion also contributes to the selective formation of six-membered phenol derivatives in the cyclocarbonylation of 2,4-pentadienyl acetates.

III. NOVEL CARBONYLATIONS BY HOMOGENEOUS MULTIMETALLIC CATALYSTS

A. Hydroformylation of Olefins by Cobalt–Ruthenium Bimetallic Catalyst

Homogeneous multimetallic catalysts have recently attracted increasing attention, because it can be expected that cooperative or successive participation of different metal complexes during a catalytic cycle might lead to more selective and efficient reactions, and in some cases, to new types of reactions which cannot be achieved by homometallic catalyst systems.[25,26]

About 10 years ago we embarked on our study toward development of unique reactions by using homogeneous multimetallic catalysts. At that time, we were very interested in the findings that the selectivity of ethanol in the homologation of methanol with synthesis gas was notably improved by addition of a small amount of ruthenium compounds to $Co_2(CO)_8$ (Eq. 11).[27,28] Detailed investigation of this reaction catalyzed by the Co–Ru bimetallic systems led us to discover that the $Co_2(CO)_8$–$Ru_3(CO)_{12}$ bimetallic system shows high catalytic activity for hydroformylation of olefins compared with $Co_2(CO)_8$ or $Ru_3(CO)_{12}$ alone.

$$\mathrm{MeOH} \xrightarrow[Co_2(CO)_8\ -\ RuCl_3]{CO,\ H_2} \mathrm{EtOH} \tag{11}$$

Table 7 summarizes the hydroformylation of cyclohexene in THF using $Co_2(CO)_8$–$Ru_3(CO)_{12}$ mixed metal catalysts (Eq. 12).[29]

$$\text{cyclohexene} \xrightarrow[Co_2(CO)_8\ -\ Ru_3(CO)_{12}]{CO,\ H_2} \underset{\mathbf{44}}{\text{cyclohexane-CHO}} \tag{12}$$

The yield of cyclohexanecarbaldehyde (**44**) with the bimetallic catalyst was much higher than would be expected by a simple additive effect based on the values observed for $Co_2(CO)_8$ or $Ru_3(CO)_{12}$ alone, respectively. As more $Ru_3(CO)_{12}$ was added to $Co_2(CO)_8$, the initial rate increased. When the Ru:Co ratio was 9.9:1, the initial rate was 27 times faster than that with $Co_2(CO)_8$ alone. 2,4,6-Tricyclohexyl-1,3,5-trioxane was also formed together with **44** in reactions with the conversion of cyclohexene higher than 50%. Hydrogenation of **44** to cyclohexanemethanol and of cyclohexene to cyclohexane were observed only to a minor extent, if any. Use of phosphine-substituted ruthenium carbonyls, such as $Ru(CO)_3(PPh_3)_2$ and $Ru(CO)_4(PPh_3)$ instead of $Ru_3(CO)_{12}$ as the catalyst component, decreased the catalytic activity drastically.

Table 7. Hydroformylation of Cyclohexene by $Co_2(CO)_8$–$Ru_3(CO)_{12}$ Catalyst[a]

Catalyst	*Ru:Co*	*Yield (%)*[b]	*Relative Initial Rate*
$Co_2(CO)_8$	—	14	1.0
$Ru_3(CO)_{12}$[c]	—	3	0.3
$Co_2(CO)_8$–$Ru_3(CO)_{12}$	0.34	32	3.6
$Co_2(CO)_8$–$Ru_3(CO)_{12}$	0.95	52	5.9
$Co_2(CO)_8$–$Ru_3(CO)_{12}$	3.2	62	8.4
$Co_2(CO)_8$–$Ru_3(CO)_{12}$	9.9	100	27
$Co_2(CO)_8$–$Ru(CO)_3(PPh_3)_2$	0.96	0	0
$Co_2(CO)_8$–$Ru(CO)_3(PPh_3)_2$[d]	0.89	10	1.1
$Co_2(CO)_8$–$Ru(CO)_4(PPh_3)$[e]	0.99	35	3.3

Notes: [a]Reaction conditions: $Co_2(CO)_8$ 0.1 mmol, cyclohexene 80 mmol, THF 10 mL, 110 °C, CO:H_2 = 40:40 atm, 4 h.
[b]GC yield based on the starting cyclohexene.
[c]0.2 mmol.
[d]130 °C.
[e]Benzene 10 mL as solvent.

Solvent effect is shown in Table 8, where the initial rate is represented as the relative value compared with that using $Co_2(CO)_8$ in THF. The initial rate of the hydroformylation of cyclohexene with $Co_2(CO)_8$ was essentially independent of the nature of the solvent employed. In contrast, the initial rate of the reaction with the $Co_2(CO)_8$–$Ru_3(CO)_{12}$ catalyst was considerably influenced by the nature of the

Table 8. Solvent Effect on the Hydroformylation of Cyclohexene by $Co_2(CO)_8$–$Ru_3(CO)_{12}$ Catalyst[a]

		Yield (%)[b]			
Solvent	*Relative Initial Rate*	*CyCHO*	*CyCH$_2$OH*	*Acetal*[c]	*Ester*[d]
hexane	5.4 (1.1)	58 (12)	1 (0)		
dioxane	5.8 (0.9)	61 (11)	1 (0)		
THF	5.9 (1.0)	52 (14)	1 (0)		
benzene	9.1 (1.1)	73 (12)	2 (0)		
ethanol	15 (1.1)	39 (2)	2 (0)	54 (14)	2 (1)
methanol	19 (1.1)	19 (0)	1 (0)	68 (15)	5 (1)
methanol[e]	32	16	2	76	4

Notes: [a]Reaction conditions: $Co_2(CO)_8$ 0.1 mmol, $Ru_3(CO)_{12}$ 0.067 mmol, cyclohexene 80 mmol, solvent 10 mL, 110 °C, CO:H_2 = 40:40 atm, 4 h. Values in parentheses are those catalyzed by $Co_2(CO)_8$ (0.1 mmol) alone.
[b]GC yield based on the starting cyclohexene.
[c]$CyCH(OEt)_2$ or $CyCH(OMe)_2$.
[d]CyCOOEt or CyCOOMe.
[e]$Ru_3(CO)_{12}$ 0.67 mmol.

solvent. Alcohols, such as methanol and ethanol, exhibited the largest rate enhancement among the solvents investigated, although the aldehyde initially formed is mostly converted to the corresponding acetal. The rate of the hydroformylation by a catalyst with a Ru:Co ratio of 10:1 in methanol was 32 times faster than that by $Co_2(CO)_8$ alone in THF.

A synergistic effect for cobalt and ruthenium was also observed on hydroformylation of other olefins, such as 1-hexene and styrene, although the enhancement of the reaction rates was smaller in comparison with that of cyclohexene. The initial rate of the hydroformylation of 1-hexene in benzene by the $Co_2(CO)_8$–$Ru_3(CO)_{12}$ catalyst (Ru:Co = 1:1) was about 3 times faster than that by $Co_2(CO)_8$ alone (Eq. 13). The normal:iso ratio of the aldehydes formed was essentially the same as in the case of $Co_2(CO)_8$.

$$ {}^{n}C_4H_9CH{=}CH_2 \xrightarrow[Co_2(CO)_8\ -\ Ru_3(CO)_{12}]{CO,\ H_2} {}^{n}C_4H_9CH_2CH_2CHO + {}^{n}C_4H_9CH(CHO)CH_3 \qquad (13) $$

Catalyst	Yield (%), (n/i)	Initial rate
$Co_2(CO)_8$	22 (3.2)	1.0
$Ru_3(CO)_{12}$	1 (5.0)	0.2
$Co_2(CO)_8$ - $Ru_3(CO)_{12}$	50 (3.1)	3.1

In the hydroformylation by $Co_2(CO)_8$, the reaction rate decreases in the following order: 1-hexene >> styrene ~ cyclohexene. However, if the $Co_2(CO)_8$–$Ru_3(CO)_{12}$ bimetallic catalyst was used, cyclohexene as well as 1-hexene was much more readily converted into the corresponding aldehydes than styrene. This shows that the synergistic effect for cobalt and ruthenium on hydroformylation of olefins occurs in the following order: cyclohexene > 1-hexene > styrene. It is of special interest that cyclohexene, which is the least reactive substrate in the $Co_2(CO)_8$-catalyzed hydroformylation, is the most effectively hydroformylated if the bimetallic catalyst is used.

In order to elucidate the mechanism of hydroformylation catalyzed by the $Co_2(CO)_8$–$Ru_3(CO)_{12}$ catalyst, we attempted a high pressure IR spectral study of the reaction solution, but it did not show the existence of cobalt–ruthenium clusters. Based on this observation, we presumed that the synergistic effect for cobalt and ruthenium might be explained by dinuclear reductive elimination of aldehydes from cobalt acyls and ruthenium hydrides (Scheme 18).

The catalytic cycle of the hydroformylation of an olefin by $Co_2(CO)_8$ is considered to include alkylcobalt formation by olefin insertion into a cobalt hydride and subsequent acylcobalt formation by CO insertion into an alkylcobalt.[30] The last hydrogenolysis step of acylcobalt intermediates forming aldehydes has been left obscure. In addition to mononuclear hydrogenolysis of the acylcobalt species, i.e.,

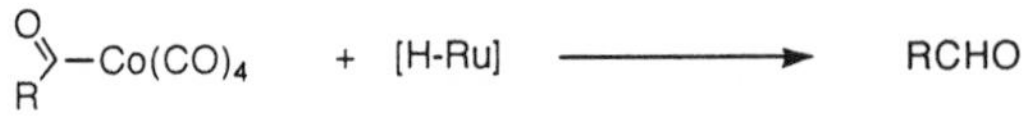

$$RC(=O)-Co(CO)_4 + [H\text{-}Ru] \longrightarrow RCHO$$

Scheme 18.

oxidative addition of H_2 followed by reductive elimination of aldehydes, dinuclear reductive elimination of aldehydes from the acylcobalt species and metal hydrides has been proposed as a possible product-forming step.[31] However, only a few reports have demonstrated that dinuclear reductive elimination reactions play an important role in actual catalytic reactions[32] despite that such reactions have been attracting much attention in connection with the product-forming steps in catalytic processes.[33–44]

Therefore, the reactions between n-$C_5H_{11}COCo(CO)_4$ and several kinds of metal carbonyl hydrides including $HCo(CO)_4$, $[HRu(CO)_4]^-$, and $[HRu_3(CO)_{11}]^-$ have been investigated as model reactions (Eq. 14).[45] Interestingly, the ruthenium hydride $[HRu(CO)_4]^-$ reacted 4 times faster with the acylcobalt complex at ambient temperature to form haxanal than $HCo(CO)_4$. This suggests that also in the hydroformylation with the $Co_2(CO)_8$–$Ru_3(CO)_{12}$ bimetallic system, a ruthenium hydride, which is generated from $Ru_3(CO)_{12}$ under the catalytic conditions, is involved in the aldehyde-forming step from an acylcobalt intermediate.

$$C_5H_{11}CO\text{-}Co(CO)_4 + [HRu(CO)_4]^- \xrightarrow[20\ °C]{} C_5H_{11}CHO \quad (14)$$

Another phenomenon which supports the above hypothesis was observed in the hydroformylation of norbornene (Eq. 15).[46] In the hydroformylation by $Co_2(CO)_8$ alone, the yield of norbornanecarbaldehyde (**45**) was low and lactone (**46**), formed from two molecules of norbornene and two molecules of CO, was the major product. In contrast, addition of $Ru_3(CO)_{12}$, which is essentially inactive for the hydroformylation of norbornene, remarkably promoted the yield of **45**, causing it to be the major product. At the same time, total increase of norbornene conversion was observed.

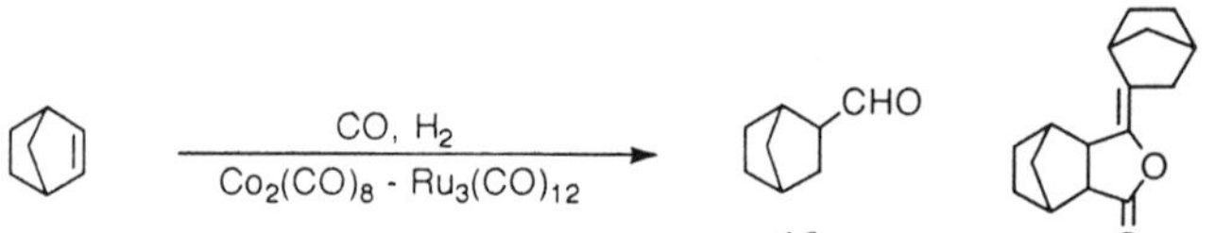

$$\text{norbornene} \xrightarrow[Co_2(CO)_8\ -\ Ru_3(CO)_{12}]{CO,\ H_2} \mathbf{45}\ (\text{CHO}) + \mathbf{46}$$

Catalyst	Conv.	Yield (%) 45	Yield (%) 46
$Co_2(CO)_8$	41	9	20
$Ru_3(CO)_{12}$	3	0.8	tr.
$Co_2(CO)_8$ - $Ru_3(CO)_{12}$	99	55	15

(15)

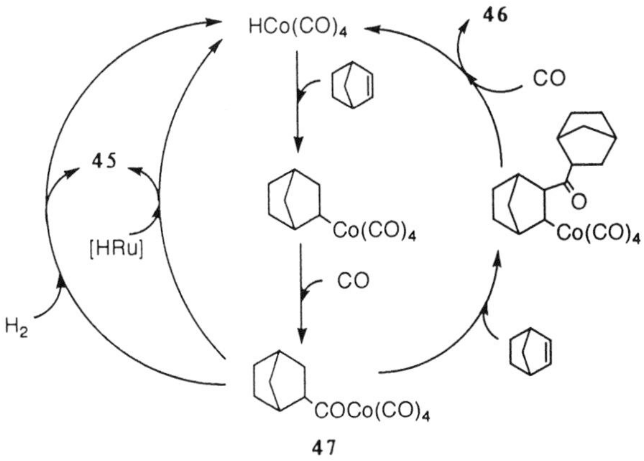

Scheme 19.

The result shown in Eq. 15 is rationalized as follows (Scheme 19). In the $Co_2(CO)_8$-catalyzed hydroformylation of norbornene, acylcobalt complex **47** is supposed to be an intermediate. Once **47** is formed, both the oxidative addition of H_2 to **47** to yield **45** and the insertion of norbornene into the Co–C bond in **47** leading to **46** proceed competitively. This is because norbornene is easily inserted into an acyl–metal bond.[22c,47] This competition decreases the selectivity of aldehyde formation. The fact that addition of $Ru_3(CO)_{12}$ increased the selectivity of hydroformylation indicates that the presence of ruthenium favors the hydrogenolysis of the acylcobalt, and this is explained by considering a ruthenium hydride species as a hydrogenolysis reagent toward the acylcobalt **47**. As a result, the total rate of norbornene conversion increased and the aldehyde **45** was obtained as the major product.

As described above, the synergistic effect for cobalt and ruthenium is understood, at least, in terms of the dinuclear reductive elimination between an acylcobalt and a ruthenium hydride species. However, Whyman[48] and Mirbach[49] proposed that the initial interaction of an olefin with $HCo(CO)_4$ might be the rate-determining

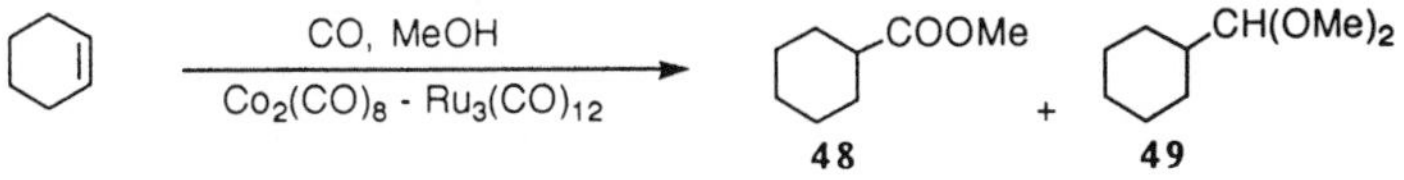

CO, MeOH; $Co_2(CO)_8$ - $Ru_3(CO)_{12}$ → COOMe (**48**) + CH(OMe)$_2$ (**49**) (16)

Catalyst	Yield (%) **48**	Yield (%) **49**
$Co_2(CO)_8$	16	0
$Ru_3(CO)_{12}$	3	0
$Co_2(CO)_8$ - $Ru_3(CO)_{12}$	53	2

step, especially in the hydroformylation of internal olefins. The fact that the synergistic effect for cobalt and ruthenium is most effectively realized in the case of cyclohexene indicates that ruthenium might also promote the olefin insertion into the Co–H bond. This is compatible with the finding that not a large but distinct synergistic effect for cobalt and ruthenium was also observed in the hydroesterification of olefins (Eq. 16), which does not involve hydrogenolysis of an acyl intermediate.

B. Formylation of Aryl and Alkenyl Iodides by Palladium–Ruthenium Bimetallic Catalyst

In order to obtain further information on the role of dinuclear reductive elimination in catalysis and to develop a novel multimetallic catalyst, we have turned our attention to a Pd–Ru catalyst system and found that the $PdCl_2(PPh_3)_2$–$Ru_3(CO)_{12}$ bimetallic catalyst exhibits a high catalytic activity in the formylation of aryl and alkenyl iodides.[50]

It is well known that palladium complexes such as $PdCl_2(PPh_3)_2$ are effective catalysts for the formylation of aryl and alkenyl halides.[51] This formylation has usually been conducted at reaction temperatures of 80–150 °C, and the reactions at 80 °C were reported to be rather slow despite that reactive alkenyl iodides were used as the substrates. Considering that the rate-determining step of the palladium-catalyzed formylation of iodides is probably the hydrogenolysis of acylpalladium intermediates, and that certain metal carbonyl hydrides effectively work as hydrogen donors in the dinuclear reductive elimination reactions to form aldehydes,[33–42,45] we expected that an improved bimetallic catalyst system for the formylation of iodides could be developed by combining a palladium complex with a metal carbonyl complex. From this viewpoint, we have examined the formylation of iodobenzene by using various $PdCl_2(PPh_3)_2$–metal carbonyl catalysts (Eq. 17).

$$\text{PhI} \xrightarrow[\text{PdCl}_2(\text{PPh}_3)_2\text{ - metal carbonyl}]{\text{CO, H}_2\text{, NEt}_3} \text{PhCHO} \qquad (17)$$

The results are summarized in Table 9. The catalytic activity of $PdCl_2(PPh_3)_2$ alone at 70 °C is low and the yield of benzaldehyde after 1.5 h was only 17%. The metal carbonyls of Cr, Mo, W, Mn, Fe, Co, and Ru show essentially no catalytic activity under these reaction conditions. Interestingly, when $PdCl_2(PPh_3)_2$ and $Ru_3(CO)_{12}$ were used in combination at a Pd:Ru ratio of 1:2, the yield of benzaldehyde became almost 4 times as high as by $PdCl_2(PPh_3)_2$ only. Obviously the $PdCl_2(PPh_3)_2$–$Ru_3(CO)_{12}$ bimetallic system showed a synergistic effect on the formylation of iodobenzene. However, other metal carbonyls did not work effectively as the second component of the bimetallic catalyst for this reaction. The Pd:Ru ratio has a large effect on the activity of the $PdCl_2(PPh_3)_2$–$Ru_3(CO)_{12}$ catalyst. The highest activity was achieved when the Pd:Ru ratio was 1:2. In the reactions where more $Ru_3(CO)_{12}$ was used, precipitation of Pd metal was observed and the yield of benzaldehyde was lowered.

Table 9. Effects of Metal Carbonyls on the $PdCl_2(PPh_3)_2$–Catalyzed Formylation of PhI[a]

	Yield (%) (Conv. (%))[b]	
Metal Carbonyl	*Pd + M*	*M*[c]
none	17 (17)	—
$Cr(CO)_6$	15 (15)	0 (0)
$Mo(CO)_6$	25 (24)	0 (1)
$W(CO)_6$	20 (19)	0 (0)
$Mn_2(CO)_{10}$	23 (23)	0 (0)
$Fe(CO)_5$	15 (14)	0 (0)
$Co_2(CO)_8$	24(25)	1 (1)
$Ru_3(CO)_{12}$	65 (66)	2 (2)

Notes: [a]Reaction conditions: PhI 10 mmol, $PdCl_2(PPh_3)_2$ 0.1 mmol, metal carbonyl 0.2 mmol as metal atom, benzene 10 mL, NEt_3 12 mmol, 70 °C, 1.5 h, $CO:H_2$ = 50:50 atm.

[b]Determined by GC. Yields are based on the starting iodobenzene.

[c]$PdCl_2(PPh_3)_2$ was not used.

Formulation of iodobenzene by the $PdCl_2(PPh_3)_2$–$Ru_3(CO)_{12}$ catalyst. $PdCl_2(PPh_3)_2$ (0.10 mmol), $Ru_3(CO)_{12}$ (0.067 mmol), iodobenzene (10 mmol), NEt_3 (12 mmol) and benzene (10 mL) as a solvent were charged in a stainless-steel autoclave under a nitrogen atmosphere. The autoclave was pressurized to 100 atm with CO-H_2 (1:1) and heated to 70 °C. The reaction was allowed to proceed at that temperature for 1.5 h with magnetic stirring. After the reaction, the autoclave was rapidly cooled to room temperature, and the pressure was slowly released. GC analysis of the liquid phase showed that the conversion of iodobenzene was 66% and benzaldehyde was formed in 65% yield based on the iodobenzene initially charged.

The $PdCl_2(PPh_3)_2$–$Ru_3(CO)_{12}$ catalyst was effective for the formylation of various aryl and alkenyl iodides. As shown in Table 10, the yields of the corresponding aldehydes were increased 2–5 times by the use of the bimetallic catalyst in comparison with $PdCl_2(PPh_3)_2$ alone. However, no positive effect was observed in the formylation of bromobenzene, probably because the rate determining step in the formylation of bromobenzene is the oxidative addition of bromobenzene to a palladium(0) complex where any ruthenium species does not participate.

In the formylation of (Z)-3-iodohex-3-ene, the Pd–Ru catalyst showed not only higher catalytic activity but also higher tendency to retain the stereochemistry around the C=C bond (Eq. 18).

When a benzene–NEt_3 (6:1) solution of $Ru_3(CO)_{12}$ was kept in contact with H_2 under conditions similar to the catalytic reactions (70 °C, CO: 40 atm, H_2: 40 atm,

Table 10. Formylation of Aryl Iodides by $PdCl_2(PPh_3)_2$–$Ru_3(CO)_{12}$ Catalyst[a]

Substrate	*Reaction Temperature (°C)*	*Reaction Time (h)*	*Yield (%) (Conv. (%))*[b]	
			Pd[c]	*Pd + Ru*
p-Iodotoluene	70	1.5	19 (20)	58 (61)
m-Chloroiodobenzene	70	3	12 (16)	56 (78)
p-Iodoanisol	70	3	25 (31)	60 (66)
2-Iodothiophene	70	3	23 (26)	51 (62)
1-Iodonaphthalene	70	1.5	22 (23)	53 (68)

Notes: [a]Reaction conditions: $PdCl_2(PPh_3)_2$ 0.1 mmol, $Ru_3(CO)_{12}$, 0.067 mmol, substrate 10 mmol, benzene 10 mL, NEt_3 12 mmol, CO: H_2 = 50: 50 atm.

[b]Determined by GC. Yields are based on the starting aryl iodides.

[c]$Ru_3(CO)_{12}$ was not used.

Et(I)C=C(Et) —[CO, H_2, NEt_3 / $PdCl_2(PPh_3)_2$ - $Ru_3(CO)_{12}$]→ Et(CHO)C=C(Et)

Catalyst	Conv. (%)	Yield (%), (*Z/E*)
$PdCl_2(PPh_3)_2$	32	32 (1.9)
$PdCl_2(PPh_3)_2$ - $Ru_3(CO)_{12}$	57	54 (3.9)

(18)

1 h), the IR spectrum of the solution clearly showed the formation of $[HRu_3(CO)_{11}]^-$ as the major ruthenium species.[52] On the other hand, in the palladium-catalyzed formylation of aryl iodides, acylpalladium complexes are assumed to be intermediates which give the corresponding aldehydes by hydrogenolysis with H_2.[51] Therefore, in the reaction catalyzed by the $PdCl_2(PPh_3)_2$–$Ru_3(CO)_{12}$ bimetallic system, it is reasonable to assume that the ruthenium hydride $[HRu_3(CO)_{11}]^-$ and an acylpalladium complex coexist in the reaction mixture under the catalytic reaction conditions. Takezaki et al. reported previously that the rate-determining step of the formylation of iodobenzene by $PdCl_2$ in pyridine is the hydrogenolysis of a benzoylpalladium complex.[53] In addition, reaction of $Pd(CO)(PPh_3)_3$ and iodobenzene readily gives a benzoylpalladium complex $PdI(COPh)(PPh_3)_2$ at room temperature.[54] Therefore, the rate-determining step of the formylation of iodides by a Pd–phosphine catalyst is also presumed to be the hydrogenolysis of intermediary acylpalladium complexes. On the basis of these things, we examined reactions of $PdI(CO\text{-}p\text{-}Tol)(PPh_3)_2$ with several anionic metal hydrides or H_2 to clarify the role of ruthenium in the catalytic formylation (Eq. 19).

$$PdI(CO\text{-}p\text{-}Tol)(PPh_3)_2 + [HRu_3(CO)_{11}]^- \longrightarrow p\text{-}TolCHO \quad (19)$$

The results are listed in Table 11. Ruthenium hydrides, especially $[HRu_3(CO)_{11}]^-$, were found to react smoothly with $PdI(CO\text{-}p\text{-}Tol)(PPh_3)_2$ to give *p*-tolualdehyde

Table 11. Reactions between $PdI(CO\text{-}p\text{-}Tol)(PPh_3)_2$ and Metal Hydrides[a]

Metal Hydride[b]	*Solvent*	*Temp. (°C)*	*Yield (%)*[c]
$[Et_4N][HRu_3(CO)_{11}]$	CH_2Cl_2	0	20
	benzene	40	68
$[PPN][HRu(CO)_4]$	CH_2Cl_2	0	23
	benzene	40	30
$[Et_4N][HCr_2(CO)_{10}]$	THF	0	70
$[Et_4N][HMo_2(CO)_{10}]$	CH_2Cl_2	0	75
H_2	benzene	40	0

Notes: [a]Reaction conditions: $PdI(CO\text{-}p\text{-}Tol)(PPh_3)_2$ 0.05 mmol, solvent 2 mL, CO 1 atm, 50 min.
[b]Metal complex, 0.05 mmol; H_2, 40 atm.
[c]Determined by GC.

under mild conditions. In benzene, the solubility of both $PdI(CO\text{-}p\text{-}Tol)(PPh_3)_2$ and $[NEt_4][HRu_3(CO)_{11}]$ are low, but the complexes dissolved gradually and the yield of *p*-tolualdehyde increased up to 68% after about 1 h. In contrast, $PdI(CO\text{-}p\text{-}Tol)(PPh_3)_2$ did not react with H_2 at 40 °C even under 40 atm and no tolualdehyde was detected by GC analysis.

The results of these stoichiometric reactions strongly suggest that in the catalytic formylation by the $PdCl_2(PPh_3)_2$–$Ru_3(CO)_{12}$ catalyst the hydrogenolysis of the acylpalladium intermediate $PdI(COR)(PPh_3)_2$ (**50**) with a ruthenium hydride (prob-

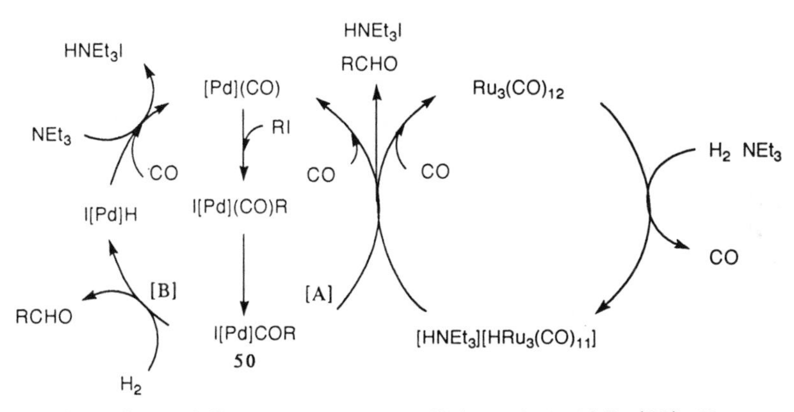

Scheme 20.

ably $[HRu_3(CO)_{11}]^-$) (Scheme 20, path A) proceeds more rapidly than the hydrogenolysis with H_2 (path B). Palladium(0) and ruthenium(0) species formed concurrently with the aldehyde are again transformed into $PdI(COR)(PPh_3)_2$ and $[HRu_3(CO)_{11}]^-$, respectively, under the catalytic reaction conditions, and this completes the bimetallic catalytic cycle (Scheme 20, path A). Since the rate-determining step seems to be the hydrogenolysis of $PdI(COR)(PPh_3)_2$, the above mechanism accounts for the rate enhancement by the $PdCl_2(PPh_3)_2$–$Ru_3(CO)_{12}$ bimetallic catalyst.

It is interesting to point out that $PdI(CO\text{-}p\text{-}Tol)(PPh_3)_2$ also reacted with $[NEt_4][HCr_2(CO)_{10}]$ and $[NEt_4][HMo_2(CO)_{10}]$ to give *p*-tolualdehyde in good yields even at 0 °C, although neither $[NEt_4][HCr_2(CO)_{10}]$ nor chromium and molybdenum carbonyl exhibited any appreciable synergistic effects in the catalytic formylation. This is probably explained by nonregeneration of the anionic carbonyl hydrides of chromium and molybdenum under the catalytic reaction conditions.

C. Carbonylation of Iodoarenes with $CO/SiHR_3$ by Palladium–Cobalt or Palladium–Ruthenium Bimetallic Catalysts

Similarity in reactivities of H_2 and hydrosilane toward transition metal complexes has been well documented, and a variety of carbonylation reactions have been developed by using $CO/SiHR_3$ system.[55] The enhanced formylation of iodoarenes by the $PdCl_2(PPh_3)_2$–$Ru_3(CO)_{12}$ catalyst stimulated us to investigate carbonylation reactions of iodoarenes with $CO/SiHR_3$ by the bimetallic catalysts composed of $PdCl_2(PPh_3)_2$ and various metal carbonyls.[56]

Table 12 summarizes the results with *p*-iodotoluene. Neither $PdCl_2(PPh_3)_2$ nor the metal carbonyls tested exhibited appreciable catalytic activity under the conditions shown in the table. However, when $PdCl_2(PPh_3)_2$ and $Ru_3(CO)_{12}$ were used together, iodotoluene was carbonylated to give *p*-tolualdehyde (**51**), *p*-$TolCH_2OSiEt_3$ (**52**), and $(p\text{-}TolCHOSiEt_3)_2$ (**53**) in 10, 40, and 4% yield, respectively (Eq. 20). The $PdCl_2(PPh_3)_2$–$Co_2(CO)_8$ bimetallic catalyst also showed a remarkable synergistic effect to give **52**. In contrast to the formylation of iodoarenes described above, these reactions smoothly proceeded in the absence of a base and $SiIEt_3$ was formed concurrent with the carbonylated products.

$$p\text{-TolI} \xrightarrow[PdCl_2(PPh_3)_2\ -\ \text{metal carbonyl}]{CO,\ SiHEt_3,\ (NEt_3)} \underset{\mathbf{51}}{p\text{-TolCHO}} + \underset{\mathbf{52}}{p\text{-TolCH}_2\text{OSiEt}_3} + \underset{\mathbf{53}}{(p\text{-Tol})(Et_3SiO)HC\text{-}CH(OSiEt_3)(p\text{-Tol})} \quad (20)$$

Interestingly, when NEt_3 was added to the Pd–Co bimetallic system, **53** became the major product. In this case, half of the iodine in *p*-iodotoluene was trapped as $HNEt_3I$ salt and the rest was converted to $SiIEt_3$. Similar results were obtained in

Table 12. Carbonylation of *p*-Iodotoluene with CO/$SiHEt_3$ by $PdCl_2(PPh_3)$–Metal Carbonyls[a]

		Yield (%) (Conv. (%))[b]							
		Pd + M				*M*[d]			
M	NEt_3[c]	**51**	**52**	**53**		**51**	**52**	**53**	
none	—	3	0	2	(10)				
none	added	2	0	0	(4)				
$Cr(CO)_6$	—	4	0	1	(13)	1	0	0	(11)
$Mo(CO)_6$	—	1	0	1	(9)	1	0	0	(8)
$W(CO)_6$	—	2	0	0	(9)	1	0	0	(8)
$Mn_2(CO)_{10}$	—	2	0	1	(12)	0	0	0	(9)
$Fe(CO)_5$	—	1	0	0	(9)	0	0	0	(8)
$Ru_3(CO)_{12}$	—	10	40	4	(79)	0	1	0	(5)
$Ru_3(CO)_{12}$	added	17	10	3	(64)	0	0	0	(0)
$Co_2(CO)_8$	—	0	76	0	(85)	0	0	0	(4)
$Co_2(CO)_8$	added	2	6	57	(70)	0	0	0	(5)

Notes: [a]Reaction conditions: *p*-TolI 2.5 mmol, $HSiEt_3$ 7.5 mmol, $PdCl_2(PPh_3)_2$ 0.05 mmol, metal carbonyl 0.05 mmol as metal atom, benzene 5 mL, CO 50 atm, 80 °C, 3 h.

[b]GC yield based on the starting *p*-TolI.

[c]NEt_3 3 mmol.

[d]$PdCl_2(PPh_3)_2$ was not used.

reactions of *p*-iodoanisole and 1-iodonaphthalene. We must await further investigation to elucidate the reasons why synergistic effects appear in these reactions and why the product distribution remarkably changes depending upon the presence or absence of NEt_3.

IV. CONCLUSION

In this chapter we have described our two approaches directed toward development of novel carbonylation reactions. The palladium-catalyzed cyclocarbonylation of allylic compounds has proved to provide a unique route to naphthyl acetates, fused-ring heteroaromatics, and phenol derivatives. On the other hand, homogeneous bimetallic catalysts such as $Co_2(CO)_8$–$Ru_3(CO)_{12}$, $PdCl_2(PPh_3)_2$–$Ru_3(CO)_{12}$, and $PdCl_2(PPh_3)_2$–$Co_2(CO)_8$ were shown to be more effective for several types of carbonylation reactions than conventional homometallic catalysts. These reactions are not only useful as an effective tool in organic synthesis, but also quite interesting from a mechanistic point of view because they are based on the novel reactivity of acyl complexes formed in the catalytic cycle. We believe that further study toward this direction will lead to development of new catalytic reactions.

REFERENCES AND NOTES

1. Mullen, A. In *New Syntheses with Carbon Monoxide*; Falbe, J., Ed.; Springer: Berlin, 1980, pp. 433.
2. Tkatchenko, I. In *Comprehensive Organometallic Chemistry*; Willkinson, J.; Stone, F.G.A.; Abel, E.W., Eds.; Pergamon: Oxford, 1982, Chapter 50.3.
3. (a) Sakakura, T.; Sodeyama, T.; Sasaki, K.; Wada, K.; Tanaka, M. *J. Am. Chem. Soc.* **1990**, *112*, 7221; (b) Tanaka, M.; *CHEMTECH* **1989**, 59.
4. (a) Kunin, A.J.; Eisenberg, R. *J. Am. Chem. Soc.* **1986**, *108*, 535; (b) Kunin, A. J.; Eisenberg, R. *Organometallics* **1988**, *7*, 2124.
5. (a) Jintoku, T.; Fujiwara, Y.; Kawata, I.; Kawauchi, T.; Taniguchi, H. *J. Organomet. Chem.* **1990**, *385*, 297; (b) Fujiwara, Y.; Jintoku, T.; Takaki, K. *CHEMTECH* **1990**, 636.
6. Doyama, K.; Fujiwara, K.; Joh, T.; Maeshima, K.; Takahashi, S. *Organometallics* **1991**, 508.
7. Hong, P.; Cho, B.; Yamazaki, H. *Chem. Lett.* **1979**, 339.
8. Bruson, H.A.; Plant, H.L. *J. Org. Chem.* **1967**, *32*, 3356.
9. Kim, P.J.; Hagihara, N. *Bull. Chem. Soc. Jpn.* **1965**, *38*, 2022.
10. (a) Koyasu, Y.; Matsuzaka, H.; Hiroe, Y.; Uchida, Y.; Hidai, M. *J. Chem. Soc., Chem. Commun.* **1987**, 575; (b) Matsuzaka, H.; Hiroe, Y.; Iwasaki, M.; Ishii, Y.; Koyasu, Y.; Hidai, M. *J. Org. Chem.* **1988**, *53*, 3832; (c) Ishii, Y.; Hidai, M. *J. Organomet. Chem.* **1992**, *428*, 279.
11. Iwasaki, M.; Matsuzaka, H.; Hiroe, Y.; Ishii, Y.; Koyasu, Y.; Hidai, M. *Chem Lett.* **1989**, 1159.
12. (a) Iwasaki, M.; Li, J.-P.; Kobayashi, Y.; Matsuzaka, H.; Ishii, Y.; Hidai, M. *Tetrahedron Lett.* **1989**, *30*, 95; (b) Iwasaki, M.; Kobayashi, Y.; Li, J.-P.; Matsuzaka, H.; Ishii, Y.; Hidai, M. *J. Org. Chem.* **1991**, *56*, 1922.
13. Friedrich-Friechtl, J.; Spiteller, G. *Tetrahedron* **1975**, *31*, 479.
14. Sargent, M.V.; Stransky, P.O. *J. Chem. Soc., Perkin Trans. 1* **1982**, 1605.
15. Matsuzaka, H.; Hiroe, Y.; Iwasaki, M.; Ishii, Y.; Koyasu, Y.; Hidai, M. *Chem. Lett.* **1988**, 377.
16. Ozawa, F.; Son, T.; Osakada, K.; Yamamoto, A. *J. Chem. Soc., Chem. Commun.* **1989**, 1067.
17. Fujiwara, Y.; Asano, R.; Moritani, I.; Teranishi, S. *Chem. Lett.* **1975**, 1061.
18. (a) Dötz, K.H.; Fügen-Köster, B. *Chem. Ber.* **1980**, *113*, 1449; (b) Dötz, K.H. *Angew. Chem., Int. Ed. Engl.* **1984**, *23*, 587.
19. (a) Anderson B.J.; Wulff, W.D. *J. Am. Chem. Soc.* **1990**, *112*, 8615; (b) McCallum, J. S.; Kunng, F.-A.; Gilbertson, S.R.; Wulff, W.D. *Organometallics* **1988**, *7*, 2346.
20. Dötz, K.H.; Dietz, R. *Chem. Ber.* **1978**, *111*, 2517.
21. Iwasaki, M.; Ishii, Y.; Hidai, M. *J. Organomet. Chem.* **1991**, *415*, 435.
22. (a) Negishi, E.; Wu, G.; Tour, J.M. *Tetrahedron Lett.* **1988**, *29*, 6745; (b) Negishi, E.; Tour, J.M. *Tetrahedron Lett.* **1986**, *27*, 4869; (c) Sen, A.; Lai, T.-W. *J. Am. Chem. Soc.* **1982**, *104*, 3520.
23. Nickel-catalyzed synthesis of *m*-cresol: Casser, L.; Foa, M.; Chiusoli, G.P. *Organomet. Chem. Syn.* **1971**, *1*, 302.
24. (a) Ishii, Y.; Gao, C.; Iwasaki, M.; Hidai, M. *J. Chem. Soc., Chem. Commun.* **1991**, 695; (b) Ishii, Y.; Gao, C.; Xu, W.-X.; Iwasaki, M.; Hidai, M. *J. Org. Chem.* **1993**, *58*, 6818.
25. Braunstein, P.; Rose, J. In *Stereochemistry of Organometallic and Inorganic Compounds*; Bernal, I., Ed.; Elsevier: Amsterdam, 1988, Vol. 3.
26. Roberts, D.A.; Geoffroy, G.L. In *Comprehensive Organometallic Chemistry*; Wilkinson, G.; Stone, F.G.A.; Abel, E.W., Ed.; Pergamon: Oxford, 1982, Vol. 6, pp. 763.
27. (a) Hidai, M.; Orisaku, M.; Ue, M.; Uchida, Y.; Yasufuku, K.; Yamazaki, H. *Chem. Lett.* **1981**, 143; (b) Hidai, M.; Orisaku, M.; Ue, M.; Koyasu, Y.; Kodama T.; Uchida, Y. *Organometallics* **1983**, *2*, 292.
28. (a) Butter, D.; Haute, T. U.S. Patent 3285948, 1966; (b) Mizoroki, M.; Matsumoto, T.; Ozaki, A. *Bull. Chem. Soc. Jpn.* **1979**, *52*, 479.

29. (a) Hidai, M.; Fukuoka, A.; Koyasu, Y.; Uchida, Y. *J. Chem. Soc., Chem. Commun.* **1984**, 516; (b) Hidai, M.; Fukuoka, A.; Koyasu, Y.; Uchida, Y. *J. Mol. Catal.* **1986**, *35*, 29; (c) Hidai, M.; Matsuzaka, H. *Polyhedron* **1988**, *7*, 2369.
30. Heck, R.F.; Breslow, D.S. *J. Am. Chem. Soc.* **1961**, *83*, 4023.
31. Pino, P.; Piacenti, F.; Bianchi, M. *Organic Syntheses via Metal Carbonyls*; Wender, I.; Pino, P., Eds.; Wiley: New York, 1977, Vol. 2, pp. 43.
32. Collman, J.P.; Belmont J.A.; Brauman, J.I. *J. Am. Chem. Soc.* **1983**, *105*, 7288.
33. Tanaka, M.; Sakakura, T.; Hayashi T.; Kobayashi, T. *Chem. Lett.* **1986**, 39.
34. Norton, J.R. *Acc. Chem. Res.* **1979**, *12*, 139.
35. Nappa, M.J.; Santi R.; Halpern, J. *Organometallics* **1985**, *4*, 34.
36. Warner K.E.; Norton, J.R. *Organometallics* **1985**, *4*, 2150.
37. Martin, B.D.; Warner K.E.; Norton, J.R. *J. Am. Chem. Soc.* **1986**, *108*, 33.
38. Kovács, I.; Ungváry F.; Markó, L. *Organometallics* **1986**, *5*, 209.
39. Bullock, R.M.; Rappoli, B.J. *J. Am. Chem. Soc.* **1991**, *113*, 1659.
40. Jones, W.D.; Huggins, J.M.; Bergman, R.G. *J. Am. Chem. Soc.* **1981**, *103*, 4415.
41. Halpern, J. *Acc. Chem. Res.* **1982**, *15*, 332.
42. Barborak, J.C.; Cann, K. *Organometallics* **1982**, *1*, 1726.
43. James, B.R.; Wang, D.K. W. *J. Chem. Soc., Chem. Commun.* **1977**, 550.
44. Renaut, P.; Tainturier, G.; Gautheron, B. *J. Organomet. Chem.* **1978**, *150*, C9.
45. Koyasu, Y.; Fukuoka, A.; Uchida, Y.; Hidai, M. *Chem. Lett.* **1985**, 1083.
46. Ishii, Y.; Sato, M.; Matsuzaka, H.; Hidai, M. *J. Mol. Catal.* **1989**, *54*, L13.
47. Fr. Pat. 1352841 (1964) to Rhone-Poulenc S.A.
48. Whyman, R. *J. Organomet. Chem.* **1974**, *81*, 97.
49. Mirbach, M.F. *J. Organomet. Chem.* **1984**, *265*, 205.
50. Misumi, Y.; Ishii, Y.; Hidai, M. *J. Mol. Catal.* **1993**, *78*, 1.
51. Schoenberg, A.; Heck, R.F. *J. Am. Chem. Soc.* **1974**, *96*, 7761.
52. Han, S.H.; Geoffroy, G.L.; Dombek, B.D.; Rheingold, A.L. *Inorg. Chem.* **1988**, *27*, 4355.
53. Yoshida, H.; Sugita, N.; Kudo, K.; Takezaki, Y. *Bull. Chem. Soc. Jpn.* **1976**, *49*, 1681.
54. Kudo, K.; Hidai, M.; Uchida, Y. *J. Organomet. Chem.* **1971**, *33*, 393.
55. (a) Murai, S.; Seki, Y. *J. Mol. Catal.* **1987**, *41*, 197; (b) Murai, S.; Sonoda, N. *Angew. Chem., Int. Ed. Engl.* **1979**, *18*, 837.
56. (a) Misumi, Y.; Ishii, Y.; Hidai, M. *Organometallics* **1995**, *14*, 1770; (b) Misumi, Y.; Ishii, Y.; Hidai, M. *Chem. Lett.* **1994**, 695.

INDEX